European Landscapes in Transition

European rural landscapes as we experience them today are the result of ongoing processes and interactions between nature and society. These are changing fast: the future landscapes will be different from those we know currently. Written for academics, policy makers and practitioners, this book is the first to explore the complex histories of rural landscapes in Europe as a basis for their sound governance in future. Tensions between the needs of agricultural spaces driven by economic incentives and a variety of non-agricultural functions are explored to demonstrate current challenges and the shortfalls in the policies that address them. Using inspiring case studies that highlight the roles of regional agents and communities, the authors go further than the usual analyses to illustrate the importance of local context. Written by experts currently working to revitalise the rural landscapes of Europe, the text concludes with suggestions for improving landscape policy and planning practice.

TERESA PINTO-CORREIA is Associate Professor and Head of the Institute of Mediterranean Agrarian and Environmental Sciences at the University of Évora, Portugal. Her research interests and record of publications span rural landscape dynamics and management of European rural landscapes at multiple scales. She develops her research mainly with a focus on the dynamics of agricultural landscapes in the Mediterranean, and how they are affected by different sets of drivers, between production, consumption and protection, creating tensions but also synergies. She coordinates an interdisciplinary research group and she is Director of a research unit with more than 100 researchers, centered on Mediterranean agriculture and environment, aiming to contribute to the sustainability of the production systems and their related ecosystems and landscapes.

JØRGEN PRIMDAHL is Professor in Countryside Planning and Management at the University of Copenhagen. With a background in landscape architecture, his main research interest is agricultural landscape (patterns, functions, and change), public policy and spatial planning. He has published extensively on rural landscapes. In 2010, he co-edited *Globalisation and Agricultural Landscapes: Change Patterns and Policy Trends in Developed Countries* (Cambridge University Press), and in 2017, he co-authored *Landscape Analysis – Mapping the Character and Potential of Space*.

BAS PEDROLI is Associate Professor in Land Use Planning at Wageningen University, Senior Researcher at Wageningen Environmental Research and Chair of UNISCAPE, the university association dedicated to the implementation of the European Landscape Convention. He is strongly involved in research and education related to sustainable land use and cultural identity. Holding a PhD in landscape ecology, he is an active member of various European scientific and NGO networks. He has published many research papers and edited several scientific books, all related to the challenge of finding new futures for Europe's precious landscapes. He is engaged enthusiastically in supporting strategies for equitable land use and landscape management and nature conservation planning, fostering community based perspectives.

Cambridge Studies in Landscape Ecology

Series Editors
Professor John Wiens *Colorado State University*
Dr Peter Dennis *James Hutton Institute*
Professor Lenore Fahrig *Carleton University*
Professor Marie-Jose Fortin *University of Toronto*
Professor Richard Hobbs *University of Western Australia*
Professor Bruce Milne *University of New Mexico*
Professor Joan Nassauer *University of Michigan*
Professor Paul Opdam *Wageningen University*

Cambridge Studies in Landscape Ecology presents synthetic and comprehensive examinations of topics that reflect the breadth of the discipline of landscape ecology. Landscape ecology deals with the development and changes in the spatial structure of landscapes and their ecological consequences. Because humans are so tightly tied to landscapes, the science explicitly includes human actions as both causes and consequences of landscape patterns. The focus is on spatial relationships at a variety of scales, in both natural and highly modified landscapes, on the factors that create landscape patterns, and on the influences of landscape structure on the functioning of ecological systems and their management. Some books in the series develop theoretical or methodological approaches to studying landscapes, while others deal more directly with the effects of landscape spatial patterns on population dynamics, community structure, or ecosystem processes. Still others examine the interplay between landscapes and human societies and cultures.

The series is aimed at advanced undergraduates, graduate students, researchers and teachers, resource and land-use managers, and practitioners in other sciences that deal with landscapes.

The series is published in collaboration with the International Association for Landscape Ecology (IALE), which has chapters in more than 50 countries. IALE aims to develop landscape ecology as the scientific basis for the analysis, planning and management of landscapes throughout the world. The organization advances international cooperation and interdisciplinary synthesis through scientific, scholarly, educational and communication activities.

Also in the Series
Sources, Sinks and Sustainability
Edited by Jianguo Liu, Vanessa Hull, Anita T. Morzillo, John A. Wiens
978-0-521-19947-6 (hardback)
978-0-521-14596-1 (paperback)

Globalisation and Agricultural Landscapes
Edited by Jørgen Primdahl, Simon Swaffield
978-0-521-51789-8 (hardback)
978-0-521-73666-4 (paperback)

Key Topics in Landscape Ecology
Edited by Jianguo Wu, Richard J. Hobbs
978-0-521-85094-0 (hardback)
978-0-521-61644-7 (paperback)

Issues and Perspectives in Landscape Ecology
Edited by John A. Wiens, Michael R. Moss
978-0-521-83053-9 (hardback)
978-0-521-53754-4 (paperback)

Ecological Networks and Greenways
Edited by Rob H. G. Jongman, Gloria Pungetti
978-0-521-82776-8 (hardback)
978-0-521-53502-1 (paperback)

Transport Processes in Nature
William A. Reiners, Kenneth L. Driese
978-0-521-80049-5 (hardback)
978-0-521-80484-4 (paperback)

Integrating Landscape Ecology into Natural Resource Management
Edited by Jianguo Liu, William W. Taylor
978-0-521-78015-5 (hardback)
978-0-521-78433-7 (paperback)

TERESA PINTO-CORREIA
UNIVERSITY OF ÉVORA, PORTUGAL

JØRGEN PRIMDAHL
UNIVERSITY OF COPENHAGEN

BAS PEDROLI
WAGENINGEN UNIVERSITY & RESEARCH

European Landscapes in Transition

Implications for Policy and Practice

University Printing House, Cambridge CB2 8BS, United Kingdom

One Liberty Plaza, 20th Floor, New York, NY 10006, USA

477 Williamstown Road, Port Melbourne, VIC 3207, Australia

314–321, 3rd Floor, Plot 3, Splendor Forum, Jasola District Centre, New Delhi – 110025, India

79 Anson Road, #06–04/06, Singapore 079906

Cambridge University Press is part of the University of Cambridge.

It furthers the University's mission by disseminating knowledge in the pursuit of education, learning, and research at the highest international levels of excellence.

www.cambridge.org
Information on this title: www.cambridge.org/9781107070691
DOI: 10.1017/9781107707566

© Cambridge University Press 2018

This publication is in copyright. Subject to statutory exception and to the provisions of relevant collective licensing agreements, no reproduction of any part may take place without the written permission of Cambridge University Press.

First published 2018

Printed in the United Kingdom by TJ International Ltd. Padstow Cornwall

A catalogue record for this publication is available from the British Library

ISBN 978-1-107-07069-1 Hardback

Cambridge University Press has no responsibility for the persistence or accuracy of URLs for external or third-party internet websites referred to in this publication and does not guarantee that any content on such websites is, or will remain, accurate or appropriate.

Contents

	Foreword by Marc Antrop	ix
	Preface	xi
	Acknowledgements	xii
1	Introduction: A Landscape in Disequilibrium	1
2	What Is the Rural Landscape About?	42
3	Conceptualising Rural Landscape Change	64
4	Evolving Activities in the Rural	110
5	Changing Relationships between the Rural and the City	156
6	Landscape Policy and Planning – Managing Conflicts and Making Places	198
7	Common Grounds for Colourful Futures	240
	Bibliography	254
	Index	283

Colour plates are to be found between pp. 146 and 147.

Foreword

Landscapes are changing. Everyone can experience this, and all landscapes seem to be affected. Landscapes are always changing, naturally. So why bother? Because there is much evidence that the current changes are not for the better and the accelerating pace and magnitude indicate that changes might be out of our control.

About three centuries ago, important social and technological revolutions began that accelerated and upscaled landscape dynamics, resulting in an irreversible transition from the past. Global driving forces transformed existing landscapes into urbanised and globalised ones in a networked society. Local forces were not able to sustain the traditional land management that had created the characteristic landscapes, in particular the rural and (silvo-)pastoral landscapes with a long history and carrying past traditions and memories. They are considered a common good and a valuable heritage that is becoming lost. Particularly in Europe, the diversity of the rural and pastoral landscapes is astonishing and characteristic of Europe's identity. The first assessment of Europe's environment in 1995 by the EEA, the 'Dobříš Assessment', formulated the situation concisely as follows:

> *The richness and diversity of rural landscapes in Europe is a distinctive feature of the continent. There is probably nowhere else where the signs of human interaction with nature in landscape are so varied, contrasting and localised ...*
>
> *Despite the immense scale of socio-economic changes that have accompanied this century's wave of industrialisation and urbanisation in many parts of Europe, much of this diversity remains, giving distinctive character to countries, regions and local areas.* (European Environmental Agency 1995)

The main driving forces of this transition are human (demography, economy, politics and technology) and natural (tectonics, climate and calamities), interacting through complex feedback loops. The main driving forces induce a variety of processes such as supplying natural resources, production (in agriculture, forestry and industry), urbanisation and communication networking. These forces act from global to local scales and are essentially not sustainable and are indifferent to the persistence of landscapes. When displacement is the mantra of global mobility to increase profit, this is not an option at the local scale, where sustainability can be attempted only by adaptation.

Although the driving forces have been identified and the general trend of the transition is known, much remains unknown about the specific processes at the local scale and the mechanisms that are involved in making the real changes on the terrain. On one hand, geographical space becomes polarised, where people, activities and infrastructures concentrate in hotspots of increasing urbanisation, and on the other hand, vast areas in the periphery of economic activities are being abandoned. Consequently, landscapes everywhere are affected. These changes also affect the attitudes people have towards the landscape. In 2014, 54 per cent of the world's population lived in urban areas, and the United Nations expects an increase to 66 per cent by 2050. Non-urban areas (i.e. rural and natural areas) depopulate and lose services that once were useful for urban places, such as providing food and natural resources. Nowadays, 'nice', 'traditional' and 'natural' landscapes attract urbanites for tourism and recreation, creating new challenges and problems for the rural countryside.

European Rural Landscapes in Transition deals with the cascade of processes that affect the complex dynamics of the rural landscapes in Europe, in particular focusing on the agricultural ones. A cascade is an appropriate metaphor here as the authors explore it from the bottom to the top. Learning from local case studies, they achieve understanding of the mechanisms involved in the transition and zoom out to the European scale as a whole. The perspective starts from the characteristics of the rural and in successive chapters the changes and the responses to them are analysed. Special attention goes to the changing interactions between rural and urban. This certainly helps to formulate realistic, optimistic visions of valuable future landscapes, visions that may inspire a landscape-planning policy that sees careful conflict management as an inherent component of equitable and balanced, integrated landscape planning.

Marc Antrop
Landscape Research – Department of Geography, Ghent University

Preface

European rural landscapes have been our common research area for years. Based on each of our research projects and involvement with local landscapes and landscape agents, we have experienced how change patterns and the policies that guide them have been transformed, although not always in ways that properly address the new dynamics. These insights formed our main motivation for writing this book together. It is the result of a three-year journey into the literature and concrete examples of landscape transitions and policy initiatives, and, not least, our joint reflections and numerous discussions on change patterns, concepts, frameworks and policy approaches. We hope that academics, policy makers, planners and students who are seeking to better understand and manage the transitions processes of European rural landscapes will find this book useful.

The European continent is highly diverse with rich natural and cultural histories, and the rural landscapes reflect this diversity. In fact, the variety and heterogeneity of the landscapes, at both the regional and local scales, to a large extent characterises the European continent. This complexity makes it difficult to generalise and identify common ground for research and the practical management of European landscapes. We have tried to balance our analyses and suggestions between generalised approaches and concrete examples and we hope that we have captured the main topics concerning Europe's current transitions and policy challenges.

We, the three authors, maintain responsibility for any errors or shortcomings.

Acknowledgements

During our research, we have benefitted greatly from critical comments, constructive suggestions, practical support and assistance from many individuals and institutions. The three of us share the privilege of working for academic institutions which provide opportunities for conducting work like this. Therefore, we are grateful to the Institute for Mediterranean Agrarian and Environmental Sciences (ICAAM), the University of Évora, the Department of Geosciences and Natural Resource Management, the University of Copenhagen and to Alterra and Wageningen University for providing the practical conditions for writing this book. Thanks also to the main sponsors of the Future Landscapes programme: Realdania Foundation, Nordea Foundation, 15th June Foundation and the Danish Outdoor Council for enabling us to start the work with a series of public lectures in Denmark on European landscapes in transition. Thanks also to Sara Folvig for preparing several of the figures, to Stuart Wright for his excellent work on enhancing our non-native English, and to Victoria Parrin, Kirsten Bot and Dominic Lewis from Cambridge University Press who have been of great help in preparing the manuscript for publication as well as being generously patient. Simon Swaffield's and two anonymous reviewers' comments on the original book proposal were also greatly appreciated.

Each of the book chapters was reviewed by colleagues and, therefore, we would like to thank Marc Antrop, Peter Howard, Gertrud Jørgensen, Søren Pilgaard Kristensen, Mattias Qviström, Simon Swaffield, Richard Wakeford, Karlheinz Knickel and Anders Wästfelt for their critical and constructive comments on the chapters.

We have included numerous examples of specific landscapes and the changes they are currently undergoing in this book. These are presented at the start of each chapter, in text boxes and in the form of eight case studies. Three of these cases were provided by colleagues, and our sincere thanks go to Janez Pirnat, University of Ljubljana, Slovenia, and Diana Surová and José Muñoz-Rojas, ICAAM, University of Évora, who generously gave us the opportunity to include the cases in this book.

Finally, we wish to thank our families, first and foremost Ole, Vibeke and Annejet, who have been disturbed by home meetings, often several days in duration, and have had to deal with extra busy partners for more than three years.

1

Introduction: A Landscape in Disequilibrium

It is still surprisingly far to the centre of the old harbour town of Edam from the lake IJsselmeer (formerly sea: Zuyderzee). But the walk is worth the effort. The town, capital of the homeland of the famous Dutch cheese, is pretty and small. Modest merchant houses with crow-stepped gables line the narrow streets. Arched stone bridges and wooden drawbridges cross the water that once formed the artery of local commerce, connecting the peat meadow 'polder' landscape with the Zuyderzee. We warm ourselves in a cosy pub by the inner port; they serve delicious smoked eel.

After lunch, we continue our walk into the polder, where there are meadows for as far as the eye can see, and here and there some sheep. The last snow makes the dug-out peaty soil along the just-cleaned ditches seem still darker. A low, winding dike, a ditch with a high water level, and on the other side another polder, slightly deeper. To the left in the distance are some farmsteads; triangles leaning heavily on the earth under the grey winter sky. To the right, the strong dike of the IJsselmeer. Unbelievable how medieval communities started digging out the peatbogs and erected new dikes, and how later, on the wet peat meadows, they kept the cattle that we know from the Edam cheese. The landscape proudly carries the traces of this history.

In the next polder – this time straight dikes and ditches – we discover a few hundred white-fronted geese, quietly grazing. One of them has a neckband and we recognise the code; it was ringed in Russia last spring, on its way to northern Siberia, 5000 km away. For many centuries, wild geese formed a welcome supplement to the scanty diet of the rural population. In winter in the Dutch cultural landscape, in spring in Russia. Many sightings of ringed geese are reported by Russian hunters. They may not know that without cows and Edam cheese, offering rich feeding grounds for wintering birds as well, spring would bring much fewer geese.

FIGURE 1.1
Polder in Holland near IJsselmeer
(photo: Bas Pedroli)

In the Edam landscape, there is evidence of at least three fundamental transitions of the landscape in the past 1000 years: the exploitation of the peatbog in AD 1000–1200; the draining of the lowland peat to form polders to allow dairy farming around 1300; and land re-allotment to improve accessibility and agricultural production capacity in the 1960s. All these changes occurred at varying rhythms in different places. In other words, the landscape developed in a highly dynamic way both in space and across time.

The underlying conditions for landscape transition are changing continuously and involve a variety of often conflicting interests, which sometimes favour the large landowners, and sometimes enhance the opportunities for the more flexible, smaller landowners and farmers' communities. Recent mainstream developments, however, have prioritised market-oriented agricultural production over community-based farming. Smallholder and family farming seem to have been neglected by these developments, although interesting new types of community-based initiatives are gaining momentum, extending farmers' communities into social communities.

Today the European landscape is in transition even more strikingly than in the past. It still exhibits the traits of the age-old history of evolving land-use systems and ownership patterns, but current land use is rarely in tune with the inherited landscape. In other words, if current land use had been practised in the past, it would never have created today's landscape. This also means that sooner or later the current land use will lead to a different landscape pattern, which is adapted to the users' demands. The landscape will follow the use, either in a consciously designed way, such as in land reclamation, land re-allotment and wetland restoration, or in spontaneous, unintentional ways such as in abandoned farmland, the gradual removal of treelines or farm paths, slope levelling etc. A stable landscape where change is not occurring does not exist – unless on a very small scale for a limited period. A museum landscape in which the traditional features are conserved can be maintained only when the traditional land use is being practised, something which can be afforded only to a limited extent. However, a well-elaborated, integrated, new management approach for the future of the European landscape is almost entirely missing in academic discourse, policy practices and public debate. In fact, although the majority of European countries have ratified the European Landscape Convention, there is a tendency to leave the landscape to the tourist brochures, and let it be covered as a low-priority dossier by sector policies on culture, environment or even economics.

There is, thus, a compelling need for reflection and debate about current landscape transitions, the value of the European landscape and its potential future and the ways we can create the proper boundary conditions and governance approaches to ensure a sustainable future for this impressive asset of European culture and identity. This is what this book is about.

1.1 The Diversity of Landscapes in Europe

Europe has an astonishing diversity of landscapes, from the extremely sparsely inhabited tundra of Sámiland (northern Norway, Sweden, Finland and Russia) to the silvo-agro-pastoral Montados of Portugal, the monoculture-like terraced traditional olive groves on Eastern Lesvos, the undulating cereal fields of the Paris basin and the extensive former state farms of Eastern Europe. This diversity is undoubtedly one of Europe's main assets both in terms of local identity, tourism and quality of life and as a development factor, which influences, for example, the location of new headquarters for multinational enterprises (Stanners and Bordeaux 1995).

Why are European landscapes so diverse? Europe is a small continent with a large range of climatic conditions, depending on a mixture of Atlantic, African, Arctic and Central Asian influences. Together with geology and geomorphology, this results in highly variable soil conditions, which is reflected in an equally diverse flora and fauna. People have settled in Europe since prehistoric times and – while adapting to their environment – they have substantially added to the outstanding diversity of landscapes, which represent an intricate expression of natural and cultural heritage. In the course of time, they developed a large variety of communities, traditions and social constructs, which made an imprint on the landscape.

A long history of diverse regions is reflected in the European landscape. Some regions are characterised by an extraordinarily high population density and a dense network of urban centres with well-developed commercial connections across the region and with the outside world. Over short distances, there are impressive contrasts in the way these complex settlement networks are structured and evolve. In previous centuries, areas of agricultural production have also developed into areas for residential use, communication, transportation and leisure. These highly dynamic regions have been in contrast to other more peripheral, much less populated or accessible remote areas until recently. The European landscape has evolved to support a multitude of societal functions, while farming has played its part in the development of what might be called the European multifunctional landscape, which combines societal functions in many different ways depending on the varying population density and urban networks. The high concentration of multiple functions provided by the countryside and the intricate small-scale pattern in the way these functions are combined is definitely a trait of European landscape diversity.

Naturally, this dynamic development of the European life world is mirrored in the landscape. The long natural and cultural history, the large biophysical heterogeneity and the highly fragmented political and cultural territorial development have created the great diversity of European landscapes.

Landscape has never been static, as is well illustrated by the description of the Dutch polder landscape at the start of this chapter. However, the pace of change has increased enormously in recent decades (see e.g. Antrop 2013; Bürgi, Hersperger and Schneeberger 2004; Hoggart, Black, and Buller 1995; Plieninger et al. 2016; Primdahl 2014; Van Vliet et al. 2015), even though there is generally a considerable time lag before changing landscape functions lead to noticeable changes in the perceivable landscape (Van der Sluis et al. 2015). The highly dynamic landscape in many European regions poses serious challenges to landscape planning and management. The changes are often fragmented and incremental, brought about by societal transformations and market mechanisms. Indeed, landscape – as it is sometimes claimed (Antrop et al. 2013, p. 1644) – can be seen as a mirror of Europe's identity and diversity, or rather "as mirror and mirage" at the same time (Hansen-Møller 2006; Lefebvre 2000, p. 189). Simultaneously, landscape-planning objectives are becoming increasingly complex and require integrated approaches. This book is about European rural landscapes in transition. The significant diversity of landscapes and their dynamic nature require careful analysis to identify the causes of the changes, and the opportunities for taking an active role in designing pathways for the future and steering landscape developments.

1.2 The Historical Roots of the Landscape Concept

The way people relate to landscape – consciously or unconsciously – is to a large extent determined by cultural attitudes regarding the human–nature relationship which develop over time. The classic reference regarding the birth of the concept of landscape is Francesco Petrarca's 1336 letter which describes his ascent of Mont Ventoux in southern France, 'moved by no other purpose than the desire to see what the great height was like' (Petrarch 1948, p. 45). It marks the point in modern times when reflecting on conscious observation of the landscape – in contrast to the pious introspection of the observer (Schama 1995, p. 450) – became possible (Ritter 1962). Since the Renaissance, ideas inspired by the Enlightenment and rationalism gradually took over the role of self-evidence in the human–nature relationship: rationally developed reclamations of forests and wetlands extended the agricultural land represented by fields farmed for subsistence. The rational attitude, characterised by taking a certain distance from nature and observing it from an anthropocentric point of view, has been at the core of modern society with all its achievements and failures and provides the opportunity (and responsibility) to govern the natural resources of the landscape with ever more drastic consequences. Because we cannot possibly administer all these consequences, at the end of the day, this rational use of the landscape almost inevitably leads to detachment and loss of identity (Latour 2012; Olwig 2008; Pedroli, Pinto-Correia and Cornish 2006; Stobbelaar and Pedroli 2011).

There is, however, another way Europeans have developed a relationship with the surrounding world, with the phenomena that today compose the landscape. All over Europe, but especially on the periphery of the continent, people erected large stones in intriguing patterns, e.g. in Malta,

Corsica, Portugal, Ireland, England, the Netherlands and Sweden. Even after Christianity had spread across Europe, people continued to worship the natural characteristics of the landscape as a source of truth, beauty and justice. In Ireland, for example, instead of impressive churches, simple high crosses were placed in the open landscape. The Celtic cross (the shape of which is much older than Christianity) includes a ring that symbolises the sun or the moon (Figure 1.2). For those living in the areas where such crosses were erected, it seems that the landscape was a sacred space, where they could worship nature as a divine creation (Pedroli 2012). Mythical stories and sagas describing the adventures of heroes in recognisable landscapes have survived for centuries in many Nordic areas such as the Edda epic in Iceland, the Saga of the dream of Olaf Åsteson in Norway and the Kalewala epic in Finland (Friberg, Landström, and Schoolfield 1998). This illustrates that, in northern Scandinavia, people's connection with the landscape was more self-evident than in Central Europe until recently, as demonstrated by the existence of many sacred localities (e.g. Ukonsaari Island, Figure 1.3).

For example, several authors (Bergman et al. 2008; Gaski 2011; Häkli 1999), especially Ingold and Kurttila (2000) and Roturier and Roué (2009), have demonstrated that the Sámi culture in northern Scandinavia has a fluid notion

FIGURE 1.2
Celtic high cross in the Irish landscape

FIGURE 1.3
The sacred island of Ukonsaari (centre), Inari lake, Lapland, Finland

of specific places in the landscape. In Sámi tradition, one has to be involved in a landscape before one can live in it as 'the work of memory, and hence people's sense of continuity with their own past, was intimately tied to their experience of inhabiting particular locales' (Ingold and Kurttila 2000, p. 187). This is in contrast to the modern perception in Western society where people and the landscape are distant from one another, as discussed earlier (Buijs, Pedroli and Luginbühl 2006; Ritter 1962). Adopting an attitude in line with that of the Sámi peoples makes it possible to learn from the landscape, reconnect to the environment and feel responsible for the actions we undertake. It is at the basis of the appreciation of places highly valued by ecotourists or by people searching for a quiet retreat away from sophisticated technological measures improving the production capacity of the landscape. Perceiving the landscape with all senses, refraining from intellectual labelling of the observed phenomena brings a sense of belonging, of identity (Wattchow 2013).

Place and space may have a different meaning for the Sámi than they do for most European people. And because the Sámi still live in a landscape in which every place may be worth taking care of depending on the season, the year, the climate, the abundance of wildlife and fish and, last but not least, the snow conditions, we can learn from their connectedness with landscape, their inner view instead of the detached consumer's view of modern society. Of course there is no point in returning to a nomadic lifestyle. Despite the achievements of modern technology, science and the dominant economic models, it seems that society has lost the ability to observe the landscape as a whole, and be aware of a sense of belonging and identity (Senge et al. 2004). Sustainable landscapes of the future will evolve only if the two mentioned attitudes towards landscape are applied in combination. This is increasingly reflected in recent developments in the concept of landscape as used in landscape research and planning.

Landscape is a complex concept with multiple meanings, the significance of which varies depending on the history of the land and the society which adopts the concept. Academic and professional disciplines are increasingly studying the meanings of the concept, and have identified ways forward to achieve responsible landscape policy, planning and management (Wylie 2007, 2013). We return to these meanings in Chapter 2.

1.3 Main Traits of European Rural Landscapes

The European landscape is very much shaped not only by the way rural settlements have evolved within their surroundings but also by public policies and state interventions. The following sections are devoted to the characteristics of Europe's rural landscapes.

European Landscape: The Land Organised around the Village

Historically, small towns and villages were central components of the rural landscape in Europe, and thus also of the organisation of the landscape as we find it today (Aalen, Whelan and Stout 2011). A powerful tradition in human culture is nucleation: people tend to gather and organise themselves in groups, and this is particularly apparent in the occupation of the European territory, from hamlets to villages and towns which have been there for centuries (Grove and Rackham 2003). Independent of population density, this agglomeration process has taken place all over the continent – in some regions of the more inland and dry Mediterranean, peasants used to live a few hours away from their fields in order to live and keep their families and social relations in small towns, often just with a small field-house in the field for staying overnight. No understanding of the European landscape can be complete without including rural–urban relationships and the influence of towns and villages on the rhythms, functions and structures of the landscape. This is the core of Chapter 5.

This rural organisation is as old as the prestigious early Christian monasteries, which performed the functions of incipient towns, as early centres with cult, market, educational and political functions (Aalen et al. 2011). Within the territory of the Roman Empire, the countryside was largely organised from the first century AD into towns, connecting routes between towns, and fields around each town, and the Roman rationalisation of the space remained many centuries later (Pitte 1983). In the south of Europe, particularly in Iberia, the Muslim influence contributed to shape economic and culturally powerful towns, but it also transformed the rural landscape. Fundamental structures were created in the countryside, centred on villages and hamlets, mainly complex and sophisticated irrigation systems, but also field organisation. Through different patterns and rhythms, settlement structure has evolved quite diversely across Europe. Still, the gathering of the population in centres, the organisation around the church, the shop, the school and the public space is at the heart of the community organisation and thus also of the organisation of the land.

Many hybrid settlement forms have also appeared through history. The single hamlet structure has in fact often been the conjunction of two or three hamlets at very short distances from each other and with a shared organisation of the surrounding fields. The form, size and functional relationships with the surroundings vary across the continent and this variation may even be significant within relatively small regions (Antrop 2004a). Traditionally, the village is typically located in the centre of relatively fertile arable land, the infield, which in turn is surrounded by more extensively used outfields (Renes 2010). Another common pattern is the location of the village on the edge between the infield and outfield, wetlands on the outfield side and well-drained arable land on the infield side (Uhlig 1961). In both cases, the outfield historically produced grass and hay for the livestock – sometimes combined with fodder from trees – while the infield produced vegetables and annual and permanent crops. In the parts of Europe where livestock was kept inside during the winter, farmyard manure could easily be transported from the stable to the arable fields (Emanuelsson 2009).

Defence sometimes played a role in the location of the village. The location of many Mediterranean villages on hilltops was for protection, although this may not have been very functional from an agricultural point of view. In Western European regions, villages were very rarely located on the coast. Even in regions where fishing was an important source of subsistence, the village was typically located a few kilometres inland as a location at the seaside was simply considered too dangerous (Aalen et al. 2011). Even when coastal areas became safe in the late Middle Ages, as was the case in the countries surrounding the Baltic Sea including Denmark, the village was moved to the coast in only a few cases (Porsmose 2008).

In general, some villages have grown into towns, while others have been abandoned; currently, some villages have been in decay for some time, while others, despite having been in decline for decades, have recovered in recent years due to the rural renaissance and urban population's quest for a new lifestyle (Aalen et al. 2011; Kovách and Kučerová 2006). Rural settlements have thus been quite dynamic through history. Still, except in more recent reclamation areas, existing hamlets, villages and small towns are not new sites as most of them are located on the sites of very ancient settlements (Pitte 1983), which implies that the organisation of the landscape around them has been in place for centuries or even millennia and bears traces of this close urban–rural relationship, which we may not always be aware of.

A Place of a Thousand Agricultures – Europe, the Continent of Cheese

When comparing European landscapes with those on other continents, the cultural history immediately comes to mind as a major factor associated with the European identity reflected in its landscapes. Much more than in the so-called New World (the United States, Canada, Australia, New Zealand), European landscapes represent deep and continuous cultural histories. Already in early prehistory, man substantially changed his environment, e.g. by introducing controlled grazing, arable cropping and iron ore extraction. Especially the Roman culture made large changes to the landscape by introducing standard parcel systems, (straight) military roads, deforestation, new cropping methods and agricultural estates. Many of these changes can be found as layers in the European landscape today.

Notwithstanding the intricate cultural history of Europe, which gives its landscapes a special character, landscape was of course also changed by man on other continents. Native Americans in the Amazon basin define to a large extent the species composition of the tropical forest and this is their landscape (Kaplan et al. 2011). Although there is ample evidence for the impact of prehistoric humans on landscape from various archaeological accounts (Goudie 2013), not much is known about the influence of man on the prehistoric landscapes of Europe as a whole. But what is European in European history is that the diversity, grounded in an extreme biophysical diversity, survived to a large extent, e.g. in languages, local markets, local regulations, local nations. Europe has grown into 'a place of a thousand agricultures' (Ventura et al. 2008, p. 149). Indeed, after agriculture started developing in the Middle East and Asia Minor long before our era (see e.g. Grigg 1974), gradually also the entire map of European landscapes was defined by the influence of agricultural civilisations, and by the associated settlement forms.

Subsequent periods in history saw many new influences on the landscape: large landowners (kings, bishops, landlords) and farmers' communities always interacted (Pounds 1990). From the Middle Ages until the nineteenth century, monks – and the rulers of their communities – have contributed to European landscape developments by reclaiming and working the land in almost every corner of Europe (Duby 1961). Wars, diseases, famines, migrations and changing climate have further added to the variation in the land use, and thus the landscapes. Notwithstanding considerable balancing of the resulting differentiation in landscapes in the past two centuries (Jepsen et al. 2015), today's Europe is still largely characterised by landscapes that represent palimpsests of a large number of landscape imprints of several historical layers (Palang, Spek and Stenseke 2011). In many other parts of the world, comparable palimpsests can be recognised in the landscape, e.g. in Japan, eastern China and the Middle East, but the scale and extent to which this is the case in Europe is likely to be unique (Stanners and Bordeaux 1995; Vos and Meekes 1999).

If we try to generalise the characteristics of today's diverse European landscape, it can perhaps be symbolised by its ensemble of walls, trees, cows and fields (Pinto-Correia and Vos 2004). Indeed, permanent grassland and the associated grazing livestock constitute unique features of the historical European rural landscape, as Yves Luginbühl convincingly describes it (Luginbühl 2012, pp. 43–46), which is typically based on the arrangement of farmsteads into villages. Europe can be characterised as the continent of cheese! And to produce cheese, grazing land, cropland, scattered trees and woodland were indispensable almost everywhere. The typical farm type to manage this array of functions was the mixed family farm, which dominated European agriculture for several centuries (Antrop 2000; Gasson and Errington 1993). Many variations of this mixed family farm system still exist. Practically all of them struggle to survive today (Van der Ploeg 2009a), but they still contribute to the diversity of the European landscape. This is as valid for nomadic grazing on tundra in the north, in mountain areas all over Europe, as much as for wine and olive oil production in the Mediterranean (Grove and Rackham 2003), and all landscapes in between.

A Landscape Shaped by Public Interventions

The small-scale and dynamic landscape pattern – often in an inherited feudal landownership system – characteristic for the continent of cheese, proved not always able to meet the demands of society for increased and efficient food and feed production. Still, it is surprising how a steadily decreasing community of farmers through history has been able to feed an urban population, which has grown dramatically during the past two centuries (McCarthy and Danta 2014). An important explanatory factor in this development has been public interventions of various kinds in the technological, political and market developments of agricultural production. Therefore, in the past two centuries, increasingly large public interventions in the management of farmland have been taken to change and adapt the land-use system to new demands. We can list a few examples:

- In Denmark, large-scale state agricultural transformations took place from the late eighteenth century onwards. First the feudal system was transformed by extensive land reforms where the property structure and overall structure of the parishes were transformed throughout most of the country. New crops and rotational systems were introduced (Kjaergaard and Hohnen 1994). Next, in the late nineteenth century, agriculture changed fundamentally in response to market changes due to cheap American grain imports, from grain and beef cattle production to dairy production organised through several thousand local cooperatives, supported by public measures and incentives. Most of the present-day rural landscape patterns can be traced back to these transformations (Fritzbøger 1998), as can the overall structure of Danish agriculture.
- In Italy, Mussolini and the Fascist movement drained many wetlands to combat malaria and open up new land for agricultural production. During the same period in the Netherlands, wetlands were reclaimed through state-initiated efforts (Renes and Piastra 2011).
- The collectivisations in socialist countries between the 1940s and the 1980s had huge impacts, when complete rural settlements were removed and rebuilt elsewhere, and landscapes were opened up for large-scale collective agricultural enterprises, especially in East Germany and Romania, but also in many areas of the Baltics, Czechoslovakia and Hungary (Blacksell 2010).
- In Portugal, the Salazar dictatorship implemented the wheat policy 'Campanhas do Trigo', with large-scale monocultures of cereals, which had devastating effects on the ecology of the rural landscapes of the Alentejo (Pinto-Correia and Vos 2004). Like in Italy in the early twentieth century, this policy appeared to be a failure (Saraiva 2010), but the impacts on landscape were largely irreversible.
- In the Netherlands, land reform was undertaken with a major impetus after World War II in which Minister of Agriculture (1945–1958) Sicco Mansholt had a substantial policy stake (Grin 2012). Land re-allotments all over the country secured larger fields, better drainage and improved accessibility and living conditions for the farmers, which led to a substantial increase in agricultural production (Van den Bergh 2004) [see Case D – Land re-allotment in Chapter 4]. Later – as the Agricultural Commissioner of the European Community (1958–1972) – Mansholt contributed to the reform of agricultural policies in the entire European Community. Being the key architect of the Common Agricultural Policy (CAP), he promoted a successful policy primarily aimed at achieving self-sufficiency in food production for the European Union (EU).

As is clear from these examples, political leadership has often played a key role in the land reforms promoted by the state, and the subsequent major changes in agricultural land use. In many cases, this has taken place without sufficient awareness of the consequences for the people affected or for the natural resources in use. The role of specific leaders is becoming decreasingly relevant, but still public policies, mainly sectoral, remain a fundamental driver of the European landscapes, as we discuss in Chapter 6.

Today the most important policy that determines the functions and management of the rural landscape is the EU's Common Agricultural Policy (CAP). The CAP is such a complex policy system that it can hardly be attributed to individual politicians, except for the naming of some reforms after the responsible Agricultural Commissioner, e.g. the McSharry and the Fischler reforms of 1992 and 2003, respectively (Cunha and Swinbank 2011). However, the impact of the CAP on the structure of rural land use in an increasing number of Member States (and indirectly also outside the EU) is undeniably substantial, although scientific evidence for this relationship is relatively scarce (Frederiksen and Vesterager 2013). Stimulating measures on forestation, the conversion of grazing land to cropland, scale enlargement etc. have led to important landscape changes across Europe, but especially in the Mediterranean (Serra, Pons and Saurí 2008) and in mountainous areas (MacDonald et al. 2000).

Changing Trends in History

It is in the nature of landscape that it is continuously evolving, whether brought about intentionally by human action or by external circumstances. In history, there have been periods of relative stability and periods of considerable upheaval, even to the extent of catastrophes, and the extent of the changes has varied from place to place (Duby 1961; Pounds 1990; Roberts 2013).

A general overview of the history of the European landscape is summarised in Table 1.1. The developments summarised should be read as generalisations and many variations exist in various parts of Europe. For example, the development of agricultural land management regimes and urbanisation over the past 250 years can be differentiated over the various parts of Europe (Mediterranean, Eastern Europe etc. (Jepsen et al. 2015)), which is further discussed in Chapters 4 and 5. All these large-scale trends have significantly affected the occupation of the European countryside and shaped its landscape in several layers of structures and elements.

If we focus on the past 100 years, more detail is available and many examples of gradually changing functions of landscape leading to landscape change have been documented. One of these has already been discussed at the start of this chapter: land consolidation in the Netherlands. The most significant types of change in the European landscape are described in the following paragraphs.

TABLE 1.1 *Phases in European landscape history (inspired by Duby (1961), Pounds (1990) and others)*

phase	developments influencing landscape	landscape
pre-antiquity	development of agro-silvo-pastoralism in the Mediterranean; nomadism elsewhere	open in some patches around settlements in the Mediterranean, closed elsewhere
Hellenic times	development of the city state, depending on crop yields from the rural; development of trade in basic food products	circum-Mediterranean opening up of landscapes
Roman times	first reclamations, deforestation, introduction of new crops and breeds in Roman Empire; nomadism and some basic agriculture elsewhere; transport of enormous volumes of food and commodities, especially overseas	opening up of the landscape in Roman Empire, annual crops, pastures and patches of Mediterranean permanent crops (olives, wine, fruit)
medieval period	decline in agriculture; regrowth of forests; islands of horticulture and fruticulture around castles and monasteries; nomadism or transhumance elsewhere	re-establishment of forests, relative closing of the landscape
late medieval period	three-field rotation → development of city culture: large-scale reclamation of forests and wetlands by monasteries all over Europe and agriculture producing surplus products; construction of a large number of new, walled cities	opening up of the landscape by farming, everywhere except northern Scandinavia
early Renaissance	disease and wars: regrowth of forests; serious population decline; decline of agriculture	re-establishment of forests, but many islands of cultural farmed landscape remain
late Renaissance early modern	gradual improvement of agricultural methods and breeds; specialisation of certain areas in specific products; increasing trade in agricultural products; new reclamations of wetlands	gradual opening up of the landscape; increasing specialisation and land-use differentiation

TABLE 1.1 (*Cont.*)

phase	developments influencing landscape	landscape
Industrial Revolution	land reforms: the end of feudalism in many European countries; large-scale deforestation for fuel purposes in industry; overgrazing because of competition with Australia over wool	probably most open landscape in history, until 1950s; diversified at the farm level though with clear regional specialisation
late nineteenth–early twentieth centuries	land reforms; introduction of new crop varieties and exotic tree species; land consolidation works	restructuring of land pattern and landownership; increased access to the land
late twentieth century	large population growth; large-scale mechanisation, intensification and specialisation of agriculture under globalising world market; increasing global transports of food and feed; abandonment of marginal lands	larger parcels, fewer small landscape elements and landscape simplification; increasing land abandonment in marginal areas
early twenty-first century	increasing dominance of European policies (especially CAP); increasing differentiation of the rural space, with commercial farming, extensive land uses, forestry and small-scale, multifunctional farming; increasing importance of biofuel	differentiation between open high-production industrial landscapes on one hand, and extensively used grazing landscapes and small-scale multifunctional, including peri-urban, landscapes on the other

Proportion of Inhabitants in Urban Areas vs Rural Areas 1950–2030

The process of urbanisation in Europe is a relatively recent phenomenon, stimulated by the Industrial Revolution and many accompanying societal changes. It is estimated that the degree of urbanisation in the world was only 1.6 per cent at around AD 1600 and 2.2 per cent at the beginning of the nineteenth century, while it fluctuated between 4 per cent and 7 per cent in the mid-nineteenth century (Antrop 2004a, citing United Nations Centre for Human Settlement, 1996 and 2001). However, during the twentieth century, people also moved in the other direction – from urban areas into the rural landscapes. Such counter-urbanisation processes are described in Chapters 3 and 5. Today, in most industrialised regions of Europe, the degree of urbanisation exceeds 80 per cent (Figure 1.4). Although population growth in these developed countries is generally decreasing (Figure 1.5), most cities and towns are still growing slightly, while the rural population is declining rapidly. The rural population is expected to decrease by 1.5 per cent annually in these more developed regions (Frey and Zimmer 2001).

The current tendency of population decline, especially in northern and eastern European regions, but also in several Mediterranean regions, and population growth in the prosperous regions of Central Europe (Figure 1.5b) indicate that associated changes can be expected in the landscapes. However, hidden behind these figures are the local changes in population density between the rural and the urban, as illustrated for the Netherlands in the years to come in Figure 1.5a. Nevertheless, large variations exist across Europe. For example, England is distinctive in that the rural population has been increasing for the past 60 years, with the increase being linked to a lifelong trajectory, which involves young people moving to cities for further and higher education, where they remain until creating a family and moving to the suburbs for high-quality secondary education, and finally later moving to market towns and the countryside or coastal locations (Lowe and Ward 2007).

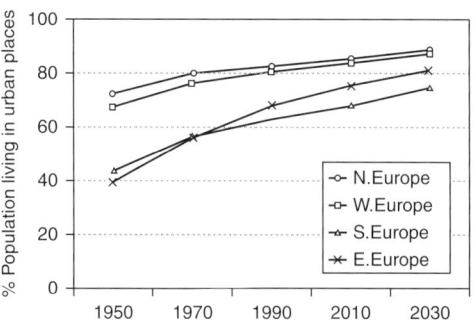

FIGURE 1.4
Evolution of the level of urbanisation in the main European regions between 1950 and 2030 (Antrop (2004a), after United Nations Center for Human Settlement, 1996 and 2001).

Farmers (i.e. land managers, rather than landowners), who have been and still are the main producers in the European landscapes, are gradually adapting their farming styles to the demands of a changing society and farming systems are changing accordingly (Van der Ploeg and Marsden 2008). Although farmers tend to be conservative – the introduction of a new crop even when subsidised may take about 20 years (Alexander et al. 2013) – these changes are definitely very substantial in the longer term.

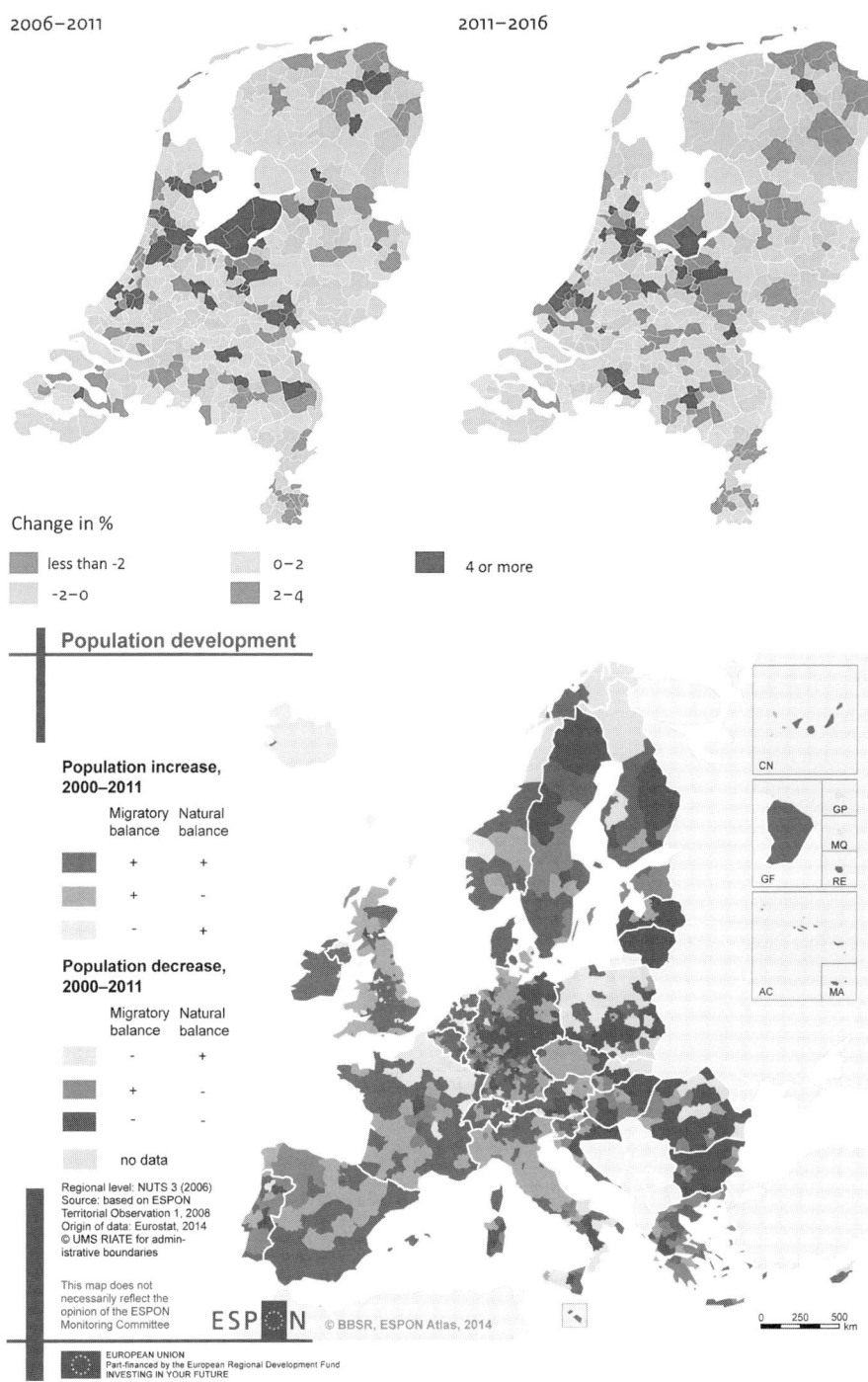

FIGURE 1.5
a. Population development per municipality in the Netherlands 2006–2016. Source CBS: www.clo.nl/nl210206. b. Population development EU 2000–2010. Source: www.Mapfinder.ESPON.eu. A black-and-white version of this figure will appear in some formats. For the colour version, please refer to the plate section.

Greening of Europe

Europe is becoming greener. However, this does not necessarily mean that agricultural practices are becoming significantly more environmentally friendly. Many marginal farming areas have been abandoned by agriculture, resulting in more natural or semi-natural vegetation. In many areas, the past 100 years has seen periods of widespread abandonment of marginal agricultural land and a gradual, but substantial increase in forest land (Fuchs et al. 2014). The transformation of traditional agricultural landscapes has had consequences for biodiversity and ecosystem services. On the one hand, the abandonment of farmland can be considered an opportunity to rewild ecosystems, particularly in highly fragmented landscapes where it could provide significant large-scale restoration of non-agricultural habitats. Many often iconic species will benefit from rewilding and natural habitat regeneration, which will provide associated ecosystem services such as carbon sequestration and opportunities for recreation and tourism development. Rewilding also provides space for ecological processes on the ecosystem scale. On the other hand, rewilding may decrease the area of semi-natural habitats that have traditionally been managed by low-intensity agricultural use (such as grazing) and its associated species of nature conservation importance, many of which are concentrated in Natura 2000 sites and high-nature-value (HNV) farmland. Once specific thresholds regarding available habitat are surpassed, biodiversity may rapidly decline. Abandonment of farmland may also result in declining aesthetic values of the traditional agricultural landscape with its varying plots and field boundaries, and the disappearance of its associated products such as traditional fruit varieties. Such decline reduces the opportunities for recreation and tourism development.

Increase in Forest Land

The area of European forests increased by about 4 million hectares between 1960 and 1990 due to afforestation of treeless and poorly forested lands, fields and pastures, which more than compensated for the conversion of forest into other land uses (Kuusela 1994). In fact, when considering the category of *other wooded land*, the increase in forest land in Europe is still considerably larger (see Figure 1.6). Forest harvesting is much more variable, as is illustrated for changing forest parameters in Europe in Figure 1.7.

Historic Land Changes 1950–2010

Fuchs and colleagues (2013) investigated the area of land affected at least once by land changes during the period 1950–2010, which is almost 14 per cent of the total area of all EU-27 states plus Switzerland (Figure 1.8). Every year, 0.26 per cent of the entire area is converted from one land use to another, an area similar to the size of Northern Ireland.

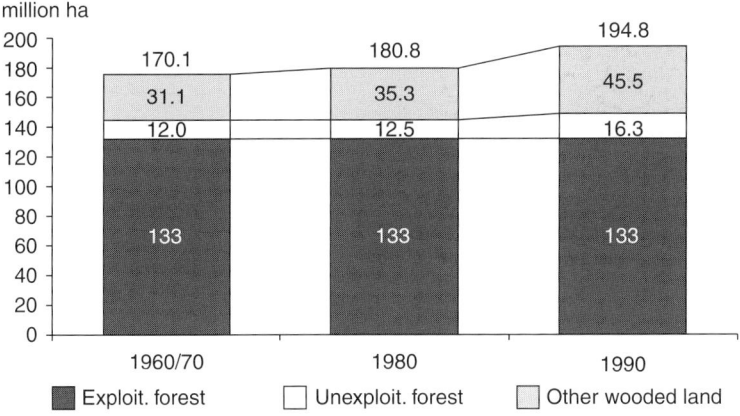

FIGURE 1.6
Estimates of exploitable and unexploitable forest and other wooded land in Europe
Source: Kuusela (1994)

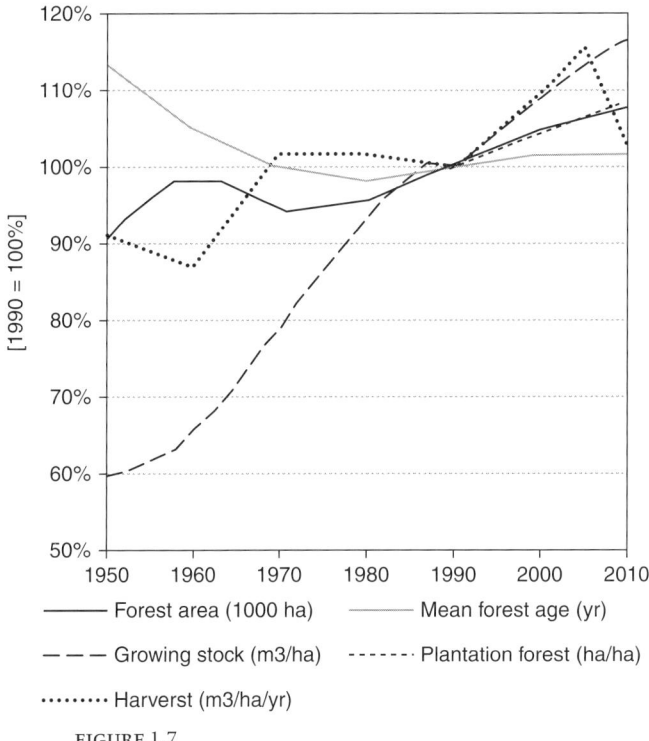

FIGURE 1.7
Area changes and management intensity changes in Europe's forests
(Kuemmerle et al. 2013)

Increasing Agricultural Production in a Decreasing Area

Figure 1.9 illustrates the change in cereal production, yield and cropped area and trends in fertiliser application and use of agricultural machinery in the EU-27 between 1961 and 2009 (Rounsevell et al. 2012). Although the total area

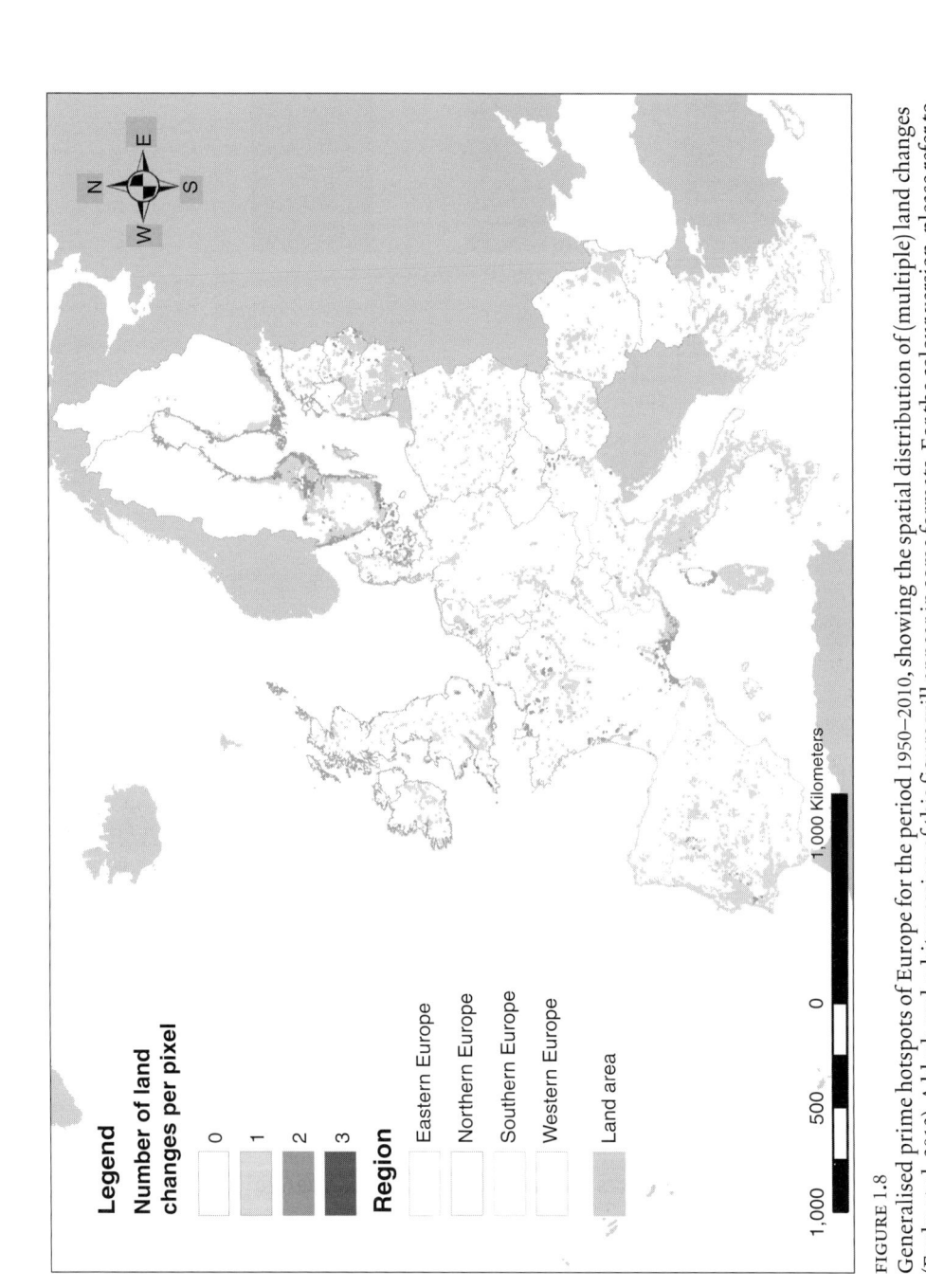

FIGURE 1.8
Generalised prime hotspots of Europe for the period 1950–2010, showing the spatial distribution of (multiple) land changes (Fuchs et al. 2013). A black-and-white version of this figure will appear in some formats. For the colour version, please refer to the plate section.

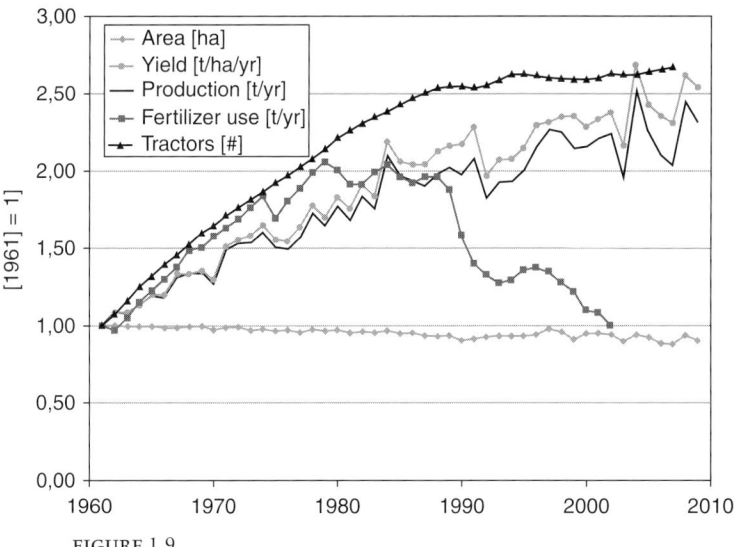

FIGURE 1.9
Land-use change: area change vs. cereal production in the EU-27 countries between 1960 and 2010 (Rounsevell et al. 2012)

under cereal cultivation decreased by about 10 per cent during the period, cereal production more than doubled due to a comparable increase in yields. Thus, production increases cannot be explained on the basis of simple relationships between agricultural area and output. Neither is the relationship between input and output a straightforward one. Understanding the mechanisms that underlie these dynamics in different contexts requires more insights.

This overview illustrates how a highly diverse set of large-scale drivers has influenced the development of the countryside in Europe, finally resulting in the heterogeneous landscape we know today. This heterogeneity depends not only on a mosaic of biophysical conditions but also multiple layers of human action and interaction, which have had an influence, one after the other, or even simultaneously, from the large scale to the village and the small farm plot. These factors or drivers have had an influence on the landscape on multiple spatial and temporal scales.

Current Developments

Figure 1.10 illustrates hotspots of land cover change for a number of change processes in the rural area between 1990 and 2006 (Kuemmerle et al. 2013): 'In terms of conversions among broad land use types, the most widespread land use change between 1990 and 2006 was cropland decline (~130,400 km²) [Figure 1.10a], followed by forest area expansion (~70,300 km²) [Figure 1.10e], and pasture area increase (~66,300 km²) [Figure 1.10d], whereas the least common conversions among broad land-use categories were permanent cropland decline

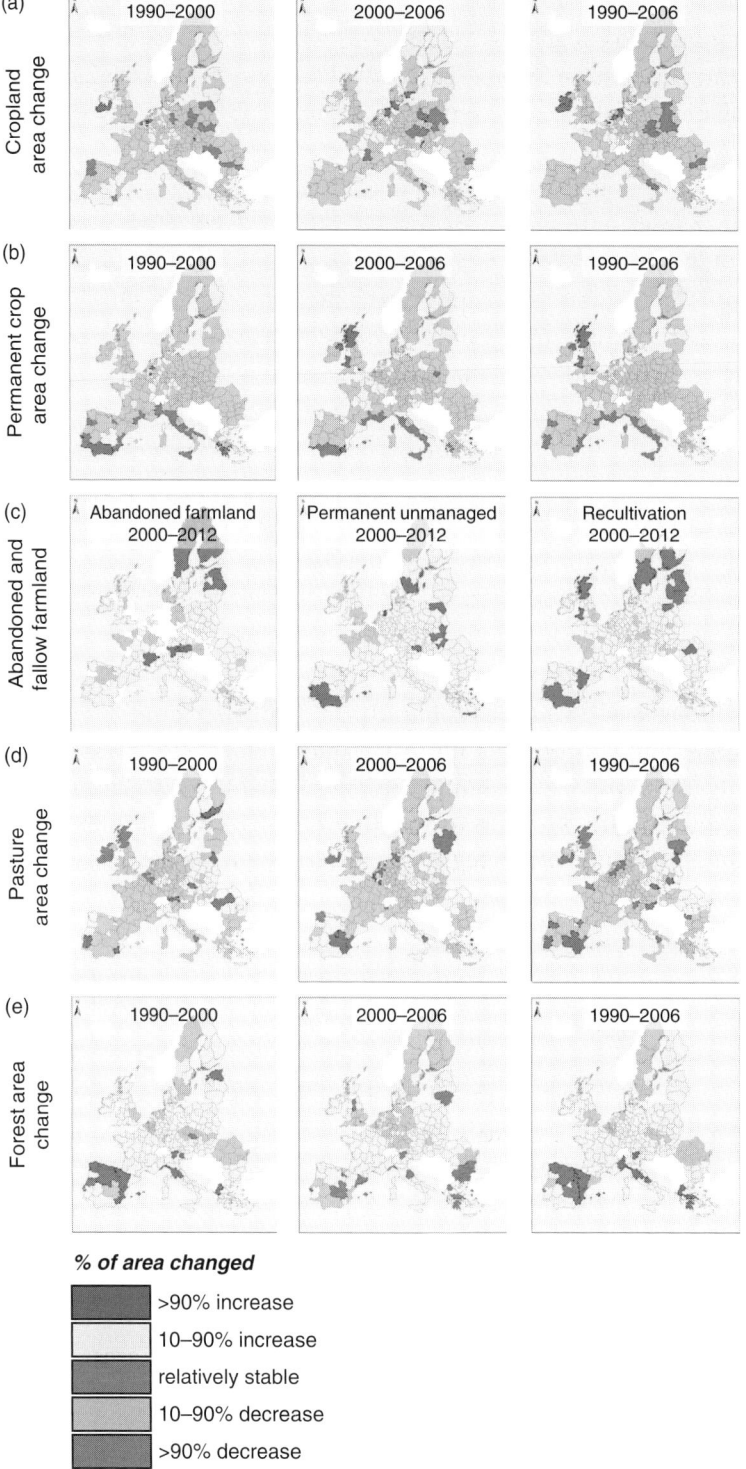

FIGURE 1.10
Hotspots of area changes in terms of broad land-use categories at the NUTS-2 level based on a range of data sources (Kuemmerle et al. 2013). Dark grey areas indicate NUTS-2 regions with area decrease, whereas light grey represents increasing area trends. A black-and-white version of this figure will appear in some formats. For the colour version, please refer to the plate section.

(~17,000 km^2) and urban expansion (~45,700 km^2).' Hotspots of cropland decline are mainly located in Eastern Europe (e.g. eastern and southern Poland, central Romania, Bulgaria) and the Mediterranean (e.g. northern Spain, southern Italy; Figure 1.10a), with annual rates of up to 1.04 per cent.

To prevent irreversible loss of landscape values, a continuous and increasingly dynamic change in time and space calls for strong anchor points of more stable land use (Antrop 1997). From a public policy perspective, it is certainly a challenge for land use and landscape management to identify which places deserve to be managed as anchor points because they have a distinctive and unique character. Such landscapes may be, e.g., visually attractive and ecologically resilient or they may have intrinsic historical values (e.g. battlefields, landscapes with a particular reclamation history).

What is also observed is the increasing attractiveness of certain landscapes for functions other than production. In particular, the landscapes that result from traditional farming and farming structures have become less competitive in terms of production, but more attractive in terms of amenities. Nevertheless, when landscapes become subject to 'consumption' of their amenity values (see Chapter 3), former land uses are often not maintained, other uses evolve and the landscapes may lose the character that made them attractive. This is the countryside consumption dilemma as described by many Anglo-Saxon geographers (see e.g. Marsden 1999). Although such developments are analysed and discussed widely in the literature, as we will show throughout this book, more coherent visions and regional strategies are generally missing (Wascher and Pedroli 2008).

1.4 Typical European Landscape Contexts

Polarising Trends in the Rural

An increasing polarisation can be observed in today's European landscape. New highly dynamic multifunctional landscapes are developing in peri-urban zones: urban sprawl in areas with poor spatial planning control is leading to an increase in the small-scale diversity of landscapes, although on a larger scale, whole regions then tend to be characterised by a significant degree of homogeneity regarding pattern and elements. Also, small-scale, high-value farming initiatives, which often produce organic products, are developing in the neighbourhood of larger conurbations, or even in urban areas in the form of urban agriculture, while former estates are getting new functions as conference centres, hotels or high-value residences. Brown fields and former industrial areas are being redeveloped into residential areas, and community initiatives are leading to new identities for local areas, including the branding of local products. Flood-prone areas are increasingly being adapted to cope with flooding, including the abolition of intensive agricultural use, the establishment of nature reserves and green open space, new housing solutions such as floating houses etc. (Mees, Driessen and Runhaar 2014).

On the other hand, over large areas, a decrease in diversity and a loss of character and identity is being reported (Antrop 2004b; Van Eetvelde and Antrop 2009). For example, land re-allotment schemes in the Netherlands have led to the disappearance of many characteristic landscape elements, buildings and parcel patterns, while peat bogs, peatlands and wetlands or land with a high groundwater table have been drained and converted into new agricultural production areas, which are often very similar to the surrounding landscape. The abandonment of land has also had a major impact on landscape character as it leads to overgrowing and the decay of historical landscape patterns, dry stone walls, terrace walls, hedges, irrigation and drainage systems, historic routes etc. Another tendency of decreasing landscape character is the construction of second houses and tourist resorts, the designers of which tend to give them specific characteristics of the local landscape, but generally introduce elements that can be found everywhere: playgrounds, lawns, golf courses, swimming pools etc. Also, the establishment of international chains of restaurants or supermarkets on the outskirts of settlements or near highways often adds to the anonymity of the landscape. Last but not least, the continuation – or acceleration – of the industrialisation of agriculture seems to be neglecting the heritage values embedded in the countryside, and this is particularly evident today in the former communist countries of Eastern Europe.

Interestingly, these examples seem to indicate that for different rural landscapes, the processes that shape landscape very often have similar results: decreasing maintenance of characteristic elements, clogging the landscape with buildings and plant material that could be anywhere and the introduction of design principles alien to local traditions. Perhaps new 'local traditions' are no longer evolving as coherent developments, and if they are, the motives are different, as are the actors and their orientations. Although localism is high on the agenda – especially from the point of view of decentralisation – and local traditions are being rediscovered in many places, the key word 'local' is being reframed as well: local people, local technology and local regulations *sensu strictu* are just disappearing. Moreover, this seems too often to go hand in hand with increased accessibility of a limited part of the landscape, whereas many country roads, dirt roads and transhumance pathways are suffering from a lack of maintenance, thereby hampering accessibility. Within one region, one can often observe areas that tend to develop into remoteness and anonymity, while others are being upgraded to international modernity, which also implies anonymity. And these trends tend to be intensified, as decaying areas are unattractive to newcomers, while central and dynamic areas can attract newcomers, who settle and may promote innovation and value creation in farming and other sectors.

Five Typical Landscape Contexts

As a reference for the discussion in this book, we present five typical European landscape contexts in terms of agricultural characteristics and dynamics (summarised in Table 1.2), community structure and interactions and the support of multiple societal functions. These are, of course, generalisations and in each specific landscape, aspects of different types will appear in unique constellations. However, the characterisations do emphasise the contrasting rural contexts of today's Europe. To develop visions for the future of specific landscapes will surely require different approaches to develop the quality of the landscape and the related well-being of its inhabitants.

TABLE 1.2 *Summary of the features of five characteristic European landscape contexts*

five characteristic landscape contexts	land use intensity	agricultural specialisation	orientation to local markets
highly productive agricultural landscape	high	high	low
nature, heritage and tourist landscape	low	low	intermediate
peri-urban dynamic multifunctional landscape	very high	very high	high
rural multifunctional slow landscape	high	very high	very high
remote solitude landscape	very low	very low	high

1 Highly Productive Agricultural Landscapes

In large parts of Europe, favourable agricultural conditions – naturally and socio-economically – have resulted in very intensively farmed regions characterised by high input/output levels. The fundamental driver of the landscape management and changes here is agri-food production. Examples are the recent polder areas in the Netherlands (see Figure 1.11), large-scale cereal production in northern France, Poland and Romania and the large-scale irrigated systems in southern Iberia.

Agricultural production in these areas is linked to globalised food networks. The landscape dynamics are therefore intimately connected – concerning both inputs and outputs – to more or less globalised flows and market fluctuations, resulting in landscape transitions (Baudry et al. 2010). Production often takes place in highly manipulated landscapes with large, regular field patterns, sophisticated drainage systems and (in some regions) irrigation systems. Some of the production systems are mainly cash crops; others are characterised by

FIGURE 1.11
Highly productive agricultural polder area in the densely populated Randstad Holland

animal husbandry or mixed systems. The environmental impacts of agriculture include the leaching of nutrients, pollution by pesticides, soil erosion and the removal or fragmentation of natural habitats. These productive landscapes are often highly dynamic, with a significant share of the active population involved in farming and agri-food-related businesses, although production has grown disproportionately through investment in machinery and innovation, still leaving relatively few people employed. Population density depends on the location and proximity of urban centres which provide jobs for other sectors, or create opportunities for other economic activities. In any case, tourism is rarely important in these areas, and the residential attractiveness for urban dwellers is limited. Ecological deterioration and social disintegration (with the immigration of farm labourers and the outmigration of residents with urban incomes) may represent mid- and long-term risks to landscape sustainability in such areas (see Buttimer 2001).

The future of such intensively farmed landscapes will most probably remain closely linked to specialised farming. From an overall resource conservation point of view, food production will naturally be more intensive in this type of landscape than in more marginal areas. However, the supply of (clean) drinking water (for human consumption as well as for livestock), a basic ecological infrastructure including habitat networks, soil conservation and recreational opportunities will probably be necessary to meet local and regional societal demands. Ecologically and socially well-functioning intensively farmed landscapes require overall designs that include regional infrastructure: green and blue networks, agricultural infrastructure, transportation, a renewable energy supply (biogas, wind energy) and residential areas (towns and villages). On the local scale, property and field patterns, drainage systems, habitat networks, recreational access and cultural heritage assets should be integrated into landscape patterns which allow for intensive agricultural production, while ensuring other basic land uses and landscape features.

2 Heritage and Tourist Landscapes

These landscapes are characterised by an intricate alternation of well-managed and branded elements which form an integral part of the local and regional identity. They are mainly 'consumption landscapes' for the benefit of either tourists or wealthy rural residents and retirees. Agriculture has, in some cases, ceased to be of economic significance, while, in others, it supports the local identity and produces specialised small-scale products valued by newly emerging markets. In such cases, the subsistence and well-being of the community and the continued attraction of tourists depend on the landscape. The community is partly composed of local people, but also many newcomers have

FIGURE 1.12
Tuscan estate
(photo: Bas Pedroli)

moved to the area due to its attractiveness or for the opportunities it offers for locally based businesses. Landscape management may be driven by preservation and aesthetic motivations and also by an acknowledgement of the landscape as a resource for the activities developed in the area. In many such landscapes, the level of income derived from hunting, horse riding or other lifestyle activities is approaching the earnings from traditional agricultural production. In a broader sense, this type of landscape reflects typical European conceptions such as sense of place, territorial cohesion, slow food and the notion that cultural diversity is one of the key values of the European continent that should be cherished. Examples of heritage and tourist landscapes include parts of Tuscany (IT), Kent (UK), Banská Štiavnica (SK), Crete (EL) and the Peak District (UK).

This type of landscape requires a focussed strategy to conserve its natural character and cultural heritage in a countryside that is under increasing pressure from consumption. Gentrification and the separation of agriculture and land management threaten a balanced landscape development reflecting heritage in a recognisable way. If there is no valued agricultural production, incentives may be needed to enhance such conservation, such as tax alleviation to compensate for the cost of managing valuable landscape elements, but also the joint branding of regional products promoted by the communities represented in these landscapes.

3 Peri-urban Dynamic Landscapes

These are landscapes which are under multiple pressures and are, thus, also highly dynamic. They are both production and consumption landscapes and are characterised by peri-urban, multifunctional land use with a dynamic mix of residential areas, outdoor tourism, specialised food production, farm shops, restaurants, drinking water extraction, care farming, bed-and-breakfasts etc. Therefore, they represent major urban-initiated functions which are highly regulated by local markets, but within the framework of the global mobility of goods and people. These landscapes represent the rural 'counter-picture' to the urban areas and exhibit comparable dynamics in landownership and land use. They are valued environments for the establishment of urban settlements and businesses on former farmland and farm structures. Examples of such landscapes can be found in all peri-urban fringes around large European cities, particularly in central and western regions.

Peri-urban landscapes are attractive to businesses and there are many service enterprises, small and highly technological industries and also small-scale farms evolving. The proximity of the urban market is important for the continued existence of these businesses. Short supply chains, farmers' markets and community-supported agriculture have developed in recent years. This type of landscape – like highly productive agricultural landscapes – may evolve

FIGURE 1.13
Peri-urban recreation landscape between The Hague and Rotterdam, the Netherlands
(photo: Bas Pedroli)

relatively independently, continuing the trend of recent decades towards dynamic land use on urban fringe areas. This may be enhanced by policy incentives for small-scale entrepreneurship serving urban demands that cannot be fulfilled in the cities, including leisure, healthcare in a non-urban environment and local – often organic or high-quality – food production.

The consequences of such development are fragmentation of the landscape pattern and the gradual disintegration of the cultural heritage in many places, while in others strong conservation of cultural heritage attracts urban inhabitants who are seeking a quiescent connection with the historical development of the landscape.

4 Rural Multifunctional Slow Landscapes

These landscapes are characterised by a high diversity of goods and services, land use and societal functions, which are still linked to traditional forms of agriculture. Conceptually, such landscapes can be placed in the centre of an interplay between production, consumption and protection drivers. They are different from the peri-urban landscapes described previously as they are still clearly rural, are located away from cities and have a low population density. Nevertheless, rural multifunctional landscapes are attractive to those who are

FIGURE 1.14
Hilly slopes with olive groves, overseeing the Lago Trasimeno, Umbria, Italy (photo: Bas Pedroli)

searching for a slow and locally based lifestyle and they are the site of increasingly innovative approaches to living, working and recreation that focus on the search for the enhanced self-sufficiency of local communities. These landscapes are home to a community of local inhabitants and some newcomers, in agriculture or in other activities. The land use integrates multiple functions, while diverse ecosystem services are generated everywhere. In some of these landscapes, a particular product is related to global markets, while all of them exhibit some potential for the production of local products or for increased tourism, particularly rural tourism and ecotourism. Examples of such landscapes are parts of Umbria and Marche (IT), some parts of the Carpathian wine region near Bratislava (SK), summer farming in the Norwegian mountains, north-western Portugal (Minho, Douro) and western Alentejo in south Portugal.

This type of landscape will probably vanish without the introduction of incentives to maintain it. There is considerable interest in these landscapes from a nature conservation perspective and from a select group of tourists and urban inhabitants, e.g. related to the production of slow food, and this may create new business opportunities and new jobs. However, in the longer term, these landscapes require support to continue the type of farming which maintains the character of the landscape as this may not be entirely compensated for by the market. In any case, state intervention is necessary to avoid changes which will definitively alter the character of the landscape. In times of decreasing economic prosperity, a counter-urbanisation tendency can be observed which replenishes these rural multifunctional landscapes, but in more favourable economic times, they may be neglected. Technology facilitates the sustainable management of natural resources.

5 Remote Solitude Landscapes

These landscapes are found in regions with very low-intensity agriculture, if it is still present, and high biodiversity values. In some cases, there is a separation in land uses, where the extension of nature areas needed for global biodiversity conservation has priority over short-term agricultural and forestry land-use benefits. In others, low-intensity farming remains, which provides the HNV land use. In this type of landscape, nature is often considered an autonomous asset of high intrinsic value, and it is linked to strong public interest. One considerable and perennial problem in these landscapes is their lack of people. Economic activity is limited and population density, therefore, has always been low, and there is a tendency for it to decrease. Revitalisation of the countryside may well take place in these landscapes because of a renewed focus on rural prosperity and new farming-nature tourism initiatives. Examples of these landscapes are found in the more remote parts of the Apennines (IT),

FIGURE 1.15
Abruzzi Apennines in Italy
(photo: Bas Pedroli)

the Scottish Highlands (UK), Sámiland (NO, SE, FI), south-eastern Portugal (Barrancos, Moura, Mértola) and large parts of Iceland.

Remote solitude landscapes can be found today in various parts of Europe: 2 per cent of Europe's surface area is considered wilderness, while 18 per cent is designated as Natura 2000 areas, and even if many Natura 2000 areas may well concern the previous landscape type (4), others belong to this type. Significant investment in the acquisition of abandoned lands may be a way of extending the wilderness areas; something various rewilding initiatives have recently promoted (Jepson 2015; Navarro and Pereira 2012). Also, buffer zones around the wilderness areas need to be established not only to prevent the undesired roaming of animal species (wolf, bear) but also to reduce the vulnerability to diseases and halt the invasion of non-native species. Protection of the living environment for local communities is a vital issue in these landscapes, including regulating accessibility for people so that they can experience the magnificent nature reserves.

The consequences of such development are that heritage values related to cultural landscapes will suffer due to the prioritisation of large wilderness areas, and that large distances will often have to be travelled to enjoy the beauty of the wilderness areas (or rather of their buffer zones). On an international level, climate change will force pragmatic approaches to other ecosystem service delivery as well: for example, a shift in forest ecosystems in the south to less drought-prone areas in the north.

The reader may recognise the landscape types presented here in specific landscapes. Other known landscapes are likely to exist as combinations of these types. The schematic types described are meant to illustrate that the constraints and opportunities in the different rural landscapes of Europe are highly diverse. A steadily less differentiated rural landscape at the local level seems to be developing and we risk losing some of the characteristics we have been so fond of all over Europe. We are facing specific challenges in each particular location. To make the picture still more complex (and more interesting!), these challenges are also rooted in the concept of landscape itself, as we explain in the next section.

1.5 Landscape as a Complex Notion

Five Inherent Tensions in the Comprehension of Landscape

So far in this chapter, we have shown how the landscape is important in the everyday life of European inhabitants, as well as in the expression of symbolic values. We have also shown how the landscape is an integrative concept, powerful for the integration of different activities and interests, which now play a role in rural Europe. Nevertheless, it is also a concept full of contradictions and tensions that need to be made explicit. In the following, we list five of these tensions which we consider crucial for defining how to use the landscape concept and approach for the future well-being of all those involved.

1 Landscape Has no Boundaries, but Management Arrangements Depend on Well-Defined Ones

Landscape has no single, fixed boundary. Instead, it may be useful to think of it as 'delineated' by a horizon (Casey 2001, p. 690). This horizon – which corresponds with the definition of landscape as 'an area as perceived by people' of the European Landscape Convention (ELC) (COE 2000) – can be understood in different ways: how the landscape is seen visually, how it is understood in a socially inclusive way and to what extent potential changes may be envisioned. However, land management on different scales, from the plot and the farm to the region, is based on clear boundaries and guided by regulations of who has the right to do what. How can such land-use and regulation rights be combined with open horizons on potential futures? This may be one of the reasons that Germany and Austria have not yet ratified the ELC: it is very difficult to translate the obligations of the ELC with its conceptual openness into practical legislation and policy measures.

2 Many Policies Have Significant Landscape Impacts, but Landscape Competence Itself Is Not Addressed

Almost every sectoral policy, be it agriculture, environment, energy, rural development, transport or climate, has consequences for the landscape. Yet in many European countries, the territory is not seen in an integrative way, and the landscape is not an explicitly addressed policy domain and is often dealt with by several ministries. At the EU level, landscape and spatial planning are explicitly regarded as the competences of the individual states. This is a crucial paradoxical relationship: landscape is a unifying or integrative concept, but this is not considered in policy making. In most national or international governance approaches where policy integration is a major issue, landscape policy does not exist. Even in the evaluation of policies like CAP, landscape functionality and landscape character are hardly considered issues that need to be accounted for. As a matter of fact, the EU could ratify the ELC, which would imply the cross-sector consideration of landscape, although this is not very likely to happen in the near future. The diversity of the landscapes in Europe, on the other hand, makes it very difficult to consider guidelines on the European level to safeguard landscape diversity for the sake of the European identity, whatever that may be, because different nations have different cultures and, thus, would value landscapes in different ways.

3 The Rural Landscape Is Increasingly Multifunctional, but, in Most Regions, It Is Still Structured as if Agriculture Were the Only Function

Until now, agriculture has managed landscapes through production activities, and the system of rural regulations and policies, the structures and the market, including representations of the rural, is grounded on the agricultural sector. This applies even though agriculture has been abandoned in some areas, while in others many different demands and actors should be considered which are actively also managing landscapes and creating a dynamic – though not acknowledged by policies or the agricultural sector. The agricultural sector (i.e. the agricultural landowners) still assumes it has the right to manage the countryside alone, while the other sectors create dynamics and induce changes, but do not represent univocally the rural. The lack of renewal of the rural voice and rural competences is a tension that does not seem to have been addressed in most of Europe. This represents a kind of permanent and increasing contradiction that remains unseen and thus creates an empty space.

4 Landscape Heritage Should Be Conserved, but a Living Landscape Is a Developing Landscape

The European landscape almost everywhere represents a palimpsest of many interacting historical layers. The legibility of these layers represents a principal asset in the value of this heritage, creating the character we value as supporting our well-being. On the other hand, the landscape is a complex and dynamic system: a landscape that ceases to evolve is not a living landscape and it will inevitably decline. Evolving landscapes will always lose part of their historic attributes, while developing new attributes that will make up the landscape heritage of the future. In any landscape strategy, there is thus an aim to preserve the past while at the same time looking towards the future: how can this best be done? To what extent should we preserve and to what extent should we let go in favour of change and innovation? What will we lose if we preserve and what will we lose if we change?

5 Landscape Is a Common Good, but It Is Managed Individually

Landscape is a common good (requiring public intervention and local governance) with shared responsibility (requiring citizens' participation), but it is managed mainly by individuals who may not take the overall landscape into consideration when they act.

Landscape is beneficial to many: to the owners and users of land and to the general public who can enjoy its beauty, wildlife and cultural heritage. Landscape represents not only our common past but also the starting point for our common future. Land can be owned, not landscape! That means that the government is responsible (in our name) for ensuring that the landscape is properly cared for. Yet sustainable development of the landscape is possible only through the active involvement and commitment of the citizens who live, work and play in it. This paradox implies a crucial challenge in landscape policy and planning, especially in a period when it is fashionable to rely on market regulations and to decentralise as many responsibilities of central government as possible, preferably without accompanying means.

Individuals, Community and Public Policy

The development of local infrastructure and local basic services in rural areas, including leisure and culture services, the renewal of villages and activities aimed at the restoration and upgrading of the cultural and natural heritage of villages and rural landscapes is an essential element of any effort to realise the growth potential and to promote the sustainability of rural areas ... In order to create synergies and to improve co-operation, operations should also, where relevant, promote rural-urban links. Member States have the possibility to give priority to investments by community-led local development partnerships, and to projects managed by local community organisations (EC 2013b, preamble 19)

This quotation emphasises the need to see the village and rural landscape as both territorial entities and development factors. It also points to local territorial governance as the way forward for rural development. Although these approaches are far from mainstream trends in EU policy, as will be discussed in Chapter 6, they do signal new approaches to public policy and how they relate to the individual stakeholder and the local community. Similar approaches have emerged in development policy and evaluation (Chavez-Tafur and Zagt 2014), as well as in urban planning (Healey 1998).

Inspired by Arler's work on 'landscape democracy' (Arler 2006), in Chapter 7 we discuss how individual landscape managers, i.e. the farmer together with the local community and policy agencies are (and could be) interacting to guide landscape change. Individuals cannot help but react to the market, but with regards to non-marketable, intangible values that ought to be taken care of collectively, who exactly represents this collective today (Spirn 1998)? As landscape management decisions are becoming increasingly interconnected as part of what is usually termed 'globalisation' (Giddens 1990), complexity is increasing and old-fashioned command-and-control regulation is no longer sufficient. Arler (2006) distinguishes between 'respect for the individual, respect for the collective, and respect for the (professional and academic) argument'. Within such a triangle, multiple policy approaches are available, including deregulation and privatisation, collective governance, steering through elections, expert judgement etc. More fundamentally, the rural landscape represents an interest shared between a system-world (of market and policy regulations) and a life-world (of autonomous individuals and collective actions (Habermas 1981). In Chapter 7, we return to the key factors in landscape transition.

1.6 Mapping the Diversity of European Landscapes

The extreme diversity of European landscapes has consequences for the reasoning presented in this book. Due to this diversity, and also considering the multiple stationary and dynamic factors which result in the complex system that is the landscape, creating a map of the European landscapes represents a considerable challenge. Many attempts have been made.

- The analogue Meeus Map of European Landscapes in the Dobříš assessment report (Stanners and Bordeaux 1995, p 176): 30 landscape types of European interest based on climate, vegetation and landscape openness, with some specific regional landscape types; minimum mapping unit of 100 km² (~1:20,000,000).
- The digital European Landscape Classification LANMAP (Mücher et al. 2010): 350 landscape types based on a biophysical approach using digital data on climate, altitude, parent material and land use; minimum mapping unit of 11 km² (~1:2,000,000).
- A number of other recent digital maps are available for further reference (Hazeu et al. 2011; Metzger et al. 2005; Van Eupen et al. 2012).

However, none of these reference maps assesses the characteristics of landscape identity for specific landscapes. For geographical reference, we therefore use a newly developed map (Figure 1.16) of European Landscape Character Areas (ELCA) (Pedroli, Van den Brink and Bakker 2018), which is based on landscapes that can be recognised by a unique toponym (see Figure 1.17 for a part of the map: Slovakia).

The map was created using maps, descriptions and a multitude of literature and Internet sources, complemented by advice from local experts. Travel guides appeared to be a rich source of descriptions of characteristic landscapes at a level that can be depicted on a European scale. Toponyms adopted refer to landscape names as used and recognised by citizens of the country in question.

The resulting landscape character map of Europe illustrates the diversity and unique character of European landscapes, and is based on a landscape classification of the units that can be univocally described and recognised by science, policy and civil society. To secure the credibility of the map, the landscape classification is, whenever possible, based on scientific evidence. To enhance the legitimacy of the map, specifications of the Landscape Character Areas have been provided by knowledgeable local experts and recorded in narratives (see Figure 1.12) and statistics.

The cases described in the chapters are included on the map in Figure 1.16.

FIGURE 1.16
The locations referred to in the introductory stories to the chapters (red triangles and numbers) and in the cases (white stars and capital letters), projected on the map of the European Landscape Character Areas (Pedroli et al. 2017). A black-and-white version of this figure will appear in some formats. For the colour version, please refer to the plate section.

1.7 Towards Future Landscapes: An Inviting Environment for Civil Society and Commercial Creativity Alike

What are society's demands regarding the development of the countryside? Several surveys of these demands have been made in recent years (Paracchini et al. 2011; Pérez-Soba, Paterson and Metzger 2015; Zasada 2011) and, although they are all subjective to a certain extent, a general picture emerges.

The generally adopted scenarios for global economic and societal development show a strong polarisation of land functions in Europe in the near future

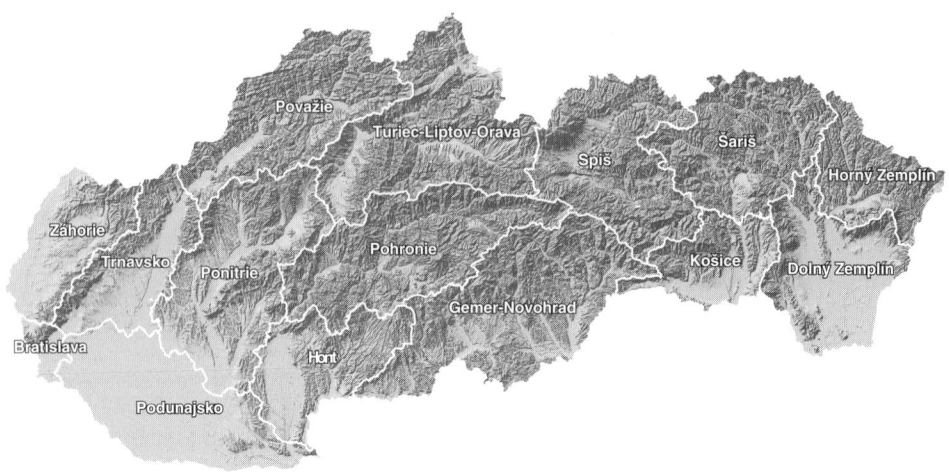

FIGURE 1.17
Slovakia on the map of European Landscape Character Areas (ELCA) (Pedroli et al. 2018). Below is the description of two of the landscape character areas

Záhorie	Danubeland / Podunajsko
The name Záhorie (transmountain region), which is based on the Latin name *Processus transmontanus*, which referred to the area beyond the Malé Karpaty Mts. or the Little Carpathians seen from Bratislava, first appeared in the seventeenth century.	The Danube region with its special landscape is situated in the south of Slovakia. The south-western and southern limits of the region follow the Slovak–Hungarian border, which follows the course of the Danube river. The lowland Podunajská nížina covers most of this region.
The region is situated in the western part of the Slovak Republic next to the border with Austria and the Czech Republic. Due to its favourable geographical position and climatic conditions, it was one of the earliest territories settled in Slovakia. The region has numerous lakes. Viticulture began to develop in the mid-fifteenth century and Záhorie has become famous especially for red wine. The most popular typical food is called Skalický trdelník – a sweet pastry with walnuts	The Hungarian language and traditions predominate at present in this part of the country (e.g. the settlements of Dunajská streda, Nové Zámky and Komárno).
	Currently, tourism is developing significantly in the Danube region, mainly centred on spa towns with hot springs and outdoor swimming pools. Especially the site Žitný ostrov has a warm climate, rich water resources and fertile soils. For these reasons, this part of Slovakia was among the first to be settled. Traditional crafts include the manufacture of mats and baskets and fishing nets.
	Water mills on the river Malý Dunaj and the associated remnants of alluvial forest are also interesting for sports fishing and ecotourism.

(Rounsevell et al. 2012; Van Vliet et al. 2015). If current trends continue, segregation of major functions ('land sparing') seems to be a much more likely prospect over large parts of Europe than integration ('land sharing'), possibly with the exception of peri-urban landscapes. Still, a large majority of the visions expressed in stakeholder consultations aims for a considerable degree of multifunctionality (Pérez-Soba et al. 2015). However, policy alternatives do not always achieve this target even if so wished: it seems that we are not getting what we want (Pedroli et al. 2015)! Apparently, it is extremely difficult to set out a structured road to a future that is substantially different from the one that could be expected from business as usual, following current trends. This planning paradox requires societal consideration and debate, especially because it seems that neglecting the long-term consequences of current trends could lead to almost irreversible land transitions, including uncontrolled land abandonment, soil degradation, biodiversity and ecosystem service loss and a decline in rural liveability.

Because landscape planning is, strictly speaking, not an EU competence, while at the same time land-use development is largely determined by a range of sectoral EU policies, accommodating cross-sectoral strategies to achieve new perspectives of land use for the future should be a priority for policy and society. At the national level, comparable situations can be observed in many countries. Creative, out-of-the-box thinking is needed to provide innovative combinations of sectoral policy targets, and to narrow down the wide range of possible futures, instead of reacting on an ad hoc basis to external developments. However, to think out of the box requires familiarity with the 'box', how it functions, how it is structured and how it changes.

The way people relate to the landscape needs to be understood here. We all acknowledge that the landscape is important for our identity, our reference system and our well-being. Nevertheless, not everyone is aware of this relationship or thinks seriously about the way he/she relates to the surrounding landscape. Consequently, when asked what is your landscape, what is the landscape of your youth, what is your favourite landscape, where do you go to lift your spirits, which landscape do you most value for recreation, people require time to answer and often find formulating answers difficult. But still, many references to one's environment, roots, cultural identity and well-being refer to landscape dimensions and assets, and when the landscape changes dramatically, and remarkable or symbolic elements disappear, people react. When someone moves from one region to another, often he/she will miss the landscape of origin, what also affects his/her well-being. We believe that the landscape is more important to people than they realise. Therefore, this book is about the future of our landscape.

1.8 The Aim of This Book

The main motivation for writing this book is to address concerns over the change processes that are affecting European rural landscapes, incomplete understanding and inappropriate policy reactions to these changes, combined with an obvious lack of vision for well-functioning and attractive landscapes.

The literature is increasingly fertile in descriptions of the particularities of different European landscapes, analyses of public demands on current landscapes and discussion of how landscape managers should be compensated for the services they are providing to society. There is, however, a surprising lack of literature that attempts to identify the processes of change that are affecting landscapes today, considers the complexity of actors and factors involved or discusses potential visions for future management approaches that integrate these different drivers and actors on multiple scales. These visions should lead to specific policy designs and institutional arrangements that support the qualities of the landscapes that society demands.

This book examines agricultural land use from a spatial perspective, related with other functions, other uses and other actors in the European landscape. This book goes further than the usual textbooks and university courses on landscape, which tend to focus on economic analyses of values and compensation schemes, by including social networks, community arrangements and people's motivations in a local context. Local context comprises the landscape functions, pattern and character, which are based on the natural and cultural history of a place, and the web of social and institutional arrangements involved in its management. This book provides a conceptual framework for understanding the drivers of landscape change and its management today, as well as grounds for supporting future management arrangements which are specific to each place and community. We hope that this understanding will help the reader to synthesise the knowledge provided in many other textbooks on landscape.

The structure of this book is presented in Figure 1.18. Methodological tools and conceptual frameworks of change for understanding European landscape are key issues in this book, including the very concept of landscape and the notions of multifunctionality, scale, driving forces and agents (Chapters 2 and 3). Based on these frameworks, we then discuss rural land-use systems in some detail, partly through selected land-use systems from various parts of the continent such as intensive crop farming, dairy production, plantation forestry and silvo-pastoral systems (Chapter 4). The urban–rural relationship is the theme of Chapter 5, which provides both accounts of

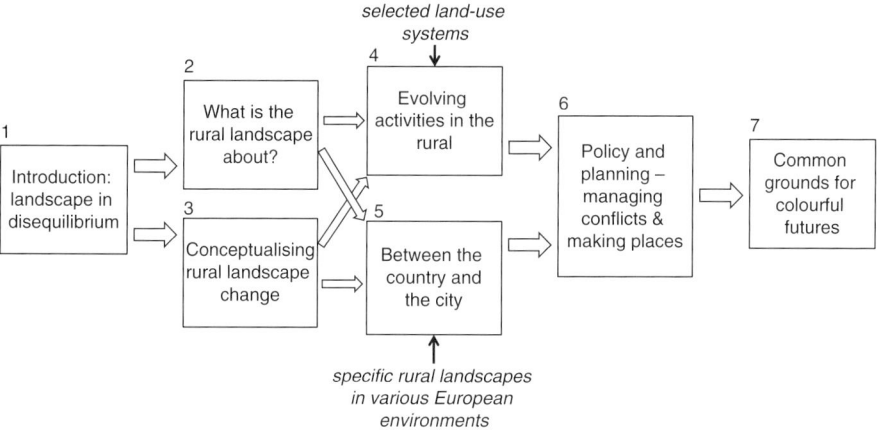

FIGURE 1.18
Structure of this book

historical developments in rural and urban interactions and current forms of urbanisation and their effect on rural landscapes. Again examples from specific landscapes located in different parts of Europe will be utilised to illustrate how urban and rural dynamics are interacting and affecting rural landscapes.

Landscape policy and planning is prioritised in this book. In the two final chapters, we first outline and then critically discuss the state of European landscape policy and planning (Chapter 6), including various conservation policies against the background of uncertainty and radical changes. This book concludes with Chapter 7, which discusses potential pathways to new paradigms of landscape policy and planning, introduces concrete examples of promising new developments in landscape governance and strategy making and, based on these, synthesises a general framework for landscape policy and planning at various levels.

2

What Is the Rural Landscape About?

'Le Climat'

It is drizzling when we walk up the gentle slope, vineyards to the left and right, the leaves shed. Further up some disorderly juniper shrubs, down in the valley a road, and simple, but well-kept houses and gardens. The Burgundy hills near the small town of Auxey-Duresses, not far from Dijon, are stony, white rocky fragments in the soil all over the place. It smells of earth and mushrooms, faintly tart, a veil of thyme. Here and there we can see someone pruning the vines; otherwise, little life is apparent in this landscape. The winegrower, Jean, picks a stone and explains that without these lime stones the grapes would not ripen properly. At daytime they absorb the heat of the sun; at night they release it to the vines. On the map the slope to the left is indicated as Climat du Val. A bit concave, nicely south exposed. This is where my most beautiful wine comes from, Jean says. The size of hardly a hectare, this piece of land has the recognition of Grand Cru. Even if he would produce a better wine on the slope to the right, that would never be a Grand Cru.

Later, in a village courtyard, we are welcomed by an unpretending man of an unmistakably important air. He leads us through a doorway down a narrow staircase. To our surprise, underground there appears to be at least as much room as in the houses above. In the long wine cellar amidst the oak-wood casks, some glasses have been put ready on a wobbly, small table. We taste some excellent wines. One red wine is really divine, rich and light at the same time, with a taste that reminds me of the autumn smells of the Climat du Val, but more in harmony. The gentleman explains that this Romanée St. Vivant – hardly affordable in the shop – obviously is of organic quality: how otherwise could one do justice to the ensemble of the soil organisms? More, the vineyard is being farmed following biodynamic principles. Not that this is indicated on the bottle; you simply do not show off with that. You do this because of dedication to your land, cherished for many generations, le climat.

What Is the Rural Landscape About? 43

FIGURE 2.1
Pruning of the vines in Burgundy, France
(photo: Bas Pedroli)

This description of a Burgundy landscape shows that the landscape has many different dimensions. When taken care of meticulously, it produces surprising qualities, not only high soil quality and wine but also other personal memories and identity, biodiversity, historical heritage, scenery and much more besides. This chapter discusses how we interpret landscape in this book, how we suggest the reader should reflect on what the landscape is and its different dimensions. As this book focusses primarily on the rural landscapes, this chapter also discusses the current meanings of rural, and how they have evolved recently and, indeed, are still evolving. Further, in this chapter we also present and disentangle the concepts which we need to be aware of and utilise when comprehending, analysing and discussing the landscape today, in particular, when working in dialogue with different scientific disciplines that deal with the rural landscape.

2.1 Landscape – an Area as Perceived by People

Landscape is about nature and culture and is everywhere. It is about what we see, but also the processes that have created what we see. And it is about working on a larger scale than just specific sites. Finally, it is about change (Countryside Commission 1998). In this book, we adopt an understanding of 'landscape' that includes all this, which, to a great extent, draws on the concept of landscape as defined in the Florence Convention (European Landscape Convention, ELC): *landscape means an area, as perceived by people, whose character is the result of the action and interaction of natural and/or human factors* (COE 2000). Landscape is, thus, seen as an 'area' and as people's perception of the area, which, when considered together, should be understood as the 'result' of natural and social processes. Although the definition is not without its shortcomings, we find it useful as a general reference point for analysing and discussing rural landscapes and their transformation. We appreciate the four words 'as perceived by people' because they emphasise the fact that the perception of the area is 'shared, valued and used by people' (Olwig in M. Jones et al. 2007, p. 215)), which reminds us that meaning and experience should not be separated from the area itself if one wishes to understand landscape. In such an interpretation, 'perception' should be interpreted as comprising more than the visual appearance of the area in question as it refers to how the area is experienced and understood more broadly. This definition is sufficiently comprehensive to capture the essential dimensions of landscape, but is still precise enough to facilitate focussed analysis and discussion, while it strengthens the human dimension of the landscape as an entity that people interact with and change, while at the same time being affected by it and the changes.

In order to deal with this complex and dynamic reality, combined and encompassing approaches are required which integrate different disciplinary perspectives, evolving into new concepts and analytical frameworks. This chapter is about the concepts that are most often applied to study and understand the rural landscapes of today's Europe and the transition process they are going through.

Box 2.1 *Landscape* as Defined in the Florence Convention (COE 2000):

* *means an AREA*: a surface of the earth in a specific location;
* *as PERCEIVED BY PEOPLE*: to be perceived by people it is local, within a geographical space; the perception of all those who relate to it, is part of the landscape itself;
* *whose CHARACTER*: what makes the landscape distinctive, what is the most determinant features and characteristics of this landscape;
* *is the result of the ACTION and INTERACTION of NATURAL and HUMAN factors*: the different components of the landscape are both natural and human-shaped elements; they change over time due to internal processes, but also due to reactions to each other – therefore there is a dimension of action but also one of interaction.

2.2 The Rural Landscape

Different Dimensions of Landscape

Many landscape-based approaches are linked to environmental processes and their spatial expression and, as such, focus on the materiality of the landscape. Landscape patterns, which express the materiality of the landscape, are the basis for the analysis of environmental processes. This spatial analysis approach can be applied at different scales depending on the elements and processes to be identified in the analysis and the scale adopted (Herod 2011). For such processes, the relationship between scales also matters, although this is not the subject of this chapter.

The landscape materiality has been the core focus of landscape ecology, including the related environment and conservation studies. Forman and Godron (1986) further developed landscape research with their seminal book on landscape ecology. They define landscape as 'an heterogeneous land area composed of a cluster of interacting ecosystems that is repeated in similar form throughout'. Landscape development is mainly considered the result of three mechanisms: specific geomorphological processes occurring over long periods, colonisation patterns of organisms and local disturbances of individual ecosystems over a shorter time. These authors' understanding of the landscape is based on the study of its structure, function and change (Figure 2.2). Many other authors have contributed by acknowledging human action as a fundamental driver of change (Bastian and Steinhardt 2002; Burel and Baudry 1999; Naveh and Lieberman 1984), and have paved the way for the human element and action to be included in all types of landscape studies. In the geographical conception of landscape (Antrop 2000; Gobster et al. 2007; Nassauer 2012), the interrelations between nature and culture are central as human action is included as a significant driver of the appearance and functionality of a landscape together with the natural conditions of the area.

As a point of departure, we understand: (a) "landscape *structure*' as the way the area in question is composed physically by patchy, linear and point elements (Forman and Godron 1986); this refers to a specific spatial pattern which includes its elements and the distribution of energy, materials and species in relation to the sizes, shapes, numbers, kinds and configuration of the different elements; (b) 'landscape *function*' as a combination of natural processes, human uses and human-driven processes, which means flows of energy, materials and species among the components of the landscape (Brandt and Vejre 2004) and; (c) 'landscape *character*' as the specific combination of the biophysical and cultural geographical patterns (Swanwick 2004). The character can be mapped as character types or as specific areas (landscapes) representing both a type and a unique character area with its own identity. Although landscape character areas are often mapped using sophisticated analyses of the different components, the areas identified often overlap with traditional regional or local areas with specific

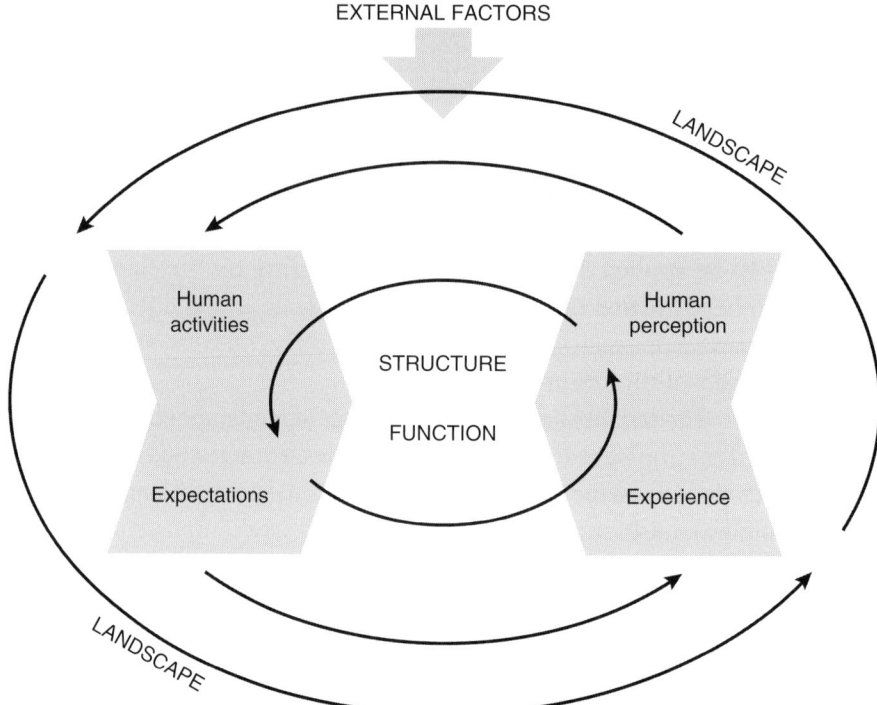

FIGURE 2.2
The landscape as a dynamic system composed of different levels of interaction between the material dimension and the human action and reaction. The material component is at the centre, and changes over time due to the interaction of structure and function. People perceive this materiality and also have experiences which influence both their action and their expectations. By acting in the landscape, they have an effect on the function and structure and, as a result, have an influence on the way in which the landscape is perceived and experienced. Activities and expectations are also influenced by many external factors, which may also have an influence on perception and experience. Therefore, this continual loop of interacting factors composes the landscape.

geographical names – as we have seen in Chapter 1. Further, we need to acknowledge: (d) *change*, which means the alteration in the structure and function of the ecological mosaic over time as a result of both natural and human-driven processes, which has an impact on the overall dimensions of the landscape.

In summary, the ELC definition of landscape as it is interpreted and elaborated here has three major components: the biophysical *structure*, which mainly refers to widespread landscape ecological terminology, the landscape *functions*, including both natural processes and human use, and landscape *character*, which refers to both the specific composition and appearance of the landscape and often also to its concrete delineation.

It is the relationship between the landscape structure and landscape functions which makes a landscape dynamic, and it is in the tensions within this relationship that we find the background to landscape change and transition. To fully understand the dynamics and what affects the landscape *drivers* from outside should also be included, as indicated in Figure 2.2.

Although the landscape should be understood as a spatial entity, it is not only about space. The landscape is also a more visually, aesthetically or artistically 'perceived area' (Selman 2012), the *paysage vécu* – lived landscape (Luginbühl 2012). Landscape pattern or structure as defined by Forman and Godron (1986) represents the landscape materiality, the basis of what people directly perceive. Further, each individual's previous personal or collective experience also leads to or influences specific perceptions of the landscape (Stobbelaar and Pedroli 2011).

In his seminal book *Symbolic Landscape and Social Formation*, Denis Cosgrove (1984) shows how landscape evolved as a *way of seeing*, as a subjective way of representing the visual dimension of landscape within a social context of power and hierarchy. The landscape has been represented by writers, painters, photographers, surveyors and geographers in ways which both reflect and produce certain views of how people and the environment are related. Landscape as a way of seeing then becomes associated with how society and space are interlinked more broadly than just a matter of visual characteristics and aesthetics. From this position, Cosgrove (1985) argues for the inclusion of the historic dimensions of landscape as a way of seeing into modern geography.

Finally, landscape has historically been understood as community or polity with its own institutions to manage conflicts and guide changes, as Olwig (2002) has shown. In fact, most political-administrative boundaries in Europe reflect historic settlement structures which, to a high degree, were intimately linked with underlying structures such as rivers, mountain ridges, drainage patterns, soil conditions etc. Today the ecological boundary conditions and the socio-economic activities even in rural communities are no longer intimately linked. At the same time, the parish boundary may very well still determine how rural residents 'perceive' their own local landscapes because it is it determines where people go to church and to which village hall they belong. In such situations, the landscape 'as perceived by people' of the ELC may be interpreted as a landscape perceived by a community, which should be included in policy and planning processes and be involved when new boundaries are constructed in landscape character mapping, in designations of planning zones, nature parks etc. (Olwig 2002; Selman 2012).

Furthermore, the perceived conception of the landscape also prompts human action, and thus, in turn, influences human intervention in the landscape (Gobster et al. 2007; Nassauer 1997). Joan Nassauer states that the concept of landscape refers to both literal settings, i.e. real places, and to a conceptual field that examines how humans affect and are affected by geographical space (Nassauer 2012). This means the word has both analytical and experiential implications (Wylie 2007), which are not necessarily contradictory as they mostly overlap and complement each other. Many authors have conceptualised these different perspectives on the landscape. For example, Han Lörzing (2001) distinguishes four layers of interaction between people and the landscape: (1) the layer of intervention, where the landscape is what we make; (2) the layer of knowledge, where the landscape is what we know; (3) the layer of perception, where the landscape is what we see and; (4) the layer of interpretation, where the landscape is what we believe it to be. Maarten Jacobs (2004) describes the true, the right and the real landscape, where the true landscape refers to its materiality, the right refers to power and decisions and the relative importance of the different roles of the landscape and, finally, the real represents the landscape construction in people's minds. When studying the landscape, we always need to: (1) acknowledge these different dimensions and interpretations and; (2) clearly identify and express our positioning to others (Stobbelaar and Pedroli 2011).

Landscape as a Place and as a Space

As we saw in the previous section, landscape is about space and it has a spatial dimension. However, it is much more than just space or a section of the Earth's surface observed and considered at a specific scale. Landscape is also about place: a more geographically 'bounded area' with its specific ecological and social functions and its own history which determines its character. This dichotomy is partly related not only to different scales of conceptualisation and analysis but also to the different understandings of the landscape and our (conscious or unconscious) attitude to it, as we saw when we explained the differences between the relationship of people with the sacred places and spaces of worship in Section 1.2. We might as well say that landscape is about space and landscape is about place.

Human intervention in the landscape is not limited to the direct action of those who make decisions regarding land use such as farmers, landowners and other decision-makers. It also includes the indirect influence of those who use and demand the landscape, and those who participate in the local communities (Nassauer 1995; Selman 2012). The landscape may be seen as the meeting point or the arena where owners and users interact in formal or informal ways, thus expressing the increasing interplay between the production and consumption of the countryside (Pinto-Correia and Kristensen 2013). Consequently, regarding human intervention in the landscape, decisions are conditioned not only by the specific biophysical and socio-economic characteristics of each place and by its location and, thus, geographical context, but also by the interaction between people, including owners and users.

These interactions create the place, which is also the local landscape (Pedroli 2000; Selman 2012; Wylie 2007). Therefore, when attempting to understand the landscape, one needs to look beyond its spatial expression or visual appearance. The place is the strongest reflection of the character of the landscape and is connected to history and the sense of belonging and engagement, care and interaction. The notion of place is related to the attributes that make a place special and distinct, and it is often derived from the activities and experiences of the people who live there and interact within a community or from external actors interacting with that community (Selman 2012). The shaping and differentiation of a place occurs when knowledge is gained from the coordinated use of the senses in carrying out various tasks in the area or by observing it with a certain purpose (Olwig 2008). At the same time, the local biophysical context is also part of the place and is reflected in its character and determines how the place is shaped by people through time (Hersperger et al. 2010; Pinto-Correia and Kristensen 2013).

Box 2.2

The difference between landscape as a place and landscape as space can be shown by the manner in which people express the way they see their own living place embedded in the landscape.

Farmers in two different Danish landscapes were asked about their own farms, their farm production and how they conceived change in the landscape (Primdahl et al. 2010a, 2010c). Half an hour into an interview, they were asked to imagine they were talking to a distant relative on the phone when this person suddenly asked: 'I have never visited you where you live. What kind of place is it? What's it like to live there? What is the landscape like where you live?' The interview included deliberately a series of sub-questions because both place and landscape can have different meanings to people.

In both landscapes, the farmers were all satisfied with their area (with one exception). However, the concrete answers were of one kind in one landscape, and of another in the other.

In one landscape, the farmers were enthusiastic because it was close to attractive places – a big river, a well-known hill and scenic areas were specifically mentioned. What they did was describe their local context by referring to (other) places, and as such their landscape remained seen as space.

With regards to the other landscape, all of them gave a very different answer. No one referred to nearby places – they all talked about the qualities of their own landscape – about the tranquillity, the dark night, the fjord, the forests and the wildlife. No one mentioned the nearby North Sea coast with its spectacular dunes – they talked about their own landscape. In contrast to the first landscape, this one was conceived as a place. And in this case the conditions for agriculture are difficult and there has been a long tradition of collaboration on landscape issues.

Edward Casey (2001), with reference to Relph (1976), argues that landscape provides the context and the fundamental conditions for the place. A place is defined by its boundaries, which should be thought of as boundaries which can be passed through rather than physical barriers which keep people out. A place may be defined at different scales depending on the context and the perspective, but it always has a local dimension. Landscape, in turn, may be defined by its horizon in a physical sense (what you see), in a mental sense (what you conceive) and in an imaginary sense (the changes you envision). The place dimension in the landscape is related to a connecting dimension, i.e. people's attachment to nature, history, culture and with each other, and, thus, relations between the material and the immaterial which occur in each place (Herod 2011; Primdahl and Swaffield 2010a; Selman 2012; Van der Ploeg 2003).

Landscape is a complex reality, but at the same time it is a powerful concept for the sustainable future of our territories. It can link creativity and social learning in spatial analysis and scientific knowledge of processes with place-specific design, community engagement and citizen empowerment. Landscape can support a broader anticipatory inquiry in landscape science, more intelligent design and more connected communities (Nassauer 2012). Landscape is no simple phenomenon to deal with in planning or policy design, but a challenging arena full of strategic possibilities for the involvement of different groups of agents. Anne Spirn (1998) has developed a coherent grammar or language to think about and work with landscape, a language which is both precise in explaining fundamental rules for landscape design and planning and open for the virtues of breaking the very same rules when this is needed.

2.3 The Rural: Between Materiality, Representation and Imagination

The landscapes addressed in this book are rural landscapes, which are in transition from being dominated by agriculture, including forestry, to being characterised by multifunctionality (Jollivet 1997; Marsden et al. 1992; Woods 2011). As already expressed in 1992 by Cloke and Milbourne, quoted by Ilbery (1998), 'there is no longer one single rural space, but rather a multiplicity of social spaces that overlap the same geographical area.' This is particularly true for Europe, where this interaction is extreme and takes many different forms, which makes grasping and characterising such diversity challenging.

What Do We Mean by Rural?

The complexity and intersecting types of spaces which characterise the rural demand a renewed understanding of the rural. For this, and considering the central role which agriculture has played so far in shaping and changing the rural space, we need to identify the relationships between the rural and agriculture, but considering the emergent demands and activities in this space, we also need to understand how this space and each place within it relate to the different activities developed and to changing representations of the rural (Woods 2007). The idea of the rural has very ancient origins related to the differentiation between the settlements where populations were concentrated and the open areas without settlements, i.e. the rural space – which commonly has been considered as the space that provides natural resources and food (Mormont 1990; Woods 2011). In this sense, the rural, until recently, has also mainly been seen by the public as the territorial basis for agriculture and forestry. In fact, most of the rural space in Europe has been under farm or forestry management which has been focussed on production and, therefore, the rural has been dominated by agrarian paradigms (Emanuelsson et al. 2009; Marsden et al. 1992). Complementary with this representation of the rural as provider of natural resources and food, other representations have increasingly spread, of the rural as nature, and of the rural as communities with close relationships. These different representations are there simultaneously, and are increasingly conflicting (Jollivet 1997; Mormont 1990). In Chapter 5, we discuss, at some length, how urban–rural relationships have developed over time and how intersecting urban and rural factors are affecting current European landscapes.

European landscapes, thus, offer a deep-rooted cultural and symbolic representation of the rural, which not only comprises an important agricultural production space but also a pleasant environment, a characteristic and specific landscape and a human dimension of social relations (Marsden et al. 1992). In recent decades, this has led to increasing focus on the rural as a space for consumption (Holmes 2006). As the European population is becoming increasingly urbanised, i.e. living in urban environments and adopting urban lifestyles, the societal demand of the rural as a consumption space has grown. The relative importance attached by society to farming as a source of food and fibre and to leisure opportunities in the farmland landscape has been changing (Buijs, Pedroli and Luginbühl 2006). The rural is increasingly seen as a space that supports various amenities, where recreation and leisure activities can be pursued, and as a source of spiritual or aesthetic inspiration. Further, the rural may also be a place to live, permanently or on a temporary basis (Primdahl and Kristensen 2011). Seeking out the rural is due to reasons related to the beauty of the landscape, the closeness to nature, the tranquillity and even remoteness of some places, but also because of its potential for social well-being, the opportunity to belong to a community and develop close relationships with other people and be more locally connected (Ilbery 1998; Lowe et al. 2003; Woods 2005, 2011). Nevertheless, the increased commodification of the rural landscape has not developed without tension and conflict with the productive sector, i.e. agriculture and forestry, which are still the dominant land uses and economic activities in many regions.

The Role of Agriculture or Agricultures

What has happened to agriculture during the same period? It has become increasingly specialised and polarised with highly intensive farming areas at one end of the scale, and at the other, increasingly marginalised small-scale farm mosaics and extensive land-use systems, or even abandoned land often located on the so-called peripheries (Lowe et al. 2003; Pinto-Correia and Breman 2009) (see Table 1.2). These trends are the reason for the increasing differentiation of rural landscapes in Europe, as we briefly addressed in Chapter 1. We return to this subject in Chapter 4.

When there are favourable biophysical and structural conditions for agriculture, a productivist paradigm often dominates with agricultural production as the main driver of land use (Primdahl and Swaffield 2010a). At the same time, the social demand for non-commodity functions related to environmental concerns and recreation and quality of life has been increasing. These trends have resulted in a new awareness concerning the different values of the physical landscapes and a need to consider other farming outcomes besides production. This has led to pressure on the farming sector, especially in peri-urban regions, while in attractive rural landscapes, intensive agriculture competes for space with other land uses.

On the other hand, in rural areas with difficult or marginal conditions for industrial agriculture, such as many areas in Southern Europe, large parts of Eastern Europe, Scandinavia and the Western islands, agricultural systems have only recently (if at all) entered the productivist phase (Ortiz-Miranda et al. 2013a; Perfecto, Vandermeer and Wright 2009; Robinson 2008). Moreover, the process is differentiated depending on the sociopolitical culture in the countries. The limitations to industrial agriculture are a function of a combination of many factors, including natural conditions, location, structural conditions, technology availability and sociopolitical history. In such areas, there is often a specific landscape character with high interest for nature conservation. Such landscapes are valued by society due to the high quality of the natural resources and nature conservation goals, but also as they support many recreational and cultural activities, although these are not integrated in the modernisation farming discourse or ideals (Pinto-Correia et al. 2013). The case of the Súľov-Hradná landscape in the Slovak Carpathians (Case A) is one example of such a landscape: highly valued today for its scenic beauty and specific character, but marginal in relation to agriculture and therefore prone to land abandonment and afforestation of former agricultural fields.

CASE A Land Abandonment and Tourism on the Súľov-Hradná, Slovak Carpathians

Landscape Appearance

The nature reserve of Súľovské vrchy in Slovakia is characterised by a rich diversity of rock formations, valleys and small intra-mountain basins (Figure 2.A.1). These hills belong to the most famous rock formations in the country. The village of Súľov-Hradná is located in a typically concave basin, the base of which consists of soft marl. The valley has soft, rounded ridges between which relatively wide shallow river valleys are incised. The boundaries of the settlement are naturally formed by surrounding rock slopes, which adds a whiff of history to the village's shape and dimension. As in numerous other Slovak villages, the dominant building of the village is a church, with next to it the local pub.

The oldest evidence of the settlement dates back to the prehistoric period. The village is located in the lowest part of the valley. Typical for settlements in Slovakia, the arable land is close to the settlement and occupying the gentlest parts of the slopes. Pastures were grazed intensively in the past, but after a transition to intensive meat production, they became abandoned in the more remote areas and gradually overgrown by forest species. Natural meadows rich in species diversity are preserved only in the more remote, extensive exploitation sites. On a landscape scale, the pastures and meadows dominate over the arable land. At the end of the nineteenth century, the average agricultural parcel had 0.5 hectares. Traditional employment of the local population was farming and sheep and cattle breeding. However, the soil is poor and rocky, and in the small farm units, it was hardly possible to produce enough to feed the large families. During winter time, when the fields were covered with snow, inhabitants were occupied preparing tools and equipment for agricultural activities. People lived in small, wooden houses with the entrance hall and one or two rooms, often several generations living together.

Already in the nineteenth century the place attracted visitors because of the beautiful scenery surrounded by rocky mountains. In the past, the rocks would impress

FIGURE 2.A.1
Location of the Súľov-Hradná cultural landscape on ortophoto and map

Case A (Cont.)

FIGURES 2.A.2A AND 2.A.2B
Landscape view of the Súlov-Hradná from the adjacent castle built in the fourteenth century, serving as an observation point of nearby roads

more than today, as they are now largely covered by forest vegetation. Progressively a transition has taken place: many local people migrated searching for better life conditions; more distant and marginal farm fields were abandoned; the agricultural area decreased; but also, new income sources were created by tourism and tourist visits, new economic activities developed, and a service based community was formed. Nowadays, the parish has eight walking trails, five biking trails and good facilities for mountaineering. The locality often appears in descriptive photos of the Slovakian countryside. There can still be found some wooden houses serving nowadays as tourist accommodations. Some local traditions are still alive in the village, as Carnival, Easter egg painting, building Mayas and various family and religious customs.

> **Case A** (*Cont.*)
>
> **Landscape processes**
>
> - Natural afforestation of the pasture area takes place by abandonment of the grazing land.
> - Due to the pasture abandonment in mountainous areas and subsequent natural forest growth, the rocks are less visible for visitors.
> - The poor and rocky soil leads to the abandonment of some crops, like potatoes and corn.
> - The wooden houses are replaced by brick family buildings.
> - The place of living is progressively changing to a place of consumption, especially for the leisure activities: e.g. agricultural exploitations are diversifying in a way to meet tourism demand by horse breeding and agro-tourism activities and cross-country skiing.
>
> **Results: Identification of Future Roles**
>
> This is a landscape where the transition has taken place, from a production countryside to a consumption countryside. By being peripheral and with hard natural conditions, it would be prone to agricultural abandonment, but the quality of the scenery has made it perfectly adequate to become a touristic landscape. The questions here are: Will it be possible to maintain the quality of the landscape as it is today? Will the abandonment of the fields closer to the village be avoided, so that an open countryside and open view to the mountain is maintained?
>
> The mechanisms which will support new land management arrangements are not yet in place, and would need to be created if this landscape and its touristic and iconic value are to be preserved.
>
> *Diana Surová*

In these regions, agriculture is most often characterised by its extensive character, due to limiting natural conditions, and is often associated with low population density. These are often high-nature-value (HNV) farming systems, which can be difficult to maintain today (Oppermann, Beaufoy and Herzog 2012) due to not only their very low productivity but also low investment and entrepreneurship associated with the farming sector, which leads to limited capacity to innovate and create new solutions (Pinto-Correia, Menezes and Barroso 2014).

Between the extremes, several possible combinations exist which reflect the differences in the agricultural sector per se, but also differences in the balance between drivers of change and regional specificities. The kind of land-use changes which have started already many decades ago in Súľov-Hradná (see Case A), as well as the growing tourist attraction to the area, may have occurred much later in other peripheral mountain areas of South-western Europe, and with other specificities. Changes occur in different places at different rates, on different scales and in different directions as a result of a complex combinations of drivers and with different impacts leading to an increasingly structural and functional differentiation of rural areas in Europe (Lowe et al. 2003; Pinto-Correia 2010; Potter 2004; Woods 2011). The paths of agriculture, space and rural society are increasingly dissociated (Domon 2011).

A New Community

The changes in the agricultural sector also mean that the importance of agriculture in the rural communities of Europe is declining as the share of the active population employed in agriculture is decreasing and large parts of the rural population are now employed in other sectors (Woods 2005, 2011). The rural community in Europe can no longer be defined by its direct or indirect connection to farming. Exceptions can still be found in Eastern Europe, perhaps mostly in Poland and Romania, although they will probably soon follow (Dannenberg and Kuemmerle 2010; Müller et al. 2009; Sikor 2004). As shown by Ilbery (1998) and Woods (2005, 2011), the increased mobility of people, goods and information has led to a radical change in rural communities, and to the emergence of new power relationships and actor networks. External linkages are often stronger than internal ones (Murdoch and Marsden 1994). Furthermore, they are combined in a new type of social positioning of the rural actors in the global society (Hedberg and do Carmo 2012; Marsden and Sonnino 2008). Just as in society as a whole, changes in rural communities are being increasingly affected by new social relations and networks around the planet, where a new network society is developing across scales (Castells 2010). These relations on the global scale and social practices which are communicated at a distance (*space of flows*) are complemented, on the local scale, by local connections and relations that persist and increase the resilience of places (*space of places*). Furthermore, as Castells describes, global networks in the networked society include some people, activities and territories while excluding others, thereby creating a new social, economic and technological inequality. As a consequence, the new community compositions in the rural are not without trade-offs: some places are characterised by significant innovation and entrepreneurship, while others are characterised by tension and conflict of a type which was previously unknown (Barbieri and Valdivia 2010; Ilbery 1998).

Many Different Representations or a Different Representation

The representations of the rural that dominate today's European society do not match with the complex and dynamic reality described earlier in this chapter and the different understandings are contested (Macnaghten and Urry 1998). The rural has been changing so fast and in such an overarching way that people have been unable to combine all dimensions in order to interpret and relate to the changes and the resulting reality (Philo 1997). As a consequence, besides the rural as a site of food production and the rural idyll, many other rurals have been created as social constructs. Today, distinct and separate representations of the rural still prevail; the rural as agriculture, the rural as nature and, increasingly, also the rural as amenity provider. Unfolding the way in which these different representations interact is extremely complex and, therefore, they remain separate. This is expressed in political discourses at different levels of governance and is apparent in public policy strategies and design, which remain largely sectorial and connected to previous paradigms, as we discuss in Chapter 6. The phenomenon is also expressed in public opinion, as shown in literature, in the media and in responses to surveys (Macnaghten and Urry 1998; Pinto-Correia et al. 2013; Surová, Pinto-Correia and Marušák 2014). Looking at the specific level of the local landscape, acknowledging the local differentiating factors may lead to a more updated understanding of what is at stake in the rural today.

2.4 Landscape Functions and Services

In order to understand the rural landscape in the context of the changes occurring regarding use and the expectations of society, we need to consider landscape functions. According to classical landscape ecology (Forman and Godron 1986), as discussed in the beginning of this chapter, landscape function refers to the interaction of the spatial elements, i.e. the flows of energy, materials and species between the component ecosystems. Functions express internal functional relations.

However, the concept of function may be used to refer not only to traditional landscape ecology but also to the capacity to support services, which has led to some confusion regarding its use.

In the context of the new expectations society has regarding the rural and, thus, regarding its multifunctionality (which is analysed in more detail in Chapter 3), function is defined as the capacity of the landscape to provide goods and services that satisfy human needs, demands and goals, directly or indirectly (De Groot and Hein 2007). Functions which are associated with habitat management, hunting, the protection of water quality, the maintenance of cultural identity, supporting leisure, recreation and quality of life are usually linked to the landscape scale even if the decisions concerning landscape change, which affect the provision of these functions, are mostly made by landowners at the property level (Vejre et al. 2007a). This mismatch creates tensions and additional problems when it comes to operationalising the concept of function.

With either perspective, landscape function remains a quality of the landscape itself. Functions are translated into services when they are valued by people, i.e. when there is a benefit that can be obtained by people (Burkhard et al. 2013; Termorshuizen and Opdam 2009). For clarification, in this book we adopt the concept of 'goods and services' when expressing what society expects and derives from the landscape, and use the concept of 'functions' to express the internal functioning of this same landscape.

Goods and Services Provided by the Farm and Demanded at the Landscape Level: A Mismatch

The debate on multifunctionality started with the need to identify the goods and services provided by the agriculture activity beyond the production of agricultural products. It was connected with an economic question: how to compensate farmers for the public goods they deliver, besides their agricultural production activity. Further, the management provision of land use to secure these goods and services is mostly dealt with at the farm level, while the overall provision and the demand are normally associated with the landscape scale, which results in a mismatch which needs to be overcome.

This mismatch creates tensions as the market value of the goods and services provided by the landscape may be outside land managers' sphere of influence as is often the case regarding the recreational or tourism use of the landscape. Furthermore, theoretically, it has been acknowledged that in, any case, the positive external outputs of agriculture cause market failures because producers may not take the benefits to society into account and, therefore, may underprovide the product that generates the benefit (Van Huylenbroek and Durand 2003), which also adds to the tension. The added value of a beautiful landscape is beneficial to the tourist companies who organise tourist activities in that landscape, while it also contributes to the quality of life of the farmers and their families who live in such landscapes and, therefore, benefit from them. Nevertheless, it often does not produce a positive economic return for the farmers, who maintain and foster the landscape character. This 'free rider' predicament becomes a significant policy issue when it comes to protecting landscape goods and services which have been linked to agricultural production, but are no longer (Hodge 2016).

The lack of an economic return for the land manager happens even if there is close interaction between the land manager and landscape user at the local landscape level (Pinto-Correia and Kristensen 2013). The land managers are the ones who manage the land and make the related decisions, while the users are those who demand and use the landscape for various functions and, therefore, benefit from it. The users may influence the managers in different ways: through daily contact in different types of networks, by actually paying for some of the goods and services they obtain from the farm, or through the pressure they may exert on the government to support these goods and services through public policies. However, except for the enjoyment they derive from living in a beautiful landscape, farmers do not receive any direct compensation for the public goods they provide to society (Brouwer and Van der Heide 2009).

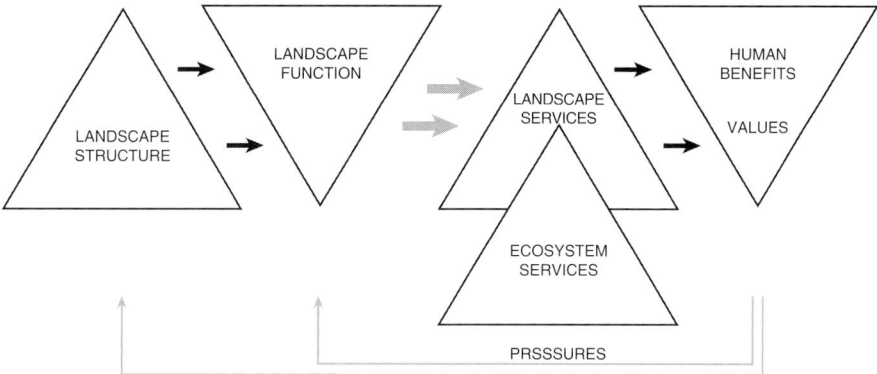

FIGURE 2.3
The relationship between landscape structure, landscape function, ecosystem services and landscape services (which largely overlap) and human benefits. The human benefits, and the expectations concerning them, lead to pressures which have an effect on both the function and the structure of the landscape.

Many of these goods and services do not always have a market, or the market does not work correctly. The non-existence or the imperfect functioning of the market are closely connected and the difference is not always clear. Private and public goods are distinguished on the basis of the concepts of excludability and rivalry (OECD 2001; Van Huylenbroek and Durand 2003). Excludability means that it is possible to exclude someone from using or consuming a product or activity, while rivalry means that the consumption of a product or service by one person or group means that it is no longer available for others to consume. In contrast to agricultural products, public goods are characterised by non-excludability and non-rivalry. Multifunctionality often applies to situations where goods have both private and public attributes, or where private and public goods are interrelated in production (Brouwer and Van der Heide 2009).

Public policy intervention may be used to ensure the provision of public goods by influencing private decisions, something which we discuss in Chapter 6.

The Ecosystem Services Approach and its Translation into Landscape Services

Parallel to landscape functions, a growing research field which has its roots in ecology and ecological economics is focussing on ecosystem services. The more widespread the concept of ecosystem services becomes, the greater the need to clarify its meaning and how it is used (Burkhard et al. 2013; Costanza et al. 2014; Crossman et al. 2013; Turner and Morling 2009; Schröter et al. 2014). There are clear parallels between the way in which ecosystem services are generally defined, even if emerging from different disciplinary fields, and the concept of landscape services. This is closely related to the functioning of the landscape, thus, the landscape functions as classically defined in landscape ecology (Potschin and Haines-Young 2013; Termorshuizen and Opdam 2009).

In other scientific traditions, the term 'ecosystem services' has become central in the debate about human–nature relationships (Fisher et al. 2009) and has emerged as a way to communicate societal dependence on ecological life support systems (Daily 1997). Ecosystem services have so far been defined in multiple ways. A simple definition has been provided by Burkhard and colleagues (2013), who state that ecosystem services are the contributions of ecosystem structure and function – in combination with other inputs – to human well-being. These authors distinguish between: (a) ecosystem function, which includes structure and processes; (b) the supply of ecosystem services including regulation, provision and cultural services and; (c) the demand for ecosystem services, which corresponds to the human benefits, which may be translated into the value people derive from these services (Figure 2.3). Seen from a landscape ecology perspective, there is, thus, a parallel between landscape function and ecosystem function as they both refer to the functioning of internal flows and processes in the system. Additionally, there is also a parallel between landscape services and ecosystem services as they are both related to the values of people – the difference seems, thus, to be a matter of scale (Potschin and Haines-Young 2013).

The concept of ecosystem services has been influential in landscape research and environmental research more generally as it has been widely used in environmental policy domains. The concept has drawn attention to the relationship between people and nature and it has been an effective way to communicate societal dependence on ecological life and support systems (Gómez-Baggethun et al. 2010; Norgaard 2010).

Yet the use of the concept is highly controversial (Schröter et al. 2014). First it links aesthetic and symbolic values of the landscape directly to specific landscape features – a specific church, for instance, will be given an ecosystem value for its contribution to the landscape scenery and its symbolic value. This means that all the benefits of a landscape are attributed to landscape features and are, thus, separated from the landscape as a whole, including the people living in it. Some of the landscape analytical methods practised in Western European countries in the 1960s and 1970s took similar approaches to landscape value (Christians 1979; Neuray 1982). Here the landscape values were measured through a combination of preference studies and the accumulation of values attributed to the individual landscape features. These approaches to landscape analysis were never established as part of landscape planning practices and are no longer in use.

Second, the attribution of values to specific features to measure the landscape's contribution to human well-being seems appealing as it addresses the problem of farmers and other landowners and managers contributing to 'common goods and services' without getting paid for it. In this way, the ecosystem service approach fits nicely into neo-liberal ideology aimed at market regulation (Norgaard 2010). Although such market creations have a role to play in promoting more sustainable practices in, for instance, farming, they reinforce the commodification of the landscape, in which the use value of the landscape is replaced by market values in various forms, which may be more likely to lead to injustice and increase the scale problem with landscape management and change occurring at the individual property and business levels and landscape services occurring at the landscape level. This does not mean, of course, that it is not possible for the public domain to pay for specific actions (or to prevent 'non-actions' beyond owner rights) through payments or other forms of compensation.

The ecosystems approach has, thus, received criticism due to its focus on increasing quantification and narrow view of nature and ecological processes (Norgaard 2010; Schröter et al. 2014). From an ecological science perspective, advocates of this approach, which started as a metaphor to support the understanding of the human–nature relationship, have reduced the analysis of ecosystems to their individual components and have ignored or underestimated the complexity of ecosystems and the complex methodological frameworks already developed within ecology to tackle this complexity (Norgaard 2010). The focus on monetary evaluation has contributed by attracting political support for conservation, but has also commodified a growing number of ecosystem services (Gómez-Baggethun et al. 2010). Having said this, the ecosystem services approach, even if the variety of concepts and disagreements between scientists may be obstructing its acceptance (Burkhard et al. 2013), has gained public and political attention, and, if used with clear definitions and goals, has the potential to support decision-making related to the people's complex uses of nature at different scales – and thus also to support landscape management from the farm to the local and regional scales.

Towards a New Understanding of Landscape Services?

In line with the understanding of landscape function as internal to the landscape, the difference between function and service becomes more straightforward: a function can be translated into services when they are valued by people. One function can provide several services. Functions continue to exist in the absence of people, while services exist because people value and use the landscape (Termorshuizen and Opdam 2009). From the landscape and particularly the landscape ecology perspective, conceptualising services may be seen as a way of including perceptions of value in the quantitative assessments of landscape change and its impacts. While the ecosystem services approach proved useful for measuring the contribution of nature to human well-being and quality of life, the landscape services approach inherently incorporates the social dimension and spatial patterns (Vallés-Planells, Galiana and Van Eetvelde 2014). Therefore, the concept of landscape service better fits the analysis of the landscape than ecosystem services as it relates directly to landscape functioning. Landscape services can be defined as ecosystem services which are provided by multiple landscape elements in combination as emergent properties (Wu 2013). When positioned in a landscape study perspective, using the concept of landscape service seems more coherent (Termorshuizen and Opdam 2009): landscape services are more easily linked with pattern–process relationships, better unify scientific disciplines and are more relevant and legitimate to local practitioners. Nevertheless, the concept of landscape service has not achieved mainstream applicability among the rapidly growing community who deal with and communicate about ecosystem services.

The scale issue is not without significance. As indicated previously, there is, in fact, a clear scale mismatch between the management of ecosystems at the field and farm level and the demand for the related services at the landscape level (de Groot 2010; Vejre et al. 2010). Therefore, land-use decisions often ignore the value of these services (Bateman et al. 2013). A common vision to guide a transition towards sustainability across all scales is still required, which could be based on improving the delivery of ecosystem services in changing landscapes under uncertainty, or better still, landscape services provided by multiple landscape elements in combination (Wu 2013). To overcome the sustainability quest, Potschin and Haines-Young (2013) defend the need for a place-based approach in the analysis of ecosystem services in order to include the context-dependent variations. They defend the assertion that the identification of the system elements that provide the service does not explain how these services change, and that only the landscape approach can provide the land manager, or other users of the landscape, consistent information on the ecosystem services provided.

Despite their increasingly widespread use by scholars and also policy makers at higher levels of governance, we do not elaborate on the concepts of ecosystem and landscape services in this book. The approaches which are based on these concepts have some operational advantages, but they also exclude an integrative understanding of the farm and the landscape systems, which results in sectoral and individual management arrangements. As shown by some authors (Gómez-Baggethun et al. 2010; Hodge 2016), the trends towards the monetisation and commodification of ecosystem services is partly a result of a gradual move away from the original economic interpretation of the benefits of nature as use values in classical economics to their conceptualisation as exchange values in neoclassical economics. This move has led not only to greater interest from policy makers but also to the commodification of a number of ecosystem services and the reproduction of a market logic to tackle environmental problems, while at the same time focussing on the single farm when many issues can be tackled only at the landscape level with the cooperation of several actors. Therefore, we do not follow this path, even though we recognise it is powerful in many circumstances.

2.5 Conclusion

In this chapter, we have discussed a number of concepts and emerging paradigms that we consider most relevant for understanding today's rural landscapes of Europe, and for combining the different disciplines that deal with the processes affecting the landscapes. This chapter has dealt with the definition of the landscape concept and the different dimensions it includes. It has discussed the meaning of place and how a local-based understanding of the landscape is linked to the issue of place. Further, it has presented and reflected on what the rural is today and how it can be described so that the multiple activities and expectations it supports are expressed. Finally, this chapter has addressed the ecosystem services approach and classification, bridging over to landscape services as a more encompassing concept, which is more useful for a transition to sustainability.

The intention has not been to be exhaustive, almost the opposite. The intention has been to address the most significant concepts and ideas and to clarify them so they become easy for the reader to understand. Many other concepts could be included, which would certainly be relevant for the study and understanding of the transitions taking place in the rural landscapes of today's Europe. They will be addressed in the remaining chapters of this book, while others will have to be the subject of future discussions.

3

Conceptualising Rural Landscape Change

Unseen Landscape Change Going up the hill to the ruins of the fourteenth-century fortress in Montemor-o-Novo, we leave the small, white town of Alentejo behind and start discovering the surrounding landscape. This hill is the highest in the area, as far as the eye can grasp. Looking south, a river runs around the hill, edged almost continuously by a thick, riparian gallery of vegetation; a few small streams run down into this river. The gently undulating terrain is occupied by a mixture of small plots of old olive groves, open pastures, vegetable gardens and woody thickets, and a dense but dispersed settlement. On the horizon there are large extensions of Montado, the savanna-like forest resulting from the extensive silvo-pastoral land use system in large estates, bordering the outer side of the small-scale mosaic.

This small-scale Mediterranean landscape shows scattered signs of modernisation: the Madrid–Lisbon highway, the renewed national road, the residential developments in the outskirts of the town, the luxury hotel and property development just a couple of kilometres away. Many of the houses, even if kept with white walls and red tiles, as the traditional style requires, are also new buildings and considerably larger than the former peasant houses. But otherwise, the landscape's appearance is very much the same as 50 years ago. There is the same small-scale mosaic and the same tiny plots, each with a single house. There are the same old olive groves, the same pasture plots, the same trees along the river and streams, the same very small vegetable gardens, the same shrubs in the small, steep slopes close to the river, the same jingle of sheep bells. Bucolic, some would say when looking at this landscape.

But does this mean the landscape is unchanged? No; the underlying functions have suffered radical changes. Farming is no longer the dominant economic, social or land-use activity. Most of these small-scale plots are now valued due to their amenity qualities. The farmers who derived their main source of economic income from these small-scale farms have now been replaced, to a large extent, by newcomers, who have settled aiming to achieve an idealised, bucolic lifestyle. They do not depend on an income from agriculture: farming is a hobby

FIGURE 3.1
View over the Alentejo landscape from the fortress of Montemor-o-Novo, Portugal

related to the lifestyle they are pursuing. A few are engaged in farming as their main activity, and these are often niche producers who are connected directly through short supply chains to the Lisbon market, or to highly specialised international markets (e.g. organic vegetables sold directly in specialised shops or in weekly box schemes to consumers in the Lisbon metropolitan area; aromatics exported directly to Germany; high-quality wine exported to Germany and Switzerland). The local community and the connection to the town have also changed. Social interaction with people in Évora, Lisbon or any European city can be as intense as the local connections, while the local bonds have changed, becoming more diverse and with new neighbour roles and dependencies.

The changes to this landscape are indiscernible. Why? Because the landscape elements and the land cover pattern remain relatively unchanged and, thus, the appearance of the landscape is very similar to how it looked 20 years ago. When observing the landscape from a viewpoint, probably only an attentive observer who is looking for signs of change and has the necessary tools to understand them would be able to detect any changes. This is because the changes we have described so far in this chapter are, in the first instance, functional and, as is often the case, there is a lag before any structural changes become apparent. There are a number of reasons for this. First, there is an inherent inertia in the land cover in that it does not always change even when it has lost the function which has created it, e.g., an olive grove may remain an olive grove for decades even when olives have ceased to be harvested and the trees are no longer being pruned. Another explanation is that new functions emerge that have an unchanged expression in certain types of land cover, e.g., an olive grove or a pasture may be maintained as such by an owner due to heritage preservation or aesthetic motivations.

An additional important explanation for the conservation of the appearance of the landscape is related to the increasing regulation of the landscape structure from the agricultural sector and especially the planning sector. A large body of rules stipulates how housing should be distributed, where new buildings may be located, how land may be divided, which land uses are permitted etc. This set of rules is becoming increasingly complex, and it inhibits change. At the same time, the responsible authority, the Portuguese Ministry of Agriculture, does not appear to consider the hidden functional changes relevant (Helming and Wiggering 2003; Mander, Helming and Wiggering 2007; Pinto-Correia et al. 2015b). In the international arena, there is a renewed focus on the role of small-scale and family farming (e.g. 2014 was the International Year of Family Farming and Smallholder Farming). However, in Southern European countries, where traditional small-scale farming coexists with new, emerging models of small farming, small farms, to a great extent, are still predominantly considered irrelevant in economic terms and, therefore, they are, in general, not targeted by policies or other public interventions in the agricultural sector. In general, the prevailing view is that traditional farms will disappear in the medium term and, thus, do not deserve particular attention, while small niche and novel farms are considered leisure rather than economic activities. Altogether, this means that the landscape structure is becoming increasingly regulated, while the way it is being managed is becoming progressively deregulated.

At this stage, it is unclear where these new and mixed changes will lead. They may very well remain unnoticed. However, there is a tension because the previous productive use of the land is becoming less important, while the new uses for these areas, although maintaining the appearance of the areas, do not involve the same land management. As the previous farming functions disappear and new ones emerge, new actors or actor networks take over. And it is uncertain how much these new actors will be able to maintain practices (the same or new, reshaped ones) which result in the landscape as we see it today.

3.1 Introduction: How to Address the Indiscernible Changes in European Landscapes?

Similar changes are occurring all over Europe so that many rural landscapes face the same predicament with a mismatch between landscape patterns which reflect former functions and the new multifunctional uses and management demands which rely on a landscape structure inherited from the past functions. The result is that the responsibility, roles and interrelations between the different uses of the landscape and the threshold for what is acceptable regarding change are contested (Selman 2012).

Although these trends have been going on for some time in different places in Europe, there is a surprising lack of literature on how to analyse and grasp the driving forces affecting today's landscapes (Plieninger et al. 2016). On the one hand, landscape analysis commonly deals with how a landscape changes and assesses the impact these changes may have on ecosystem services and the goods and services valued by society. On the other hand, drivers of change at multiple scales are mostly analysed from an economic perspective in relation to market-driven processes and the role of public policy intervention. There is a tradition for scientific literature on agriculture change and the transition from production-oriented farming to multifunctional farm management, and vice versa, to relations between farming and rural communities and activities (Ilbery 1998; Knickel and Renting 2000; Potter 2004; Robinson 2008), but these analytical constructions are rarely linked to other processes, such as those associated with various forms of urbanisation, or to the landscape dimension. As such, the territorial dimension of farming is rarely considered, i.e. the functional relations and flows between the rural and the urban, the potential for farming to be integrated with other rural services or the benefit from new initiatives initiated by other sectors (Primdahl 2014).

In order to understand the transitions in rural landscapes and to produce knowledge that is relevant for their management, we rely on theoretical and conceptual frameworks. To improve in the understanding of the changes taking place and to grasp their societal relevance, theoretical reflection, empirical observations and analytical models which address the changes in the landscape itself (land use and land cover) as well as the changes to the driving forces, considering the actors involved, are essential (Hersperger et al. 2010).

This chapter is an attempt to understand the landscape transitions that are occurring. It extends the concepts introduced in Chapter 2 and discusses some of the conceptual frameworks that explain changes in the 'rural', their linkages to urban-generated processes and transitions in landscapes. This chapter also discusses which dimensions of the landscape need to be acknowledged if the changes in the rural landscapes are to be understood and it presents the drivers and the way they affect the landscape.

3.2 Analysing Change: Intentional, Functional and Structural Explanations

Landscape changes are due to a combination of natural and social processes, which in turn occur at various scales in time and space from annual changes at the level of the arable field to long-term climate change which affects landscapes on a global scale (Brandt and Vejre 2004; Emanuelsson, Arding and Petersson 2009; Wylie 2007).

Natural processes causing landscape changes include geological processes such as plate tectonics (causing earthquakes and volcanic activities), erosion (caused by water and wind), deposition and glaciation; ecological processes linked to the establishment and succession of plants and animals; soil developments caused by climate, vegetation and decomposition by small animals and microorganisms. These processes occur continuously, but they may also be caused by extreme events – disturbances – such as volcanic eruptions, floods, tornados and fires (Bastian and Steinhardt 2002; Forman and Godron 1986). The climate and climate change are also included in natural processes.

Through temporal cycles, the climate has caused radical environmental changes. However, the climate is no longer solely affected by natural processes as human activities are also affecting the climate, firstly through increased emissions of greenhouse gases (CO_2, methane, N_2O and others) (www.ipcc.ch/publications—and—data/ar4/syr/en/spms2.html/). Global warming has already caused immense changes, resulting in an increased risk of extreme weather events such as storms, torrential downpours, droughts etc. These natural processes are connected through complex feedback linkages. For example, warmer and drier summers in southern Iberia affect the natural regeneration of oaks in silvo-pastoral systems, which leads to progressively less dense tree cover in these areas. Sparser tree cover means that the local albedo increases, which has consequences for the micro-climate in that it leads to higher summer temperatures, which in turn exacerbates the effects of drought and hinders natural tree regeneration. A silvo-pastoral system with a low density of trees is, in turn, highly susceptible to climate change and would be adversely affected much sooner than a more dense forestry system. Furthermore, as a consequence, the climate affects all the other natural processes while it is also, to some extent, affected by them. Feedbacks also occur within other components of the natural system such as vegetation development which affects the soil, which in turn affects further vegetation development (Forman and Godron 1986).

When social processes are included in the understanding of landscape change, the cause–effect relationships in landscape change become dramatically more complex. On the one hand, landscapes are changed by primary agents such as farmers who continuously manage and sometimes change their farms as a result of a combination of their values and adaptation to changing conditions. Historically, farmers within the same region often tended to adopt the same or similar ways of farming, building styles and landscape management practices over time. Therefore, they established particular local landscapes which were sometimes very different from other local landscapes with similar natural conditions. Today, on the other hand, each individual farm is becoming increasingly linked to other farms through more or less globalised networks of various types, which means that, to some degree, they are converging. Danish dairy farmers now compete with Irish and New Zealand farmers, and they are adopting similar technologies and production systems, which means that today, the source of inspiration and the overall economic conditions go far beyond the regional scale (Primdahl and Swaffield 2010a). Furthermore, each farm is also influenced by other uses and demands from the countryside. This means that actors besides farmers and demands other than for food, fibre and energy production are influencing farmers and other landowners' strategies.

In sum, this complexity means that it is often necessary to combine different approaches when analysing rural landscape change. Michael Jones (1988) has suggested distinguishing between three types of explanation each of which is linked to specific research questions concerning landscape change (Figure 3.2). As shown in the figure, the three modes involve the agent, the functional and the overall socio-economic aspects of society and, in general, they refer to the following three spatial and temporal scales: micro (intentional), macro (structural) and meso (functional). According to Jones (1988), these three sets of explanations are the following:

1. *Intentional explanations*
   - They can be identified by a chronological-biographical method, linking specific parts of the landscape to individuals or groups. Changes in properties, specific plantings, single buildings, location of a camping site etc. are explained through this approach involving interviews and archive studies.
   - The questions asked are: who did what, where, when and why?
   - The limitations of this approach are mostly that these analyses are mainly suitable for small areas and short time spans. Generalisations of such studies may be limited.

2. *Structural explanations*
   - In this approach, the focus is on how underlying structural driving forces – technology and the market for example – lead to major

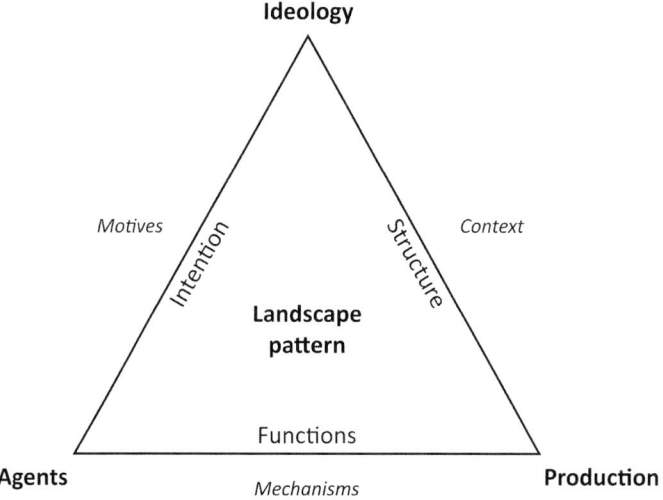

FIGURE 3.2
The three types of explanation for the changes taking place in the landscape, according to Jones (1988): intentional, structural and functional. These explanations are often all present, but due to their underlying complexity, analytical approaches tend to focus on only one.

changes. This is also where the impact of public policies, at different scales, is assessed.
- The question asked is: what is the socio-economic context for the change?
- The limitations of this approach are linked to a tendency to neglect agents or underestimate their significance. For example, the heterogeneity which characterises different local landscapes within the same region, which is derived from different decision processes and local contexts, cannot be explained.

3. *Functional explanations*
    - Here the analytical approach focusses on how the different parts of the landscapes function in relation to each other and how and why the landscape changes according to the interplay of factors. The specific adaptation of production to natural conditions can be explained through functional adaptations, although this is of less significance today as discussed previously in this chapter.
    - The question asked is: how is the landscape functioning? Common patterns are mapped and described.
    - The limitations here are that the structural dimensions of the landscape and changes to these may appear self-evident when referring to functional explanations and may lack links to personal stories or structural trends in society. This approach provides simple explanations such as 'the strong winds and open landscapes "resulted in" the new "wind power landscape".'

The insights Michael Jones (1988) provides are crucial in helping us to understand the perspective we are using when analysing landscape change. However, they also provide us with a framework to combine different explanatory approaches, to analyse and understand from different standpoints and to bring these together in more comprehensive studies.

A widespread approach to understanding landscape change is to specifically focus on the drivers and agents of change. In a landscape context, driving forces can be seen as processes which directly affect landscape change (such as the aforementioned natural forces) or as forces that influence an agent's decisions and actual changes in the landscape. The driving forces interact in complex ways with various kinds of agents almost always being involved. Hersperger and colleagues (2010) present four analytical models that link driving forces to landscape change. In three of them, driving forces are distinguished from agents (termed actors). Here the way driving forces affect the agents is at stake and the different models vary with respect to the number of driving forces analysed and whether any interaction takes place between the driving forces and the agents. In the fourth model, the driving forces are assumed to directly cause the change. According to Hersperger and colleagues (2010), correlations between the driving forces (e.g. environmental factors such as terrain and soils) and the changes (e.g. agricultural marginalisation) are statistically analysed to explain the changes. Although this fourth model represents a common approach to analysing landscape change, the explanatory power of such an analysis can be questioned. As values, intentions and practical management and change actions are crucial to understanding almost all changes to rural landscapes – they usually represent the so-called proximate causes of change (Geist and Lambin 2002; Van Vliet et al. 2015) – we find it useful to distinguish between such types of change factors and the driving forces that underlie or directly affect agents' decisions and actions.

The DPSIR Model

The OECD (1993, 2001) developed a conceptual model of relationships between driving forces, environmental states and responses to develop a framework of indicators to be used when evaluating agri-environmental policy. The general logic behind the model is that driving forces of various kinds affect the environmental state, which then changes and, therefore, causes various responses (including policy inventions), which in turn affect the driving forces. The model represented a step forward in developing more precise indicators of environmental change, and the EU Commission has adopted a widely used modified version of the model which has been employed in e.g. the European Environmental Outlook published by the European Environmental Agency (EEA 2005). In this version, two more dimensions have been added – a pressure dimension which represents changing factors caused by underlying driving forces (e.g. increased use of fertiliser caused by higher grain prices) and an impact factor representing consequences of changes in the environment (e.g. biodiversity loss affected by increased nutrient levels in soil). The driver-state-response model was, thus, extended to a driving force-pressure-state-impact-response or DPSIR model. Essentially, this conceptual model more precisely elucidates what is causing the changes and their result. Increased grain prices, for example, may represent a Driving force which causes Pressure for arable land expansion, which in turn may result in a change in the environmental State in the form of the reclamation of semi-natural habitats with negative Impacts on biodiversity which thereby results in a policy Response in the form of a new habitat-protection law (see Figure 3.3). For a discourse analysis of the DPSIR model, see Svarstad and colleagues (2008).

The DPSIR model is useful when the results of a certain policy must be evaluated (*ex ante* or *ex post*), while it can also be used to better understand how specific actions affect the landscape and how policy measures could be designed to cope

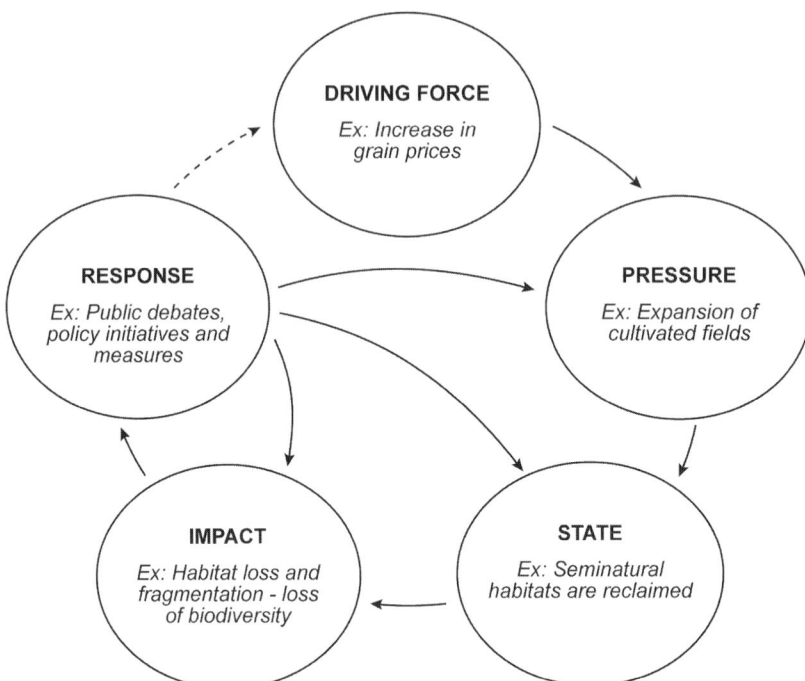

FIGURE 3.3
The expanded OECD driving force-state-response model, transformed into driving force-pressure-state-impact-response model (DPSIR)

with the consequences of such actions. However, we identify two shortcomings of the model. First, it is weak when it comes to the role of agents in landscape change; it is simply too 'macro' for this. Using Jones' modes of explanations, the model performs best when it comes to structural and functional modes of explanation, which also means that it is best when it comes to large-scale, i.e. regional-global types of change. Second, the model tends to mainly interpret policy measures as 'responses', which means that proactive and vision-driven types of policy action are mainly ignored.

An Agent-Based Framework

The DPSIR model helps us to understand the policy process at higher governance levels. Still, the maintenance and management of the landscape depend on farmers' or other land managers' everyday decisions, at the farm scale (Primdahl and Kristensen 2003; Van der Ploeg and Marsden 2008). These decisions are conditioned by policy mechanisms and depend not only on the farming systems in question but also, to a greater extent, on the farmers' strategies and the farmers and their families' relationship to the farm and the local landscape. This means that the role of the decision-maker, at the farm level, is crucial to understanding the changes that take place at the landscape level. All analyses of rural landscape change processes need to state whether the complete set of actors and actions is being considered.

The close interplay between the policy system, the farm system and the landscape system has been widely described by Kristensen and Primdahl (2016), Primdahl and Brandt (1997) and Primdahl and colleagues (2003) and is at the heart of the analysis and dynamic systems modelling of Bryden and colleagues (2011a).

An agent-based framework for analysing landscape change, inspired by the work of Primdahl and Brandt (1997), is shown in Figure 3.4. According to this model, the agent – in this case the farmer – affects the landscape from three positions as a landscape manager.

As owners, farmers manage their properties as a family or as a personal issue, taking into consideration the history of the family and of the farm, gender and age dimensions and reasons of the personal sphere. Usually the owner makes major decisions such as the establishment or removal of major landscape elements, the sale or purchase of land, investments in new buildings etc. The owner is legally responsible with regards to most, but not all regulations concerning land-use and landscape features, while it is often the owner and not the leaser who is responsible when land is leased. For these reasons, the farm owner is an important agent concerning landscape change, although this

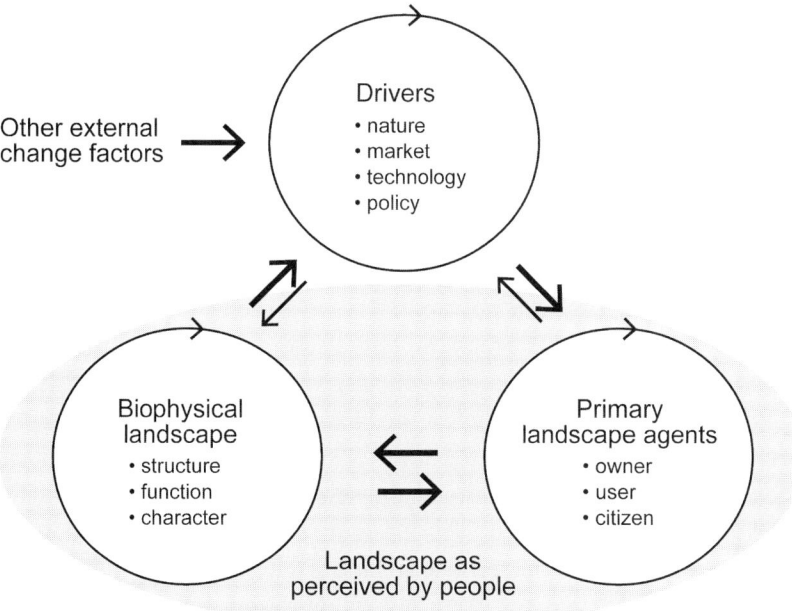

FIGURE 3.4
An agent-based framework for analysing landscape management change. After Primdahl and Brandt (1997).

is often underestimated or even overlooked in research and landscape policy making (Primdahl et al. 2013b).

The farmer, of course, also affects the landscape in his role as a producer of food, fibre and energy. All aspects of the landscape, including natural processes, social functions, physical structure and character, are affected by farming practices such as tillage, fertilisation, chemical spraying etc. (Knickel 1990). On the one hand, the farm producer has a relatively significant effect on natural processes associated with soil development, water resources and biodiversity. On the other hand, the agricultural practices maintain a certain type of landscape pattern, resulting from the production system. As a producer, the farmer sees himself or herself as a skilled professional, and it is this role in which the pride of being a farmer and the farmer's identity are nested. It seems meaningful to associate variations in farming practices within the same agricultural system with the farmers as agents even if the same practices are shared by groups of farmers within the system. A central concept in this case is 'farming styles' (Van der Ploeg 1994), which has been used to analyse differences even within specific systems (e.g. dairy farming). Accordingly, a farmer may decide to adjust his production, but any adjustment will be within the confines stipulated by his particular values and understandings of what constitutes good farming practice. Each farming style may or may not be more economically attractive than another, but the farmer will justify his decisions and farming practice based on his set of values.

Finally, a farmer is usually also a citizen within a community, and from this position the farmer shows consideration for his neighbours (when spraying for instance), and it is often in his or her role as a citizen that the farmer cooperates with other farmers or with the village community (Primdahl et al. 2013b).

Thus the primary agent, that is the landscape manager, maintains and changes the landscape through different roles, which are affected differently by the various drivers which underlie the practices. From a public policy point of view, this is important as policies always affect the landscape through affecting agents' decisions and practices. This also means that the implementation of policies is crucial for success. Policies, which will be further discussed in Chapter 6, are designed and redesigned based partly on responses to perceived problems and visions articulated at various levels. The agent-based framework as presented in Figure 3.4. acknowledges the role of the individuals who make decisions and the multiple drivers affecting them. The framework is valuable when attempting to understand local landscape change and the role of agents, which also means that it is useful when analysing policy and planning processes where the agent is crucial. On the other hand, this framework does not provide much information about overall changes at the regional level over periods longer than a few decades.

Case B

Pivka area, Slovenia, can biodiversity conservation be the driver for the control of natural forest expansion in former agricultural fields?

Landscape Appearance

The site described lies in the Pivka municipality around the village of Palčje, south of Lake Palško in the Notranjska Karst landscape as indicated on the ELCA map with nr 273 (see inside back cover). This is a typical karst region with predominating limestone and dolomite geology. Land use is a mixture of the traditional farmland, forests and extended areas of abandoned farmland with natural regrowth. It is a hilly landscape with gentle slopes, yet rocky and harsh, especially when strong winds and winter come along. North of the village lies the intermittent lake Palčje, which changes during the year from a flat meadow in the dry season into a lake after heavy rains. The area harbours a rich mixture of different habitat types ranging from large sedge communities to Mediterranean montane grasslands, from low-altitude hay meadows to wet eutrophic and mesotrophic grasslands, with a mixture of Mediterranean-montane broadleaved deciduous thickets and species-rich hedgerows. Areas of abandoned agriculture are being covered by early-stage natural and semi-natural woodlands. Various forest types are present in the area, ranging from thermophilous deciduous woodland to *Pinus nigra* woodland.

Landscape Processes

Spontaneous afforestation is one of the landscape change processes affecting landscape functioning. It is the result of a combination of structural, social and demographic reasons. In Central Europe, abandoned agricultural land has already often

FIGURE 3.B.1
Landscape appearance in the Pivka area

Case B (*Cont.*)

FIGURES 3.B.2 AND 3.B.3
Landscape appearance in the Pivka area

Case B (*Cont.*)

been transformed entirely into forest. Degradation of high-nature-value (HNV) farming systems frequently results in homogenisation of land use and loss of agricultural landscape elements, both natural and cultural ones. In cases of forest expansion, an additional dimension of this process must be taken into account: traditional land-use practice is a potential guarantee of biodiversity conservation, as opposed to land abandonment. The latter again results in homogenisation of land use (of a less valuable vegetation type, since vegetation succession has replaced the HNV farmland) and vanishing of landscape elements (natural and man-made) in a forest matrix (which is the end point of vegetation succession under Central European conditions). In Slovenia, the rate of forest cover expansion has been dramatic, starting as early as the nineteenth century and continuing well into the end of the twentieth century (Kobler, Cunder and Pirnat 2005).

Results: Identification of Future Roles

The number of large forest patches (more than 200 ha) increased due to forest encroachment and merging of smaller patches. The average patch area of the largest patches has increased as smaller patches from 1975 have now often merged with the

FIGURE 3.B.4
Ortophoto of the area described
Source: 'Sinergise, Geopedia.si' (www.geopedia.si)

> **Case B (*Cont.*)**
>
> forest matrix. Former medium-sized patches (between 50 ha to 200 ha) have practically disappeared, whereas the number of smaller patches (less than 5 ha) has not significantly changed (Kobler et al. 2005).
>
> Karst rocky pastures are gradually being overgrown and their future depends partly on grazing and fires. Remaining grasslands (pastures or meadows) or cropland are extremely important for at least two bird species, the Barred Warbler (*Sylvia nisoria*) and the Red-backed Shrike (*Lanius collurio*), as mentioned in the newest proposal for the revision of Special Protection Areas (SPA) in Slovenia. Both species are sensitive to forest expansion, preferring a mixture of meadows with shrubs and hedgerows as their optimal habitat. It is of key importance that small living forest patches and intertwinement of intensive and extensive meadows are preserved within the farming area from a biodiversity point of view.
>
> Agriculture is still practised in a range of about 1 km around the edge of villages: the open space between villages and forest are covered with a mixture of grasslands and hedges representing extremely important habitats. However, it is only through European agriculture policy subsidies that such landscapes can be kept intact, as maintaining farming here depends heavily on the recognition of its role as provider of public goods. Only a few farmers remain in the villages, which continue the traditional farming practices.
>
> **Conclusions**
>
> Besides natural characteristics (slope angle and exposition, soil types and remoteness from settlements), it is crucial to consider social conditions (small plots, ageing ownership, uncertain future of agriculture in marginal agriculture areas), in order to identify possible mechanisms to prevent a radical change in the landscape. Biodiversity payments are not enough; understanding farming decision-making processes is of crucial importance here.
>
> The preservation of these areas should be additionally supported by the CAP, which recognises the need 'to enhance the overall environmental performance of the CAP through the greening of direct payments by means of certain agricultural practices beneficial for the climate and the environment'. As far as deforestation, areas used for farming before 1975 could be suitable if in accordance with the interests of habitat preservation. The issue is whether landowners can be motivated to maintain farming in these conditions, and for how long.
>
> *Janez Pirnat*

In fact, farmers' decisions, values and motivations are not static as they may vary over time due to changes in the overall norms of the farming community or society, while they may also change in line with the life cycle of the famer and the farm. The rationale for this type of thinking is that over a lifetime, all farmers go through different stages regarding their agricultural production: firstly, a build-up and consolidation phase, during which many changes occur, followed by a phase of stabilisation with few changes and ending with a downscaling phase, which involves the extensification of production as the dominant trend (Potter and Lobley 1996). The impact of farming on the landscape is likely to differ during such phases. The presence of a successor to take over the farm may, however, result in different patterns: a high level of investment and expansion throughout a longer life cycle with other implications for the landscape (Potter and Lobley 1996). If there is no successor in the family, an older farmer may start downscaling earlier or may change focus from production values to amenity landscape values. Such processes need to be considered if processes of land-use change at the farm level are to be tackled by policies, as shown in the Pivka area, Slovenia (see Case B).

To summarise, the landscape is changing as a result of driving forces and agents' concrete actions. Various explanations and change models may be used depending on the type and scale of the changes. Having reviewed the perspectives usually taken when analysing changes in the landscape, we now present the major conceptual frameworks which support the understanding of these changes. We first address the structural dimension and thereafter the intentional.

3.3 Multifunctionality: Between Farm-Level Management and Landscape-Level Demands

The Emergence of the Concept of Multifunctionality

Multifunctionality emerged as a concept linked to agriculture during the 1990s (Helming and Wiggering 2003). In forestry, it had been in use since at least the 1970s (Mander et al. 2007). The use of the concept can be traced back to a number of societal and political transformation processes which attempted to introduce integrated rather than separate approaches, and which have influenced scientific and policy approaches in different ways in different countries and disciplines in both the natural and social sciences (Renting et al. 2009). The emergence of the notion of multifunctionality was partly due to attempts to describe current developments in land use and land-use policy (Parris 2004; Selman 2008) and partly due to a need to justify farming activities beyond the production of food and fibre and thus also the legitimacy of policy support for agriculture (Tilzey and Potter 2008). Within the context of the liberalisation of commodity markets, the OECD (2001) adopted the neoclassical approach, which recognises the production of commodities and positive externalities, as well as market failure regarding public goods, as key notions (Cairol et al. 2009; Renting et al. 2009). In this understanding, an economic activity may have multiple outputs and, by virtue of this, may contribute to several societal objectives at once. The multifunctionality of agriculture is then defined as the joint production of commodities and non-commodities by the agricultural sector. Joint production means there is a fixed or quasi-fixed relationship between two outputs of an economic activity (Van Huylenbroek and Durand 2003). Besides food, feed and fibre production, agriculture encompasses other values related to conservation, recreational and leisure activities, non-use values and other existing and emerging societal concerns, i.e. both market and non-market goods (Brouwer and Van der Heide 2009).

Productivism and Post-productivism

The notion of multifunctionality is particularly related to the debate about a shift from a dominant productivist to a post-productivist paradigm. This shift is related to the multiple expectations society has for the rural today. Productivism has been broadly conceptualised on the basis of industrial agriculture, which maximises production and farm modernisation. In contrast, post-productivism emerged in the late 1990s and represents a much fuzzier concept (Kristensen 2001; Mather et al. 2006; Selman 2009; Wilson 2001), which reflects a move away from the productivist paradigm (Hediger and Knickel 2009). The concept is related to the need to defend agriculture in the context of a more environmentally aware society, and to the emerging societal demands for public goods and services that agriculture can provide or support. Lately, this move away from the productivist paradigm combined with pressure for food security has led to the emergence of the new concept of bio-economic productivism, the aim of which is sustainable intensification (Marsden 2013). Bio-economic productivism is partly a reaction not only to the global food crisis of 2008 but also to widespread acknowledgment that food and nutrition security cannot really be achieved or maintained without sustainable production systems, which respect the limits of natural resources and preserve their balance (Knickel et al. 2013; Maye and Kirwan 2013).

Agriculture can be considered multifunctional when it provides several other goods and services besides production that society values and is, thus, not solely focussed on production and can instead be placed somewhere on the continuum between productivism and post-productivism in a broad sense (Barbieri and Valdivia 2010; Oreszczyn, Lane and Carr 2010). In fact, there is no clear divide in time and space between the two absolute orientations, but rather a mixture of various orientations in farm management, both at the local and at higher scales of analysis, which is reflected in different pathways of transition towards increased multifunctionality (Pinto-Correia and Kristensen 2013). Even in the same location, divergent processes may have been occurring side by side, leading to greater complexity in changing patterns (Buller 2005; Short 2008; Van Berkel and Verburg 2011). At present, several processes of transition, which coexist spatially, temporally and structurally, are occurring in rural areas in multiple combinations, resulting in a more complex, contested, variable mix of production, consumption and protection goals (Holmes 2012; Robinson 2008; Tilzey and Potter 2008).

Multifunctionality from the Farm to the Landscape

The debate on multifunctionality has become broader. As presented previously, it started with the need to justify agricultural subsidies and to match this need to the increased societal demand for goods and services provided by the rural. However, a broader debate has also been taking place on multifunctionality related to rural development more generally and the viability of rural communities, including their social, economic and environmental aspects (Clark 2010; Van der Ploeg et al. 2002a; Van der Ploeg et al. 2008). A more comprehensive understanding of multifunctionality as an attribute of rural space, which is available to be exploited by a much larger community of stakeholders, is expressed by Helming and Wiggering (2003) and Potter (2004). This understanding is rooted in a reinterpretation of the contribution of agriculture to rural development, its engagement in broader market processes (Van der Ploeg et al. 2002a) and the changing role of farmers and a larger community of land managers in the so-called consumption countryside, as seen in the previous section.

Nevertheless, a broader understanding of multifunctionality needs to acknowledge the fact that many of the environmental goods and services that society expects from the rural can be provided only at the landscape level. In consequence, the demand, as well as the provision of such goods and services,

---

Box 3.1

Multifunctionality: From the Plot to the Landscape Scale

Research on conservation biology has shown that the abundance and distribution of bird species related to the complex landscape in the small-scale farm mosaics of Mediterranean Europe cannot be related directly to a change in land use or land cover on small-scale farms, but only to a pattern of change over the whole mosaic area; or more precisely, to the whole landscape area, including the forest patches and their spatial relation with the agricultural area (Moreira et al. 2001; Stoate et al. 2001). Considering recreation and tourism certainly leads us to the same conclusion: the interest and attractiveness of these small-scale landscapes for leisure activities can never be assessed at the farm level – only at the landscape level, where each farm is a patch in a complex landscape mosaic (Müller, De Groot and Willemen 2010).

can be definitively assessed only at a scale larger than the farm itself (Bastian et al. 2014; Fang et al. 2015; Knickel and Peter 2005).

3.4 New Roles for Agriculture in Differentiated Rural Spaces

The multiple trends of change in the rural space, caused by the interaction at multiple scales between productivist and post-productivist drivers, have been widely analysed in the two past decades (Buller 2005; Ilbery 1998; Knickel et al. 2004; Marsden et al. 1992; Short 2008; Wilson 2001, 2007; Van der Ploeg et al. 2000). The combination of the associated transition processes has resulted in increasing spatial variation in rural areas (Lambin and Meyfroidt 2010; Marsden 2003; Murdoch et al. 2003; Van Berkel and Verburg 2011): some landscapes have become more homogeneous due to large-scale specialisation regarding land use and land cover, while others have developed into or have preserved more complex landscape patterns. The understanding of rural differentiation in literature began with an analysis which centred on agriculture, but it was soon broadened to include the rural space as a whole. Therefore, studying rural differentiation is a useful approach to improve understanding of the drivers of change in European landscapes.

In 2003, Terry Marsden conceptualised these processes as a separation between different paths: the *agro-industrial dynamics*, which is based on productivist thought and action; the *post-productivist dynamics*, which views the rural as a consumption space where nature is commodified; and the *rural development dynamics*, where multifunctionality is becoming dominant (Marsden 2003). This conceptualisation does not separate rural areas into one or the other profile, but presumes that rural landscapes are affected differently by these dynamics at different scales so that in some cases, they may be dominated by one of them, while they may be more broadly affected by all of them in others.

Murdoch and colleagues (2003) distinguish the following four types of parameters which shape the development trajectories of rural localities in the British countryside: economic, social, political and cultural. These socio-economic processes, which interact at the regional, national and global scales, are regionalising the rural, i.e. generating new constellations of socio-economic relationships which gain coherence at the regional level. The interaction of these different parameters thereby creates different types of countryside, which can be categorised into the *preserved* countryside, the *contested* countryside, the *paternalistic* countryside and the *clientilist* countryside. The preserved countryside is dominated by protectionist attitudes and decision-making and is linked to the consumption interests of a neo-rural population. The contested countryside lies outside the main commuter zones. Such areas are dominated by agricultural interests, although commercial and development interests are also at play and these are often hard to combine. Furthermore, there are also newcomers looking for a residential environment, and this leads to increased conflicts between different groups. The paternalistic countryside corresponds to areas where large, private estates dominate and the dynamics are shaped by large landowners, who may diversify their farm-based activities when their traditional extensive production fails to generate sufficient income. In the clientilist countryside in remote, rural areas, agriculture and its traditional institutions are still dominant, and the management of rural dynamics is dominated by state agencies with farmers highly dependent on public subsidies, while the socio-economic dynamics depend on state institutions. Countries with a tradition for strong rural policies at the national level such as Norway and Switzerland are clear examples of the latter. There is thus mainly an issue of changing balances between production and its conventional or new roles, and the emerging consumption interests.

This is in fact close to what John Holmes (2006, 2012) has demonstrated as new modes of rural occupancy, which are related to the societal use of the rural areas. According to the rationale in this book, this societal use can be said to correspond to drivers: what society expects from the rural space drives the rural space to respond accordingly, even if there are failures or mismatches at the spatial or temporal scale. The conceptual framework Holmes proposed is helpful for understanding and characterising today's rural space as well as the territorial role that agriculture may have in each type of space (Figure 3.5). John Holmes partly developed his conceptual framework based on empirical studies in Australia, but mainly on reflections developed by European authors. Holmes (2006) argues that the different transitions taking place in rural areas today imply a radical reordering of the three basic purposes underlying people's use of rural space, namely production, consumption and protection. A shift away from the previously dominant production goals towards a more complex, contested and variable mix of production, protection and consumption goals is occurring. Even with the recent trend for increased productivism which has been driven by food security concerns worldwide, there are many areas where drivers other than productivism are shaping the rural today. In the areas where the hegemony of agricultural *production* is declining, both in the occupation of space and in the occupation of people, amenity demands for the countryside are growing (*consumption*), while societal values are changing and turning the environment into an issue (*protection*):

- *Production*. In the industrialised world, the use of the rural space for agricultural production has been decreasing, while yields have increased in the past 100 years (Rounsevell et al. 2012), which has been made possible by expanding food markets, technological innovations and agricultural structural developments. In recent years and following the global food crisis of the early 2000s, these trends have been observed with increasing concern and food security featuring high on the European agenda once more (Maye and Kirwan 2013b). However, the belief that not all rural areas will be needed for food production in an increasingly globalised world remains.

- *Consumption.* The agricultural overcapacity Holmes (2006, 2012) described results from farm intensification in many agricultural areas and leads to farm redundancy elsewhere due to reduced viability, which farmers and land managers attempt to resolve through pluri-activity, extensification, disinvestment, conversion to non-farm uses and, finally, land abandonment. At the same time, rural areas are increasingly being 'consumed' by market-driven urban interests related to housing, tourism or well-being (Gardi et al. 2015).
- *Protection.* Finally, in addition to alternative market-driven uses, certain societal concerns – mainly those linked with the quality of natural resources and nature conservation – can be pursued effectively only in rural space (Lomba et al. 2015; Tscharntke et al. 2012).

Placed in the interplay between the three sets of drivers, rural areas are positioned differently today compared to some decades ago. The case of the Great

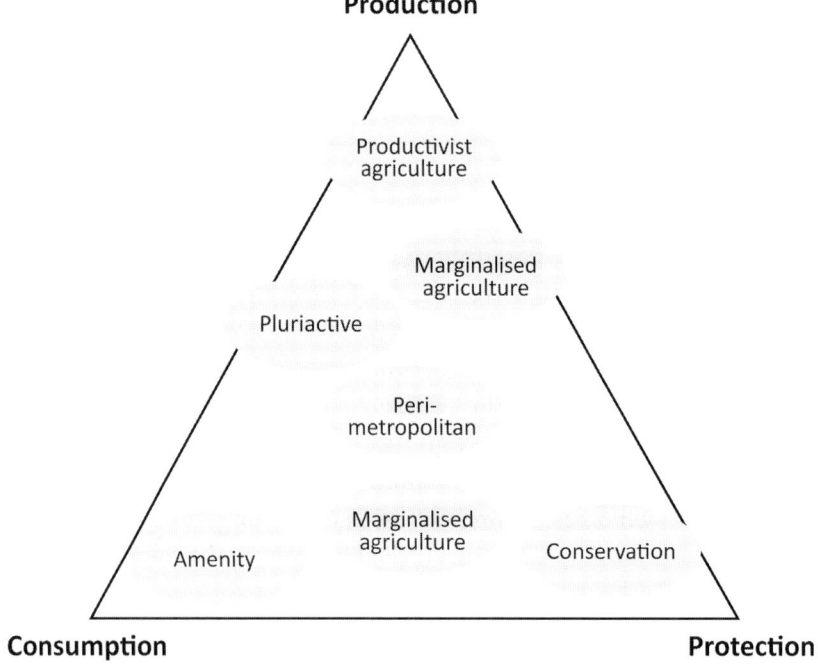

FIGURE 3.5
Modes of rural occupancy
(adapted from Holmes (2006))

Trossachs Forest (Case C) demonstrates how the social expectation for a landscape has changed from extensive production to nature protection, and how this change is causing tensions and may cause even the disappearance of acknowledged values in this landscape. This means the new trajectories of the rural space and the related landscapes are still to be defined, and the pathway for wise management is not obvious. Furthermore, changes are occurring at different intensities and scales and a review of the literature reveals that there is a spatial, temporal and structural complex of transitional processes taking place in Europe which is resulting in the increased diversification of rural areas (Domon 2011; Knickel 2011; Marsden and Sonnino 2008; Pinto-Correia and Kristensen 2013; Renting et al. 2009; Van Berkel and Verburg 2011; Wilson 2007). Bryden and colleagues (2011a) discuss the transformation of agriculture-related ecosystem services into socio-economic development and quality of life.

These three different approaches all acknowledge the assertion that there are rural areas in Europe today where market-based agriculture is still the major driver of land use and the economy, but that there are many other areas where agriculture has ceased to be viable in a global context, or has been pushed back by pressure related to urban uses and other land uses and activities which are leading to other types of community composition. These conceptual frameworks were proposed in 2000. With the recent changes in the world economy and the push towards increased food production and bio-energies in recent years, changes have occurred in the pressure for intensive land uses, which has altered the balance between the different sets of drivers (production, consumption, protection), although the overall differentiation of rural areas remains as the different drivers also persist.

Wilson (2007) has contributed to this conceptual debate by highlighting how the multifunctionality concept is better at illustrating the complexity of contemporary rural change and explaining such change than the polarisation between productivism and post-productivism. He states that there are no single, isolated decisions, but a large complexity of processes in one direction or another at different scales. Multifunctionality helps to explain the transitions taking place in the rural through their non-linearity, heterogeneity, complexity and inconsistency (Cairol et al. 2009). According to these characteristics, pathways of change in the rural can be linked to a weak, moderate or strong multifunctionality, which is ultimately assessed through their territorial expressions (Wilson 2009); something which the aforementioned authors also highlight.

In their review of conceptual approaches to multifunctional agriculture, Renting and colleagues (2009) illustrate how the multifunctionality, which is increasingly expected from farm management and is increasingly driving the decisions of a number of farmers, is resulting in larger-scale transition processes which end up leading to differentiated rural regions. These authors and Bryden and colleagues (2011b) thus also link multifunctionality directly to processes of rural change. Renting and colleagues (2009) contribute to a clarification of the analytical complexity by distinguishing the different approaches to the analysis of multifunctionality:

- *market regulation* approaches, which focus on the economic aspects of new markets for non-commodity outputs;
- *land-use* approaches, which focus on spatial or territorial changes and neglect the particularities of each land manager or land management unit;
- *actor-oriented* approaches, which look at the social constructions and decision-making processes and;
- *public regulation* approaches, which centre on the way institutions and policies support or hinder multifunctionality.

Each of these separate approaches identifies different drivers and processes of change, but they all attempt to clarify how and when rural areas are affected differently by the changing demands of society and individual actors.

The Problem of Data Availability

When considering the rural space and the landscape impacts of agricultural change, the application of such conceptual constructions to empirical analysis poses considerable challenges as discussed by Knickel and Renting (2000). Thus the previously mentioned approaches have been developed by their authors into analytical models only to a very limited extent (Knickel et al. 2009). In addition, it is extremely difficult to find data on farming and the other uses and activities in the rural to support the required analysis or to maintain consistent levels of detail, to facilitate spatially explicit combinations (Marsden and Sonnino 2008; Pinto-Correia and Breman 2009; Slee and Pinto-Correia 2014; Woods 2011).

One example is the data used in national statistics and by Eurostat, which are essentially agricultural relating to agricultural production and farm economics, while the sampling is based on the farm as an economic unit. This holds two direct consequences for land-use and landscape change. First, it means that the size of the unit (area, livestock, intensity) determines whether it is included in the sample. A farm property of 30 ha, for example, of which 26 ha are leased out would not be included in the sample because the 'farm business unit' is too small. Second, as farms owned by hobby farmers are typically much less intensively farmed than those managed by commercial farmers, they become underrepresented (with respect to land size) in the statistics simply because they are relatively small units in economic terms. This means that a large number of farms and farmers is excluded from the surveys when such surveys are used to analyse land use. Consequently, such statistics do not work well when it comes to land-use management or practices related to the landscape such as new trees, hedgerow plantations, the creation of ponds and wet areas or the maintenance of small, rural paths. Even though small (and frequently hobby) farms may in reality occupy the largest share of the area (as they often do in less-favoured and the most attractive peri-urban areas), they are underrepresented in the statistics (Primdahl et al. 2013a).

3.5 Proceeding towards Classification: Spatially Explicit Rural Types

The use of selected and especially designed indicators to produce typologies has so far been the most straightforward way to approach the differentiation of rural areas beyond conceptual discussions. The analyses, which represent a step forward in the understanding of rural change, draw upon the existing theoretical and descriptive studies combined with expert knowledge and analytical experience.

Conceptual frameworks that produce categories of rural areas do, in fact, produce conceptualisations of ideal types of rural in the Weberian sense. The types proposed by Marsden or Murdoch are first approaches to a new classification of the rural. They propose well-defined types, while reality is always much more complex and the situation in each rural area, independently of the scale, will most often be a hybrid of the different types. That is also what Holmes and Wilson aim to highlight, rural areas as the result of different and constantly changing drivers, which are now much different from the drivers of 50 years ago, and will most probably again be different in 50 years.

Building on the conceptualisation of ideal types, Pinto-Correia and Breman (2009) propose a methodological approach to classify rural areas. This classification is based on the identification of ideal types and is supported by an analysis of the current situation combined with recent trends so that the static picture of the present situation is complemented by the ongoing dynamics. Three dimensions of the rural are considered: (1) the land cover and, thus, the landscape pattern; (2) agriculture and the farm sector, and; (3) the community. The classification procedure was applied to Portugal (excluding the islands of Madeira and Azores). Data were gathered from different sources: Corine land cover, agricultural statistics and socio-economic statistical data. The analysis was done at the municipal level, with values in each indicator for each municipality. The data were collected for two different moments in time: 1990 and 2006. The most relevant indicators in the three dimensions considered (the land cover, the agriculture and the socio-economic dynamics) were selected through an initial analytical step, composed by a principal component analysis. This was followed by three separate cluster analyses, each with the selected variables in each dimension. This led to the classification of the municipalities and resulted in a typology in each dimension: land cover, the agriculture sector and socio-economic dynamics (Table 3.1). The comparative assessment of

TABLE 3.1 *A typology of rural areas driven by a conceptual framework and implemented through data analysis; this typology results from an analysis of Portugal municipalities 1990–2000*

A four-step typology of rural areas – application to Portugal			
1. Land cover types	2. Agriculture types	3. Socio-economic types	4. Rural ideal types
Dense urban	Small farm units tending to intensification	Urban and dynamic	1. Production and specialised agriculture with high profitability
Urban expansion + agriculture under pressure	Declining agriculture + progressing forest	Peri-urban dynamic	2. Extensive agriculture with high environmental quality
Forest dominant	Extensive agriculture in decay	Young and industrial	a) In homogeneous landscapes
			b) In diversified landscapes
Mosaic of agriculture + forest, dynamic	Very large farm units with extensive farming	Extremely strong decay	3. Agriculture combined with planning and conservation
Stable agriculture + forest	Diversity and dynamism in small farm units	Strong decay	a) in forestry landscapes
			b) in mountain landscapes
Mainly agriculture	Stability and important social role of agriculture	Stagnant and unqualified	c) in urban areas
Polarised dynamics	Medium and expanding farm units with extensive farm systems	Problematic but with some potential for resistance	4. Rural services agriculture
			a) in peripheral areas
			b) in dynamic and disturbed areas

these three typologies led to the identification of the ideal types, where the existing and potential role of agriculture for the dynamics of the rural was in focus (Table 3.1). Ideal types are here considered in the Weberian framework: they refer to a mental image or conception rather than a material object, and therefore they are seen as a model. An ideal type can be conceptualised as a kind, category, class or group with a particular character that seems to be the best example of it. Weber defined an ideal type as a mental construct, like a model, for the scrutiny and systematic characterisation of a concrete situation, and as such it can be used as a methodological tool to understand and analyse social reality. Following the identification of the ideal types of rural situations in Portugal, each municipality was then placed in one type through selected indicators and related thresholds. Each municipality could be placed in two types, the most dominant type and the secondary type, acknowledging the hybrid situation of most rural areas today.

The aim of this classification was to highlight the potential of each area and the role that agriculture could play in a territorial perspective. Such an approach is strongly driven by a need to assess what can support the maintenance of agricultural activity in marginal areas, and is highly relevant for the marginal regions of Europe (Breman et al. 2010). In any case, the step from conceptual framework to grounded classification was taken and the resulting classification started a debate in the relevant bodies about the resources that could potentially be mobilised in the rural areas of Portugal.

Other analytical efforts, with similar concerns, but more centred on the multifunctionality of the rural areas than on the role of agriculture, have produced typologies which have been applied to rural areas in Europe. In the same way as the analysis described earlier in this chapter, such analyses have often assessed and combined different dimensions of the rural.

Van Berkel and Verburg (2011) have used the concept of territorial capital to ground an analysis on the spatial distribution of the potential for rural development across Europe. These authors have identified and collected data for indicators expressing: (a) the intensification of agriculture; (b) off-farm employment; (c) rural tourism and; (d) nature conservation. With a combined analysis, a map was made of the capacity of European rural areas to support multiple functions, i.e. a level of multifunctionality, and also the potential for rural development options given the current conditions.

Van Eupen and colleagues (2012) have searched for the specific attributes or characteristics of rural areas to support policy targeting and have grounded their analysis on existing environmental classifications (Hazeu et al. 2011). Building on this and on new spatial and statistical analysis, they have considered three main dimensions: (a) the environmental conditions; (b) accessibility and; (c) the economic density. These three layers have different spatial expressions. By combining the three, they have produced a rural typology at different levels of aggregation, which enables the consistent identification of comparable rural areas and interrelations with urban areas in the European territory.

Fertner (2012) has conducted an interesting study of four different urban-rural typologies (OECD, DG Regio, Eurostat and ESPON) and has downscaled the typologies to local, municipal levels using Denmark as a case study. Although the four typologies vary regarding the categories used, the results – rather detailed patterns of urban, rural and intermediate municipalities – are relatively consistent across the four typologies. A typology which considers the urban and the rural and the gradients in between the two represents an important tool in analysing social changes in a spatial context, but, as the authors point out, it may also be a useful tool in landscape policy and planning, for example, in differentiating between restrictions on housing permits. In this case, downscaling is often needed.

These typologies mainly represent data-driven approaches which are undertaken by combining data layers. They show how different types of areas are more or less comparable, but they fail to identify the determining characteristic of each set of areas (Carvalho-Ribeiro, Madeira and Pinto-Correia 2013; Verburg et al. 2013). Often, they do not address the new modes of rural dynamics and occupancies, which requires a stronger conceptual background and more detailed and powerful data sets including the social and cultural dimensions, which often lag behind.

Copus and colleagues (2011b) attempted to structure different ruralities in the EDORA project which challenges outdated generalisations about the nature of rural areas. The overarching aim of EDORA was to examine the process of differentiation in rural areas in order to better understand how EU, national and regional policy can enable these areas to build upon their potential to achieve (in the words of the EU 2020 strategy) 'smart, sustainable and inclusive growth'. The aims of this approach were: (a) to describe the main processes of change which are resulting in the increasing differentiation of rural areas; (b) to identify development opportunities and constraints for different kinds of rural areas; (c) to consider how such knowledge can be translated into guiding principles to support the development of appropriate cohesion policy (Copus and Hörnström 2011).

The EDORA framework (Copus and Hörnström 2011; Figure 3.6) recognises both locational issues and structural characteristics in creating diversity and builds on the work of Dijkstra and Poelman (2008). Unlike the attempts of Marsden and colleagues (1992) and Murdoch and colleagues (2003) to simply discern different types as categories of use of space, Copus and colleagues (2011a) have actually formally mapped them. They build on a similar conceptual framework to that used by the previously mentioned

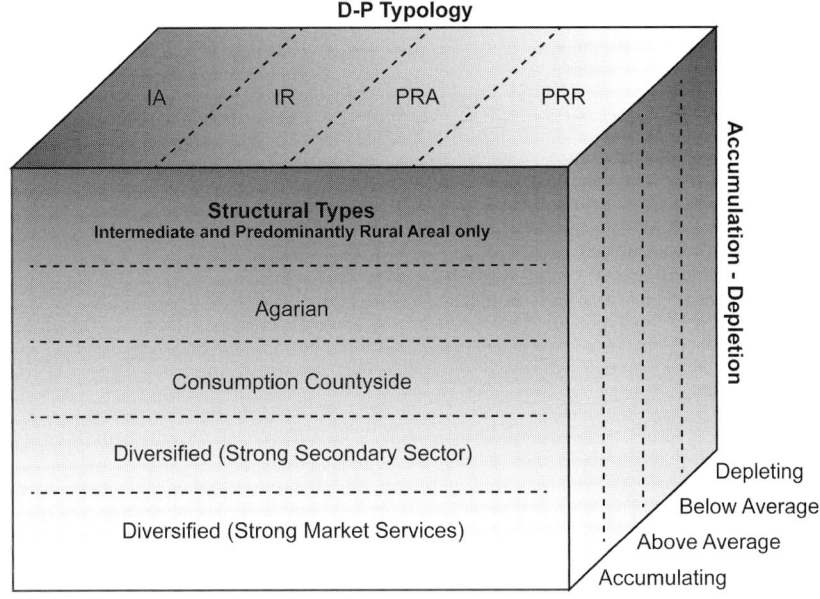

FIGURE 3.6
The EDORA Cube, as produced by Copus and Hörnström (2011)

authors, but they apply it to a typology of European rural areas through a meta-scale analysis paired with detailed case study observation. A review of the literature was the first step, followed by story lines from exemplar regions, which led to the production of three meta-narratives: (a) agri-centric; (b) urban-rural, and; (c) globalisation. These were complemented with the existing data and the creation of indicators. The combination of the complementary conceptual and data-driven paths resulted in the construction of the EDORA cube (Figure 3.6), which defines the frame for the analytical work which leads to the spatial differentiation of the NUTS 3 regions in Europe according to three dimensions: (1) rurality and access to urban areas; (2) the degree of economic restructuring; (3) socio-economic performance. These three typologies form a kind of 'triangulation' which was then used as the basis for a statistical 'portrait' of rural Europe as well as an exercise in 'foresight' which considered the key dimensions of future change over the next 20 years and described alternative scenarios and their likely policy implications.

The EDORA approach certainly represents a step forward in the possibilities to describe and identify what is going on where in rural areas besides agriculture, but without neglecting its central role (Slee and Pinto-Correia 2014).

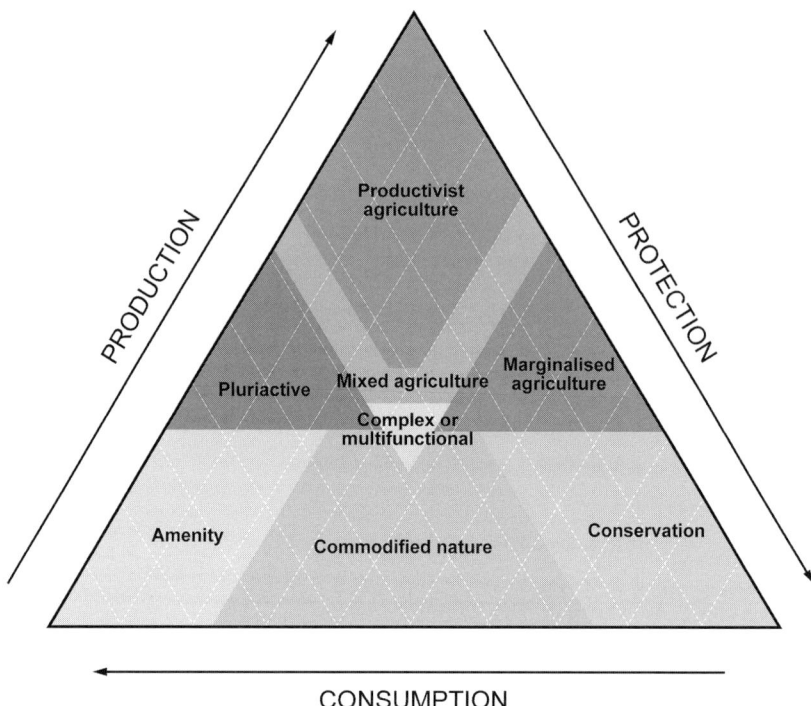

FIGURE 3.7
Three-dimensional plot used to assign a specific rural type to each spatial unit (adapted from Holmes (2006, 2012), in Pinto-Correia and colleagues (2015c))

Holmes' conceptual design (2006, 2012) still leads to a slightly different approach as it is an attempt to grasp the societal uses of the rural, e.g., how society is using the rural space. The resulting classification of the rural areas is grounded on the relative importance, in each area, of production, consumption and protection drivers, and on possible changes to this balance over time. An initial application of this model was tested on Portugal (Carvalho-Ribeiro et al. 2013), and then, with stronger analytical tools, on the European territory (Pinto-Correia et al. 2015c). In this work, two analytical pathways are described: one based on a clustering analysis of selected variables for each dimension, and another based on expert judgement of the relative weight of the variables using an Analytical Hierarchical Process (AHP) pair-wise comparison procedure, followed by a rescaling which leads to the classification expressed in the model in Figure 3.7. As these two methodological approaches have been applied to the whole of Europe, there are obvious limitations related to the availability of data from the whole of Europe or comparable data sources. Regardless of these limitations and the different results of the two approaches, the produced typologies can be seen as a step forward in addressing rural change in Europe today. They show a high degree of spatial differentiation that illustrates the hotspots of each dimension: production, protection, consumption.

With the linkage to a conceptual framework, the poor explanatory capacity of data-driven typologies is surpassed: the resulting classification of the studied areas acquires a more clear positioning in relation to the processes in focus in the analysis, and the questions addressed to the data can be predefined.

Case C The Great Trossachs Forest, the changing character of a moorland landscape in Scotland

Landscape Appearance

The Great Trossachs Forest is a 16,500 hectare area set in the heart of the Loch Lomond and the Trossachs National Park (LLTNP) in central Scotland (Figure 3.C.1), and accounts for 9 per cent of the national park area. According to Scottish Natural Heritage's Landscape Character Assessment (2009), the area is made up mainly of Highland landscape types comprising hills, upland glens, forested hills and glens and transitional landscapes, including open and forested moorland. The area is within a National Park and includes National Scenic Areas, Sites of Special Scientific Interest and Special Areas of Conservation, having itself become a National Nature Reserve in 2015. All of these designations directly contribute to the dynamics of landscape change in the area, with forest landscape regeneration being favoured on the lower slopes for both the near and the distant future. A substantial proportion of these new woodlands are the so-called wood pastures, which are being created using traditional techniques of grazing cattle in woodlands. The main focus of recent and ongoing actions on the area has been creating a mosaic of habitats, including the restoration of ancient forests, woodlands and moorlands, which are therefore prioritised over other traditional and identitary landscapes, such as highland pastures dedicated to sheep grazing.

Landscape Processes

More than 70 per cent of the population of Scotland lives in nearby cities, and therefore the forest has great potential to attract a wide range of visitors. However, the touristic, educational and conservationist potentialities recently being developed in the area are triggering subtle changes in the landscape which might remain unseen to the common eye. This is particularly the case of a series of actions that are planned within the scope of the Great Trossachs Forest Project, which has set objectives for the regeneration of woodland and other HNV landscapes for the following 200 years. The time-scalar mismatch that thus takes place between the project's lifetime and the perception spam of the end-users of the forest result in a series of landscape changes that may easily remain hidden to the visitor.

The project's vision has been presented as 'a spectacular landscape, stretching from loch shore, through pasture and wooded glens to open moorland,

Case C (Cont.)

FIGURE 3.C.1
Location of the Great Trossachs Forest, Scotland (www.thegreattrossachsforest.co.uk). A black-and-white version of this figure will appear in some formats. For the colour version, please refer to the plate section.

with high peaks in the far distance. At the woodland edge, black grouse display on a Spring dawn. Cattle graze among the trees and butterflies frequent the plentiful wildflowers. An eagle soars high above the mountains' (www.thegreattrossachsforest.co.uk). This vision is aligned with a series of concrete actions and plans related to the more concrete objectives of restoring habitats, involving people, life-long learning and partnership working. Concrete actions planned to achieve these goals include the creation of around 4,400 hectares (44 km^2) of native woodland, ultimately resulting in an area of 16,650 hectares (166 km^2) of

Case C (*Cont.*)

forest and open ground containing a mix of habitats, returning ecosystems which have been damaged by overgrazing and human exploitation to a more natural state. This objective is aligned with the overall aims of the Scottish Rural Policy which aims at expanding the woodland cover from the present 17 per cent to 25 per cent in 2020, but goes clearly beyond these objectives both in timescales and in consideration of landscape multifunctionality (Forestry Commission Scotland 2006; Scottish Government 2009, 2011).

Results: Identification of Future Roles

Ultimately, the strong efforts and objectives planned for the afforestation of the area might end up shifting the current character of its landscapes, which recently and throughout the past few centuries have been dominated by the grazed vegetation that epitomises the romanticised image of the Scottish Highlands. However, the ecological and historical analyses of long-term changes in the local and regional landscape provide a reality that clearly differs from the romanticised vision of the area, as promoted by the Victorians and as is still evident in much of the contemporary tourist literature. Whereby according to this vision, barren moorlands strongly characterise and add value to the landscape of the region, in reality, the moorland habitats encountered today are to a great extent the product of sheep overgrazing for various centuries, a process that resulted in high levels of ecological degradation and that may be now in the process of reversing through long-term visions and related action plans (see Figures 3.C.2 and 3.C.3a and 3.C.3b).

FIGURE 3.C.2
View over Loch Katrine to the West from Brenacholie Point (see Figure C.3)

Case C (*Cont.*)

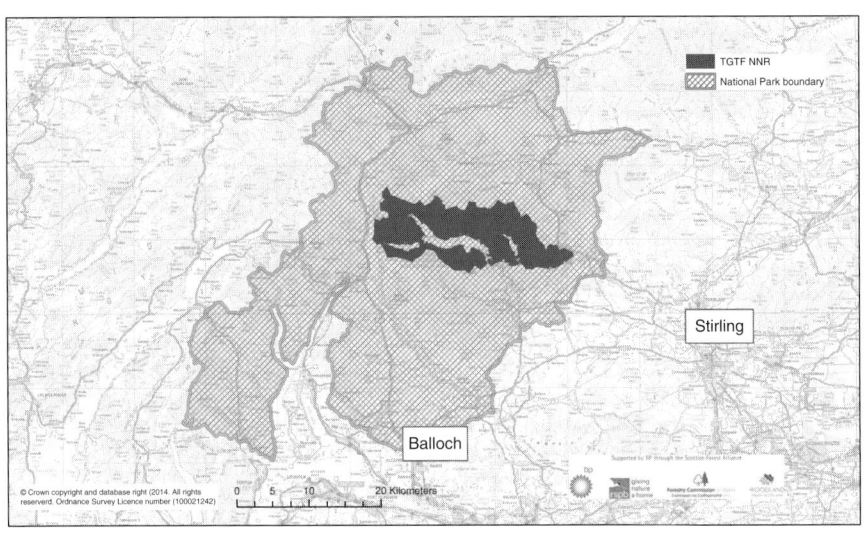

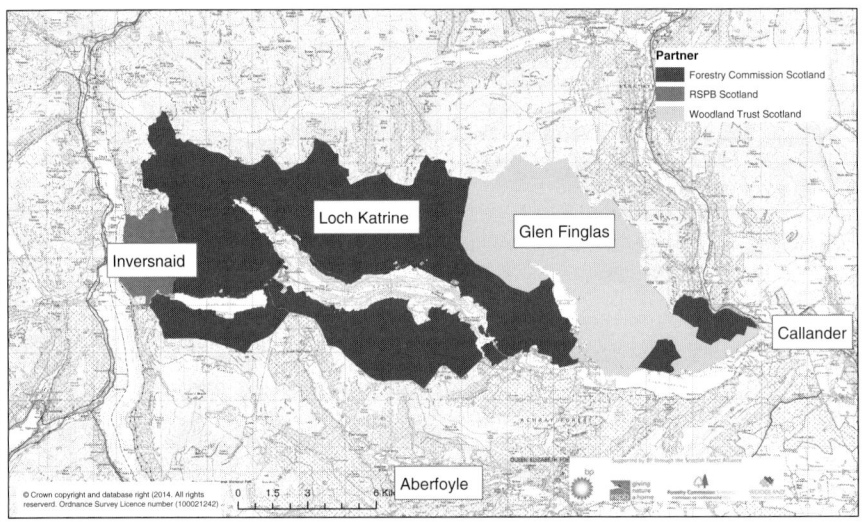

FIGURES 3.C.3A AND 3.C.3B
Local area of the Great Trossachs Forest R: Source: www.thegreattrossachsforest.co.uk. A black-and-white version of this figure will appear in some formats. For the colour version, please refer to the plate section.

Conclusions

Thus, key challenges remain for this landscape, and while a clear vision, plan and project exist that intend to shift its current character and to restore its ecological status, the objectives set may be seen by some as controversial. These include tourist-related actors profiting from a romanticised image of the area, and hunters

> **Case C** (*Cont.*)
>
> expecting to maintain in the future grouse habitats managed as they have been over the past century. Consequently, the ambitious afforestation and habitat restoration objectives may interfere with what some still perceive (e.g. tourism and the grouse-hunting industries) as a distinctive and widely acknowledged mountain landscape character, linked to the moorland and hills of the Scottish Highlands. Whilst it is fair to acknowledge the number of controversies potentially arising from such changes, it is equally important to point out the social inclusiveness, diversity and fairness that landscape transitions such as those triggered by this project might foster. Ultimately, long-term visions such as those set for the landscape plans recently approved in this area may enrich any further debates on the key challenges associated with the inherently complex nature of landscapes. Such complexity stems from the acknowledgement of landscapes as spatial units of encounter for the multiple aspects, both objective and perceived, short-term and long-term focussed, that jointly define their character, identity and uniqueness, ultimately driving differences in their degree of sustainability and resilience at different spatial-temporal scales.
>
> *José Munoz-Rojas*

But still, finding the right method and identifying the right indicators dealing with the available data remains highly challenging. The case of the Great Trossachs Forest (Case C) and its change in dominant drivers and management strategies is a good example of how adequate indicators would help clarify how exactly the area is seen today and what goods and services it is providing – to inform coming management decisions.

More simple classifications focussing on dynamic processes can also be enlightening. As shown in Chapter 5, Primdahl and Swaffield (2010b) propose a matrix of two dimensions framing current change patterns, namely the natural conditions for agriculture and the degree of urbanisation.

3.6 People in the Landscape: Farmers, New Actors and New Communities

Different Strategies of Farmers

Most farm enterprises in Europe continue to follow the productivist paradigm and the related management models. Specialisation, the global circulation of production factors and the long-distance transportation of products dominate the agricultural sector. Even if other concerns such as aesthetic values and well-being also play a role, the main income is still generated by production connected to agricultural markets. However, there is an increasing number of farmers who are introducing innovations and combined activities on the farm (Knickel et al. 2004) (see also Figure 3.8). The deepening, broadening and re-grounding of conventional processes

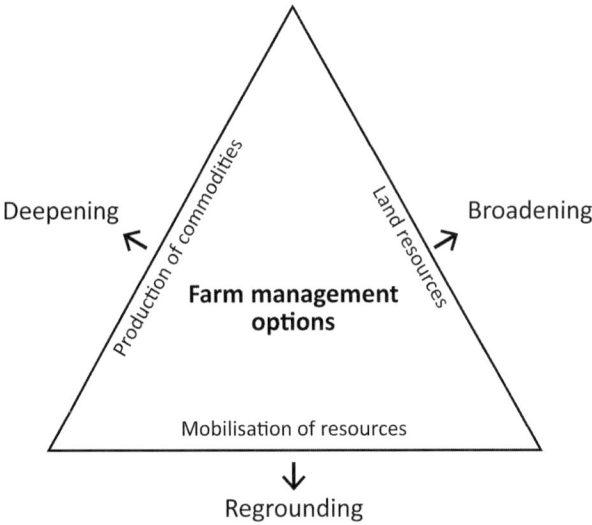

FIGURE 3.8
Deepening, broadening and re-grounding as the different options for restructuring activities at the farm level (adapted from Kitchen and Marsden (2006)). Deepening can be done through shorter value-added chains, turning organic or other special production forms etc. Broadening can be done through new on-farm activities, such as sport activities, but also nature conservation goals and management, agro-environmental measures etc. Re-grounding is done by creating structures for energy production in the farm, organising events, doing social care etc.

at the farm level jointly support more multifunctional farms that deliver a broader range of products and services than conventional farms and thereby transcend conventional sectoral and spatial boundaries (Van der Ploeg and Marsden 2008).

'Traditional' agricultural production activities represent just one aspect of these farm enterprises, which are characterised by pluri-activity or broadening, deepening and also re-grounding. Pluri-activity refers to the phenomenon of one or more members of the farming family deriving a considerable proportion of their income from outside the farm enterprise (Bryden 1993). Broadening refers to new on-farm activities which are still related to the agricultural use of the land, but which focus on other associated activities such as nature conservation or the processing and direct sale of products. Re-grounding takes place when part of the income of the farming family is derived from non-agricultural activities on the farm. There are old forms of re-grounding (the farmer having, e.g., a small construction or repair shop located on the farm), as well as a quickly expanding range of new forms, strongly associated with what is now rural development, as recreation or social care (Van Elsen, Günther and Pedroli 2006). Finally, deepening refers to the transformation of agricultural activities in order to realise more value-added per unit. Organic farming, the production of high-quality products and regional specialities, on-farm processing and marketing, as well as the creation of new short supply chains, are expressions of this new, deepening process. Importantly, all three processes imply that the borders of the classical, highly specialised agricultural enterprise are becoming blurred. That is, the newly emerging rural farm-based enterprise is becoming a multisectoral enterprise which, through pluri-activity, broadening and deepening, is also strongly linked to the non-agricultural parts of the rural economy.

Identifying Various Types of Farmers

As previously discussed, farmers and those connected to the agricultural sector are no longer the only actors in the rural or the only ones who make decisions regarding land uses. A central dimension of change in the rural landscape is the change in the characteristic attitude and behaviour of the actors involved. Until a few decades ago, communities in rural areas focussed on farming, but today, in some areas, the rural landscape is increasingly a place where diverse actors engaged in complex, new and heterodox activities meet and interact. There is, thus, a hybridisation of actors and the traditional categories are no longer sufficiently explanatory. Hybrid classification models which combine different dimensions are also required in order to understand this diversity of actors.

Fewer and fewer people are employed in large-scale, specialised and competitive agriculture. The farm sector may therefore not contribute much to a dynamic rural community. The same is happening with low-intensity farming, albeit for different reasons. In areas where the natural conditions limit farming, rural actors are increasingly focussing on alternative landscape management such as hunting, forestry, nature conservation etc. which are often undertaken by other types of actors (Ilbery 1998; Wilson 2009). This does not mean that no one lives in the rural areas where these two types of 'farming' (large-scale specialised and low intensity) dominate. But those who live in the rural are decreasingly related to farming, while the activities are increasingly decoupled from the functions that have formed the landscape. Case F in Chapter 5 illustrates the increasing significance of non-agricultural activities in the landscape.

Small-scale farming has a different pattern and is dominant in many regions of Europe, especially mountainous and hilly regions, peri-urban areas, and also some areas along the coast (Kempen et al. 2011). It still is the way of living for a considerable proportion of the rural population in some countries and regions of Eastern and Southern Europe. Traditional small-scale farming in peripheral regions is often associated with low incomes, farm decay and a tendency towards land abandonment (Shucksmith and Rønningen 2011). However, small-scale farming has also proved highly resilient to global changes, and small farms are now, in Western and Southern Europe at least, increasingly attracting neo-rural inhabitants, who turn them into hobby or lifestyle farms, produce food and create a new dynamic in the rural (Ortiz-Miranda, Moragues-Faus and Arnalte-Alegre 2013; Marsden 2013; Pinto-Correia et al. 2010; Pinto-Correia et al. 2014; Primdahl et al. 2013b).

With the differentiating trends of change in the European rural areas, the people who today are part of the rural communities are also much more diverse than a few decades ago. In particular within farming and the management of the land, thus directly influencing the landscape, many different categories can be identified.

Full-Time Farmers

Full-time farmers are economically dependent on their farm income. They are commercially oriented and work full-time on the farm. These farmers are likely also the ones with the highest level of formal professional training, but they are progressively decreasing in number in Europe as farms are becoming larger with both structures and production becoming more concentrated. Full-time farmers are mostly associated with large-scale farms or highly capital- and input-intensive, output-oriented farms. Increasingly, full-time farms are managed as an enterprise and less as a family occupation. However, full-time farmers may also make decisions regarding the use of the land which cannot be fully explained by a production rationale. Farmers are almost always inhabitants of the farm and production place they use and, therefore, their decisions concern both: (1) decisions made from a production-oriented perspective and; (2) decisions made for non-production reasons, e.g., to improve the characteristics of the farm as a living place, as a wildlife hotspot, to preserve the family heritage (Primdahl et al. 2013b). Therefore, the local landscape structure is not always the sole result of production decisions, even on highly specialised, commercially oriented farms.

Part-Time Farmers

In contrast, there are increasing numbers of part-time farmers (Van der Sluis et al. 2015). In the period 2000–2012, 4.8 million full-time jobs in EU agriculture disappeared, 70 per cent of them in the new EU Member States and 93 per cent corresponding to non-salaried workers (Eurostat 2013). Part-time farmers are those who own a rural property and use the land for farming, but are not primarily farmers by occupation or income and in many cases do not have a formal farming education, though they may be strongly commercially oriented and consider themselves professional farmers (Primdahl et al. 2013b). In general, such farmers own and manage medium-sized farms.

Part-time farmers may be less focussed on production output and, therefore, more attentive to the landscape they are producing with their farming practices.

'New Peasants', Hobby and Lifestyle Farmers

Besides full-time and part-time farmers, where farm management and production is commercially oriented, there are other farmer styles and profiles, some of which Jan Douwe van der Ploeg has described as 'new peasants' (Van der Ploeg 2009b) within the context of more peripheral countries and emerging economies. However, what we observe in Europe are farmers who are adopting a peasant livelihood, who can be classified as new peasants, although their motivation is not to make a living solely through peasantry. In a number of parts of Europe, some of these new farmers can be conceptualised as lifestyle farmers, i.e. rural landholders who farm or live on the land principally for lifestyle reasons and not for financial reasons related to farming (Pinto-Correia et al. 2015b). This does not mean that production is not central to their lifestyle or that the concept entails an affluent countryside standard of living as the term may suggest. The term 'lifestyle farmer' encompasses different types, but they all manage their land and, thus, also the landscape, whether the primary goal be farming or otherwise. Such farmers are also more likely to have a landscape and amenity concern incorporated into their management strategies as enjoying the countryside and the landscape is part of their motivation for farming. The estimated value of agricultural operations is low because the primary income is more often than not derived from non-farm sources, and especially because the income generated from agriculture is rarely the main driver of land use or occupancy. Hobby and lifestyle farmers are typically associated with small-scale farms, usually below 20 ha as observed by Bohnet (2008).

At the same time, both hobby and lifestyle farmers are often 'back-to-the-landers', i.e. migrants from the urban to the rural, who adopt agriculture as a vocation having previously worked in another sector (Halfacree 2007; Trauger 2007; Wilbur 2013), but not necessarily. Lifestyle or hobby farmers may also originate from rural areas and may have started farming with motivations other than to generate a full-time income. This profile of rural people conducting land management with different degrees of attachment to farming and no dependency on the farm income may be most common in Southern Europe due to the complex and intense relationships that characterise rural and urban communities and the close connection to the land that most families, even urban families, still have (Ortiz-Miranda et al. 2013). This situation has been intensified by the deep and long-lasting economic crisis which has been affecting these countries for a number of years.

Therefore, today, farmers in many parts of Europe have opted for new farming styles as a form of a reinvented peasant-like lifestyle which combines different characteristics of the more traditional farmer. More hybrid types of farmers appear now, combining in one single person different types of profiles or motivations and goals. In the transitional trajectories as defined by Van der Ploeg and Marsden (2008), the consumption driver in rural areas is currently associated with 'deactivation', which is characterised by a decline in agricultural activities in rural areas and a shift towards leisure, nature reserves, rural dwellings and bioenergy production. However, Van der Ploeg has conceptualised another trajectory, which he terms 're-peasantisation', which is characterised by the active construction of new levels of autonomy on commodity markets (i.e. major farm resources are produced on the farm itself) and the presence of new actors in farming. Re-peasantisation has been shown to be highly relevant in areas where the demand for commodities is increasing (Pinto-Correia et al. 2015b). Countryside consumption (*sensu* Holmes 2006) is, thus, also a driver of farm and farmland management grounded in quests for a rural lifestyle, healthy food and leisure, which may or may not be closely linked to production. However, autonomy on commodity markets surely changes the way in which farmers deal with farming or involves new types of actors entering the farming sector.

The New Contours of Rural Communities

Today, as we have seen, farmers represent a much broader and diverse group of actors than a few decades ago. The multiple profiles also respond to the adaptive capacity required by the rapid pace of changes occurring today with regards to farming technologies, markets, and society in general (Milestad et al. 2012). These hybrid and diversified farmer profiles are making a significant contribution to the diversity of the population in the rural in today's Europe.

Today, a variety of actors, who have not been present before, reside in the rural besides those involved in farming who have a direct role in managing the land. In many places in Europe, the rural community is now diverse, with farm-related actors living side by side with others engaged in a variety of activities such as back-to-the-landers and people who stay just for a while in the area before moving again – as occurs in urban societies.

A consequence of this is that the importance of agriculture in the rural community is, in general, declining, or at least changing: the share of the active population involved in agriculture has decreased to a minimum and a large proportion of the rural population now have another type of occupation, as they may also have moved to the rural area from another location and origin (Woods 2005, 2011). The data may hide those who we have previously described as lifestyle farmers as they do not appear in the statistics as farmers. However, even though farming may be returning in some small-scale farming areas, in general, the rural community in Europe can no longer be characterised by its direct or indirect connections to farming. The exceptions are in Eastern Europe, perhaps mostly Poland and Romania, although a similar trend is also expected to occur there in the near future (Dannenberg and Kuemmerle 2010; Müller et al. 2009; Sikor 2004). As shown by Ilbery (1998) and Woods (2005, 2011), the increased mobility of people, goods and information has led to the radical change of rural communities and to the formation of new power relationships and actor networks. Often, external linkages can be stronger than internal ones (Murdoch and Marsden 1994). Furthermore, these external linkages are intermixed in a new type of social positioning of the rural actors in the global society (Hedberg and do Carmo 2012; Marsden and Sonnino 2008). This change is not without trade-offs: while innovation and entrepreneurship may be stimulated, tension and conflict, which were not present previously, may also emerge (Barbieri and Valdivia 2010; Ilbery 1998).

The new complexity of actors and actor communities in the rural is reflected in the local landscape. The landscape in a given location is the meeting arena for all these types of actors: the farmer, the land manager and the landscape user (Pinto-Correia and Kristensen 2013). As such, the increasing interaction between production and consumption of the countryside is expressed at the local level through close interaction between managers and users of the landscape.

Pinto-Correia and Kristensen (2013) propose a framework which expresses the different drivers affecting the landscape today, but also the different actors interacting in landscape management and change (Figure 3.9).

Where the two axes meet in the model in Figure 3.9 is the local landscape, which is constructed and maintained through local decision-making. This decision-making is affected by the chain of factors in the social-economic and cultural axis, but also by the natural and structural set of factors, which create the context in which to act. The local level is where the landscape is constructed, but also where the landscape is 'consumed' both by individuals who manage the land and make the related decisions – the managers – and by others who demand and use the landscape for various functions – the users. The users influence the managers in different ways, not only through local networks as they may also pay for the goods and services they demand in the

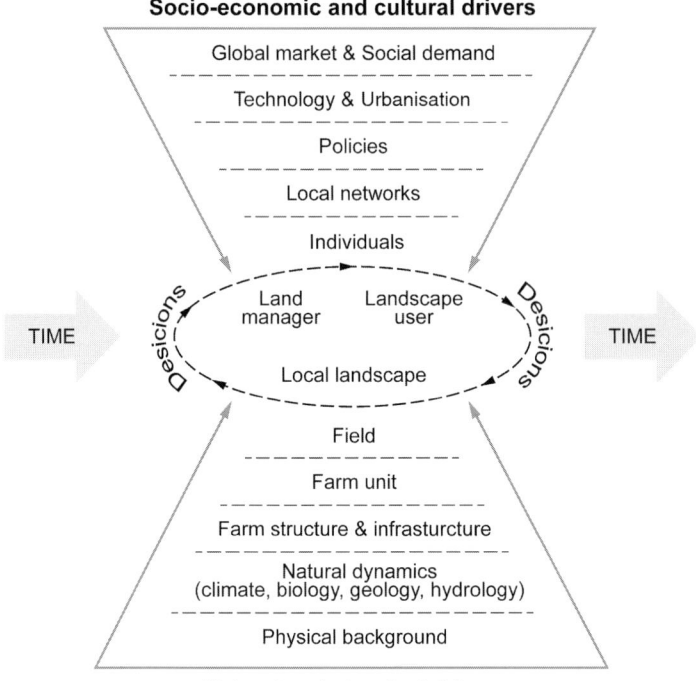

FIGURE 3.9
The local landscape as an expression of two sets of drivers (socio-economic and cultural drivers, and natural and structural drivers), but also as the arena for interaction between different types of actors, both managers and users of the land (Pinto-Correia and Kristensen 2013)

landscape or exert pressure for these goods and services to be secured through suitable public policies.

3.7 Contributions from Transition Theory

Between Structure and Agent, Understanding Change Processes

The frameworks of analysis presented in this chapter can often provide us with the linkage between the agency of landowners, farmers, inhabitants of the rural and the structural factors which we know also affect them and are transferred to landscape change through their actions. However, understanding how individuals' actions or the actions of small groups relate to the structural factors such as markets, public policies and regulations requires different approaches.

The concept of transition has been used to better understand the changes in agriculture, mainly related to the reconfiguration of activities at the farm level and the development of niche productions and activities (Knickel 2011; Van der Ploeg and Marsden 2008; Wilson 2007) and is related to the spatial, temporal and structural coexistence of several processes of transition taking place in rural areas. As observed earlier, these processes in multiple combinations have resulted in a more complex, contested, diverse mix of production, consumption and protection goals and, therefore, new modes of rural occupancy (Holmes 2012).

But much more broadly than this, transition studies rely on a large range of theoretical foundations, from evolutionary economics to sociology, innovation as well as science and technology studies (Geels and Schot 2010). They have been developed and their relevance for improving understanding of changing processes in industry, including the role of actors at multiple levels, has been demonstrated, while they have also revealed the complexity and contingency of technological change (Darnhofer 2014). Transitions in this context are seen as renewed alignments between developments of the concerned regimes, at multiple levels (Geels and Schot 2007). Transition studies consider societal systems as complex, adaptive systems, i.e. open systems that interact with their environment following constantly adaptive rules (Rotmans and Loorbach 2010).

The Multilevel Perspective depicted in Figure 3.10 has been developed by, among others, Arie Rip and René Kemp (Rip and Kemp 1998), and further developed by Frank Geels and Johan Schot (Geels and Schot 2007). The perspective view transitions as non-linear processes that result from the interaction of developments at three analytical levels: niches (the *locus* of radical innovations); socio-technical regimes (the *locus* of established practices and associated rules that stabilise existing systems); and the external socio-technical surroundings (termed the socio-technical 'landscape' by Geels (2011)). Each level refers to a heterogeneous configuration of elements with the regime more stable than the niches in terms of the number of actors and degrees of alignment between the elements. This perspective emphasises that for a transition to be successful, processes at the niche, regime and surrounding levels need to be aligned. In other words, the successful development of a novelty into a niche will not lead to transition unless, e.g., pressures from the socio-technical landscape open up a window of opportunity by exerting pressure on the regime.

Transition is a radical change at the regime level. Such a transition emerges from a succession of incremental changes over a long time period (e.g. 25–50 years) and is surrounded by great uncertainty and complexity. A transition implies a system innovation (as opposed to a series of technical add-ons). Systems innovations not only involve new technologies but also new markets, user practices, regulations, infrastructures and cultural meanings.

Socio-technical regimes are defined as the *locus* of established practices and rules (i.e. mainstream activities and their supporting institutions). The

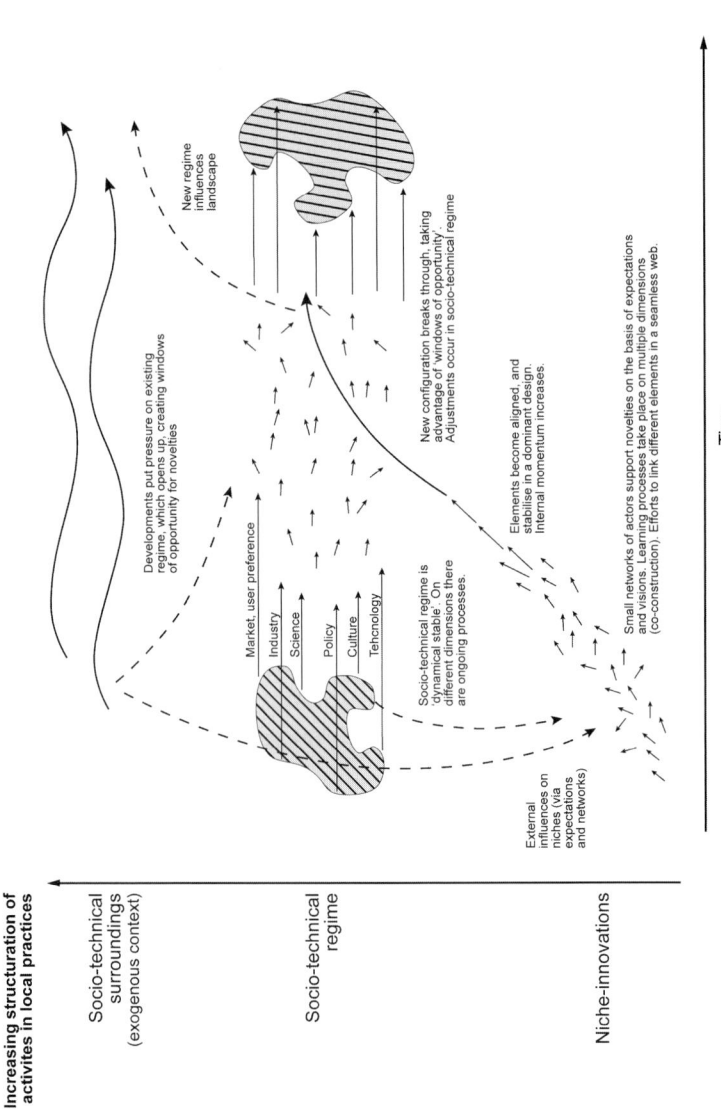

FIGURE 3.10
The Multilevel Perspective in transition studies: interactions between niches, regimes and the landscape. The niches are represented by the small arrows on the bottom of the figure, the regime is in the central layer and represented in its adaptive form in the dark patches, the societal surroundings, expressed in transition literature as 'landscape', are represented by the thin arrow, while the exogenous fluctuations in natural processes are indicated by superposed grey waves (adapted from Geels and Schot (2010)).

regime accommodates a broader community of social groups and their alignment of activities, stabilising existing trajectories in a diverse number of ways (Geels and Schot 2007). Regimes are oriented towards fulfilling specific social functions.

Niches are conceptualised as being created and protected by small groups of actors, working on innovations that deviate from existing regimes. These actors develop norms and expectations which are less stable than those for regime actors, but nevertheless guide and constrain the development of the niche. The socio-technical system – the technologies involved in the niche as well the material and social context for action – also constrains or enables development in specific directions. These niches can lead to regime change when a shift occurs in the external context (i.e. the landscape), which destabilises the regime and opens a 'window of opportunity' for niche expansion. Then the niche may or may not 'engage' with the regime in an anchoring process through which the niche can become mainstream. The anchoring process is typically both contested and negotiated, often with hybrid actors between niche and regime playing a key role.

The Multilevel Perspective in the Rural

Dealing with farming and farm management, and recognising that the agricultural sector has multiple functions, the most significant regime is the broad agricultural regime, even if others may also be involved (Pinto-Correia et al. 2015b). Studies of transitions in agriculture grounded in the Multilevel Perspective have most often addressed large-scale, national processes with a historical perspective (Grin 2012), or changes related to technical developments in the agri-food sector (Elzen et al. 2012). The existing literature on endogenous rural development provides a deep insight into the start and development of novelties which can be conceptualised as niches, as well as into the drivers of successful collective action that may lead to their subsequent transformation into innovations (Oostindie and Broekhuizen 2008). Similarly, earlier research on transition in agriculture has focussed on understanding the factors influencing the launch and growth of novel initiatives such as organic farming, short supply chains and direct marketing. The Multilevel Perspective has recently been applied to analyse processes of change in agriculture at the local and regional levels (Sutherland et al. 2015b). These show that there are significant particularities in relation to other sectors, due to the spatial nature of farming systems and the growing multiplicity regarding uses of the land. Due to the multiple interests in the rural land today, the push-and-pull effects from niche actors and regime actors which lead to transitions are from the agricultural regime as well as from other regimes, framing the non-productive uses of the land (Darnhofer 2014). Examples studied by Sutherland and colleagues (2015b) show that in farming, niches often differ from the ideal type defined in the Multilevel Perspective because the aim is not to transform the regime, they may be co-created by niche-entrepreneurs and regime actors, and they may emerge from atomistic decision-makers without any formal coordination. The anchoring capacity of such niches in the regime, or regimes, and, therefore, their transformative potential is still to be assessed. Furthermore, there is evidence for a need to distinguish between regimes given the diverse social expectations related to the use of the land, and the multifunctional nature of farming (Darnhofer 2014).

The Multilevel Perspective provides a useful analytical tool to: (1) understand the new landholders and their networks as potential drivers of change in the rural, i.e. as a niche; (2) understand the way these new landholders relate to, or challenge, existing policy and institutional frameworks which govern rural land use, e.g., how they as a niche relate to the regime (or several sub-regimes) and; (3) identify opportunities and threats to the changes promoted in the niche and how they may be overcome. When considering the contributions of such findings to targeted policy design, such an approach may reveal the niche, the niche-regime interaction and the consequent challenges for policy formulation and institutional arrangements (Selman 2012; Woods 2011).

Considering rural landscape, the added value of the Multilevel Perspective over a simple cause-effect approach lies mainly in the understanding of the interactions between the different levels of action and decision and between different actors and groups of actors at the same level (niche, regime and landscape). The fact that the rural landscape contributes to several crucial functions in society for which farmers and other land managers across half of Europe's territory are stewards and that they produce many public goods, all contribute to a high level of policy involvement (Darnhofer 2014), which highlights the need to understand these relations.

3.8 Conclusion

Understanding rural landscape change is an extremely interesting and challenging task. Furthermore, it is necessary if scientific knowledge is to support decision-making and the tailoring of public interventions. The agricultural landscape in the European Union is subdivided into 12 million farms, each of which has its own farming style, ownership structure, family tradition, subsistence level, community attachment and future perspective. All these farms are being combined at the spatial level with other uses of the space and at the social level with many different actors than farmers, which increases the complexity of the rural community and rural agency. Inevitably, this leads to a very diverse landscape, which characterises the majority of Europe's rural areas. Further, any analysis and understanding produced at the European level about the changes needs to be complemented with local evidence, and vice versa, and should acknowledge the differentiated change trajectories occurring at multiple scales.

Departing from the indiscernible landscape changes in Montemor, Portugal, in this chapter we have presented various methods of analysis to the reader, which can – to a limited extent – aid our understanding of European landscapes in transition. In any analytical effort, only a few of the presented approaches can be used at each time, otherwise the analytical complexity seriously risks becoming too high. Nevertheless, these different perspectives and methods of analysis *can* be used, and different conceptual frameworks can be helpful. In the next chapters, we will look at how the rural landscapes are in fact shaped, in between agriculture and the complex and evolving relations of the rural with the urban drivers and agents.

4

Evolving Activities in the Rural

Community Spirit

It is a bright winter day. The Alps – almost within shooting range – form a circle of watchtowers around: the snowy slopes of the Gran Paradiso on the left, the Monte Rosa colouring up in the late afternoon sun on the right. On the lower slopes, less snow appears and here and there a castle towers above deciduous woods. Beneath, where the rivers have reached the foot of the slopes draining to the Po, the flat landscape is brown with open fields and poplars, fine-meshed silhouettes of oak trees. So much the more striking is that really everywhere there are buildings, small industry, textile and shoe workshops of doubtlessly famous fashion brands and many scattered warehouses.

We are looking over the banks of a small river, the Cervo, from one of the round towers on the walls of the Ricetto di Candelo, a well-ordered ensemble of some 200 brick storehouses. The narrow streets – shelving down to the river – are paved with large cobblestones and also the bulwark is made up of flat pebbles in fishbone masonry. The roofs lean far over; in places beams jut out halfway up the facades where balconies must have rested. The local guide, Giuseppe, explains that the inhabitants of Candelo decided, in the end of the thirteenth century, to jointly store their precious harvests in a protected (ricetto) warehouse. The ground level of the stores was for grapes and wine making, the upper storey for cereals. In case of emergency – an alternation of rulers ravaged the region – the farmers' families could withdraw to the Ricetto, but they never lived here.

A starker contrast can hardly be imagined; the randomly erected temporary warehouses and chaotic urban sprawl in the Po plain, and this intact example of 700 years of community spirit, defying plague, wars and privatisation. Until a few years ago, some of the buildings were still in use as a granary. And now? Is conservation as a museum the only future?

FIGURE 4.1
Ricetto di Candelo, Piedmont Region, Italy

We are situated in the upper Po plain of the Piedmont region of Italy. The Po plain (*pianura padana*) shared by the Piedmont, Lombardy and Veneto regions in northern Italy is known as an agricultural area *par excellence*. The plain was reclaimed ages ago; field patterns from the Roman era are still recognisable in many places. Its production of rice, dairy products (*grana padano* cheese), meat (Parma ham), maize and cereals, and its trading and cultural connections with all parts of Europe and beyond have been emblematic for many centuries, forming the basis of prosperity for a large number of historical towns of eminent cultural value. Today, the landscape of the Po plain is characterised – apart from rice fields and fodder crops – by urban sprawl, land abandonment at the margins and a search for a new identity. Here the urban conversion of land is considered a territorial 'disease' (Romano and Zullo 2016, p. 109), resulting from complex economic dynamics and population growth and ineffective spatial planning regulations. Many agricultural buildings are abandoned. Paradoxically, newly built storehouses or factory halls rise along many of the roads, but they are often not in use either because they were mainly constructed as an investment opportunity rather than for a specific purpose.

The history of agricultural use that profits from the natural context helps us to understand the landscape we observe today. Lombardy, the most important region in the Po plain in economic terms, has almost 10 million inhabitants, 16 per cent of the population of Italy (www.istat.it). The Lombardy plain is one of the most urbanised areas in Europe with Milan alone having more than 5 million inhabitants. However, it is also a region with particularly favourable conditions for agricultural production and is located in a central position in relation to the European market. The particular geomorphologic characteristics have endowed this area with a natural and stable supply of water from the Alps in the north. The water is not only derived from run-off. The geological structure enables the water to percolate deeply into the earth in the Alpine zone to emerge in the plain as a series of natural springs (*fontanili*). In the past, many efforts were made to manage the abundant water supply in order to support agricultural production on the plains (Lassini, Monzani and Pileri 2007). The abundance of water influenced agricultural practices by favouring certain types of land use – e.g. grazed pastures – and by maintaining the permanent vegetation along field boundaries ('Italian *bocage*') and along the numerous irrigation channels on the plain. The consequence of this was a landscape rich in natural elements with a high degree of biodiversity and with an integrated rather than diffuse ecological network, which was also characterised by a large number of small and large forests. This pattern persisted more or less continuously from the late Middle Ages until about 1950.

The Po plain represents an agricultural system in an area of limited space. Expansion in one type of land use causes reduction in another. Thus, the high degree of urbanisation, especially on the plain, has resulted in a substantial reduction in the amount of land available for farming. In north-western Italy, the proportion of soil surface which is sealed is estimated to have more than doubled from 1956 to 2006 (Munafò, Salvati and Zitti 2013). For example, in Lombardy from 1990 to 2000, the amount of farmland decreased by more than 63,000 ha due to urban sprawl about 6 per cent of the useable farming land, UAA, while 14,000 ha were abandoned as farmland (www.istat.it). This trend continued in the period 1999–2007 with an annual decrease in farmed land of almost 4 per cent. At the same time, the area of woodland managed by farms greatly decreased: more than 90,000 ha disappeared between 1990 and 2000. Therefore, in the space of just 17 years, the landscape of the Po plain changed dramatically. A rich agriculture is still present, which is highly specialised, market-oriented and focussed on high-quality and high-value products. Yet agriculture hardly influences the structure of the rural landscape any longer or maintains the diversity of landscape elements which used to be paradigmatic of the Po landscape – this structure is increasingly lined with urban buildings which have no relation to agriculture.

4.1 The Rural Landscape in Transition

Land-Take and Rural Dynamics

The development in the Po plain is no exception in a European context. The fertile land has been reclaimed and used for farming production for centuries in many European countries. Multiple farm systems developed, often combined at the landscape level with forestry systems occupying the less fertile areas. The most fertile land is often also attractive for human settlement and was, thus, frequently the location where urban centres evolved. The presence of urban centres in turn has led to favourable conditions with market proximity for the maintenance of a diversified and, at the same time, highly intensive agriculture in these regions. However, progressively valuable agricultural land has been and is still being consumed by urban and industrial land use and by infrastructural developments.

Gardi and colleagues (2015, p. 907) estimate, based on modelling, that the impact of land-take on the agricultural production capacity for the period 1990–2006 for 19 EU Member States corresponds to a loss of more than 6 million tonnes of wheat. They conclude that Europe's intense urbanisation has a direct impact on its capability to produce food (Gardi et al. 2015, p. 898), while it is clear that the impact on landscape character is also substantial (La Rosa et al. 2013). In response to this impact, the European Commission proposed a zero land-take target for all EU countries by 2050 (EC 2011a; Prokop, Jobstmann and Schönbauer 2011). Only then, it is presumed, will it be possible for agriculture to continue to fulfil its role as the main producer of food, feed and fibre. This initiative, which relates to the global concern for food security, means future trends can be expected in the form of protecting farmland and maintaining production, which in turn means agriculture will still be the main driver of the rural landscape in the future.

This chapter focusses on the rural dynamics which are driven by farming and forestry and by demands and constraints which have an impact on rural landscapes and land-use change. To understand the origins of the trends, we explore the historical roots of European agriculture and subsequently analyse the major land-use systems in agriculture and forestry that have been shaping and still shape the landscape of Europe. As a basis for further study of the dynamics in European landscapes, we attempt to characterise the main farming systems in Europe and the current challenges they face.

Historical Roots of European Agriculture

The agricultural history of Europe started long before historical sources began to document the European civilisations of the sixth century BC (Pounds 1990, p. 9). Agricultural practices gradually spread into Europe from the Middle East through the Balkans and, depending on the physical environment and the sociocultural conditions, sooner or later began replacing hunting-gathering practices with animal husbandry and agricultural cropping. However, nomadic cultures have persisted almost until today, e.g. in Sámiland in northern Scandinavia, while transhumance, i.e. the long-distance migration of livestock from summer to winter pastures and vice versa, even if much less frequent than in former times, is still alive in various forms, mostly in Southern Europe (Grove and Rackham 2003; Roturier and Roué 2009; Terwan et al. 2004). It is evident that natural landscapes (in the sense of 'untouched by man') no longer exist in Europe and, with the exception of remote islands, have indeed not existed for the past many thousands of years.

Since prehistoric times, human influence has shaped the rural landscape in Europe at a gradually increasing pace. According to Pounds (1990, p. 7), the main vehicles of change of rural land use in history are: innovation, diffusion, specialisation and migration. Innovation brings the discovery and use of new materials and processes or the refinement in an established technology. The second agent of change is diffusion, i.e. the spread of innovation to neighbouring regions. Secrecy and reluctance to accept new ideas often play a role and may lead to differentiated rates of diffusion and adoption. Thirdly, specialisation arises as soon as a community begins to produce more of a certain product or commodity than it needs for its own consumption. This breeds experts and often induces innovation and organisational change. Fourthly, migration brings existing technologies to new places at a much faster pace than the other processes. These processes allowed people to better cope with the challenges posed by their environment.

The main phases of rural land use through history can be described as follows:

1. hunting/gathering
2. early agriculture and animal husbandry: subsistence farming
3. early market agriculture/establishment of larger settlements
4. three-field system, use of the heavy plough, larger cities and increased urban demand for food
5. era of forest and peat reclamations: monasteries
6. feudal farming including use of commons
7. independent family farming, often mixed or specialised farming
8. cooperative farming
9. industrial farming

In the past, these phases did not at all coincide across Europe. Even in the same region, several differing farming systems could coexist depending on the physical conditions and the landownership structure and also on the proximity to towns. In general, hunting/gathering was replaced by subsistence farming and early market agriculture which spread from the eastern Mediterranean in classical times, finally reaching the western and northern extremes of Europe in modern times (Pounds 1990, p. 27 seq.). After the decline of the western Roman Empire and the invasion of Germanic tribes, agriculture was reinvigorated in Carolingian times in Western Europe. The invention of the three-field system and the heavy plough and later the light plough (Pounds 1990, p. 195) resulted in major improvements in agricultural production in regions with suitable soils (Duby 1961). The late Middle Ages saw a large increase in arable land, which was made possible by an unprecedented reclamation of forests, peatlands and wetlands, mainly by Benedictine and Cistercian monks (see e.g. Lambert 1985, p. 112). The food production made possible on these reclaimed lands led to urban development in Italy, France, England and the Low Countries in the thirteenth century and in adjacent countries to a lesser extent (Duby 1976). The investments by the monasteries were generally institutionalised in the form of privileges ceded by the feudal rulers, which laid the foundation for the Church's huge land acquisitions, which still exist today in several countries. Yet secular land properties also developed into well-functioning feudal farming systems during the Renaissance, including industrial crops such as flax, hemp and mulberry (for silkworms) (Pounds 1990, p. 231). Livestock farming also increased crop production through the use of manure from the animals to fertilise soils (Luginbühl 2012, pp. 46–49).

In modern times, family farming developed into specialised enterprises in Western and Northern Europe as a consequence of the privatisation of land. With the abolition of serfdom in the eighteenth and nineteenth centuries came the right for tenant farmers and peasants to buy land which, in combination with new technologies, led to an increase in productivity (Jepsen et al. 2013). The commons gradually disappeared and more and more land became enclosed. The introduction of mineral fertilisers – though different in time across Europe (see Figure 4.2) – caused significant increases in yields, but later unseen environmental problems. In contrast, the disintegration of the Soviet Union in the twentieth century also led to the (re) privatisation of land and state and cooperatively run farms leading, in this case, to declining productivity and land abandonment (Jepsen et al. 2015; Jepsen et al. 2013).

The distribution of land management regimes over various parts of Europe in the past 250 years is summarised in Figure 4.2. It shows how the development has been uneven in Europe, although it has to a great extent gone through similar phases. These unequal changing patterns through time have had an influence on the way the landscape has and still is being changed by agricultural regimes. Fundamental changes have taken place,

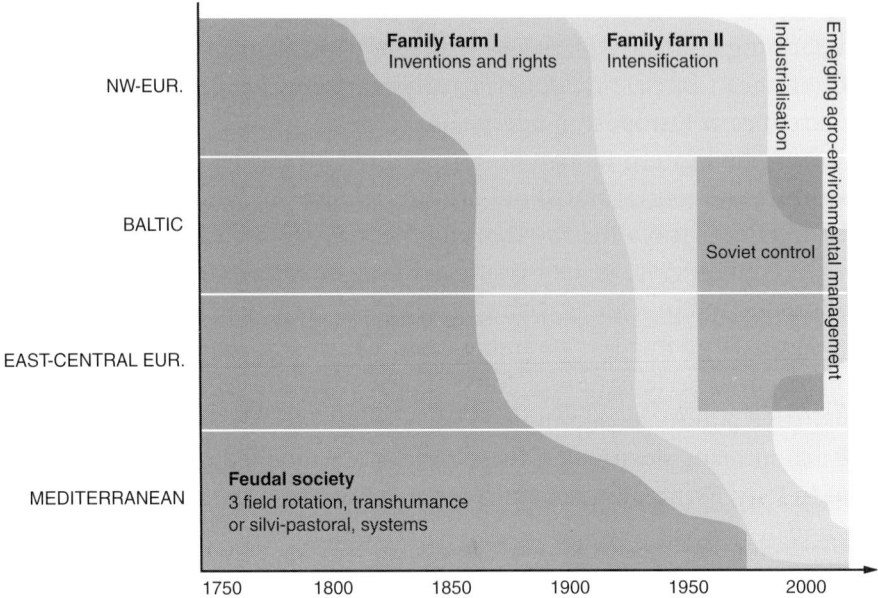

FIGURE 4.2
Large trends in the spatio-temporal distribution of land management regimes
Sources: Jepsen et al. (2015); Jepsen et al. (2013)

starting in Northern Europe and progressing later to Central-Eastern and Southern Europe. The nineteenth and a large part of the twentieth centuries have been characterised by the dominance of the family farm, though with widely differentiated characteristics in size and specialisation in different regions. In some regions, family farming has resulted in highly specialised and market-driven units since the middle of the twentieth century with the consequent simplification and rationalisation of the landscape. In others, it has meant extensive large-scale production, combining forest, grazing and crops, which has maintained the natural elements in the landscape, while in other regions, it has resulted in small-scale and diversified farming, which is closely connected to the functioning of the rural community and within traditional landscape structures. Gradually, during the 2000s, industrial/corporate farming has replaced family farming in certain areas. It is already widespread in Northern Europe and increasingly so in the Baltic States, Eastern Europe and the Mediterranean. Thus land use has become strictly oriented towards production systems and market demands, while concern for the appearance and character of the landscape has been limited to residential rural areas.

Case D Land Re-allotment/Land Consolidation Schemes in the Netherlands

Landscape Appearance

After World War II, the Netherlands government found itself confronted by the need to modernise agriculture for the sake of agricultural self-sufficiency and food security. Until then, agriculture in the Netherlands had been based on family farming, which was hampered in many places by a very fragmented field pattern due to the prevailing inheritance system of dividing up the land among the children. On the sandy soils in eastern and southern parts of the Netherlands, more than 80 per cent of farms were less than 10 ha in 1930 (Van den Bergh 2004, p. 28), which were generally fragmented into several parcels, sometimes located several kilometres from the farmstead.

Landscape Processes

The opportunity to re-allot the land to form coherent new farms with larger parcels was used to improve farming conditions by redesigning roads and water courses, improving drainage, building new farms and buying out farmers who did not want to continue. More than 200,000 ha were re-allotted in this way in less than 50 years' time (Figures 4.D.1–4.D.3). In these projects, the average number of parcels decreased by about 65 per cent (Van den Bergh 2004, p. 168). From the 1970s, recreational aspects and nature conservation issues were also taken into consideration in specific cases even in a leading role (Doevendans, Lörzing and Schram 2007; Van den Bergh 2004).

All over the Netherlands, this process was conducted in a highly organised way. The re-allotments were decided in a voting procedure among the stakeholders, with farmers who owned more land having a larger influence on the result. Besides agronomists, soil scientists and groundwater hydrologists, landscape architects

Case D (Cont.)

FIGURE 4.D.1
Maas en Waal area (Province Gelderland, the Netherlands) before and after land re-allotment. Source: RCE (2015). A black-and-white version of this figure will appear in some formats. For the colour version, please refer to the plate section.

Case D (Cont.)

Land reallotment in the Netherlands

Location of the reallotment projects requested, in preparation, in progress and completed as per 31 December 1985.

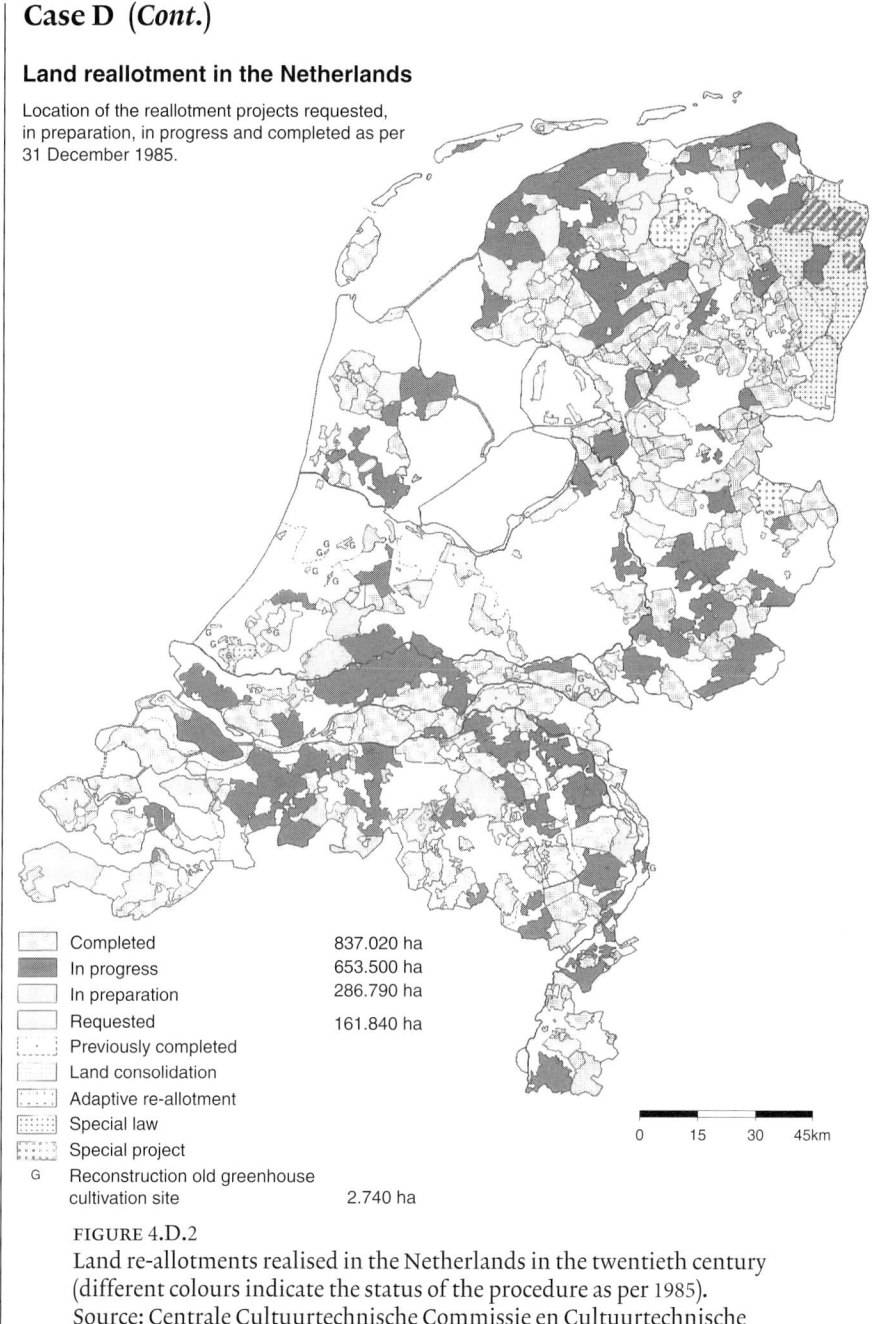

	Completed	837.020 ha
	In progress	653.500 ha
	In preparation	286.790 ha
	Requested	161.840 ha
	Previously completed	
	Land consolidation	
	Adaptive re-allotment	
	Special law	
	Special project	
G	Reconstruction old greenhouse cultivation site	2.740 ha

FIGURE 4.D.2
Land re-allotments realised in the Netherlands in the twentieth century (different colours indicate the status of the procedure as per 1985).
Source: Centrale Cultuurtechnische Commissie en Cultuurtechnische Dienst, Jaarverslag (1985). A black-and-white version of this figure will appear in some formats. For the colour version, please refer to the plate section.

Case D (Cont.)

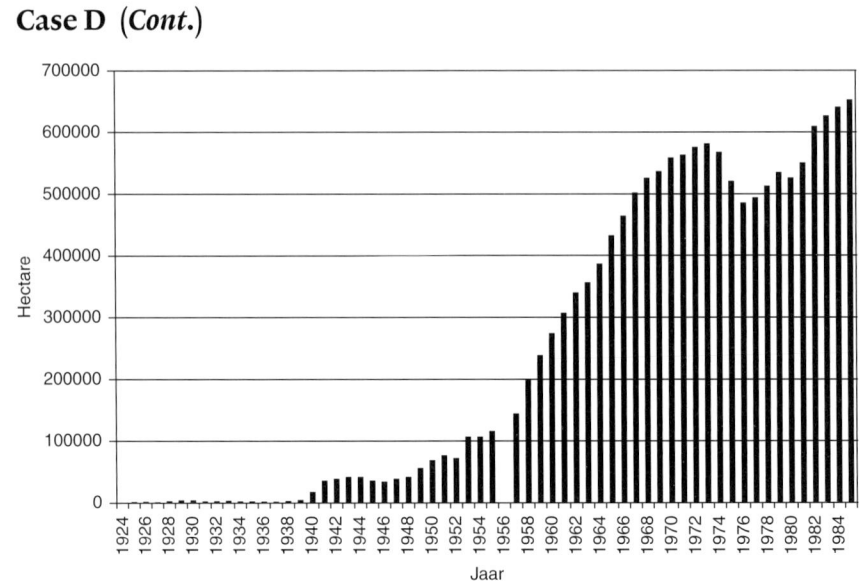

FIGURE 4.D.3
Total area (ha) under land re-allotment procedures in the Netherlands from 1924–1985 (Van den Bergh 2004)

were also always involved in designing the new landscape pattern. This process could take many years of preparation, sometimes more than 15 years. In total, more than 500 land re-allotments took place between 1956 and 1998, with the cost of preparing and implementing the plans largely financed by the government. Interestingly, some of the early finished land re-allotments proved outdated just 20 years later so a new land re-allotment was agreed upon.

In the meantime, the principle of the agricultural self-sufficiency of the Netherlands was given up for European self-sufficiency in the 1970s, and even that was given up soon after. The increasingly globalised and liberalised market would also not allow production solely for the national market.

Case D (*Cont.*)

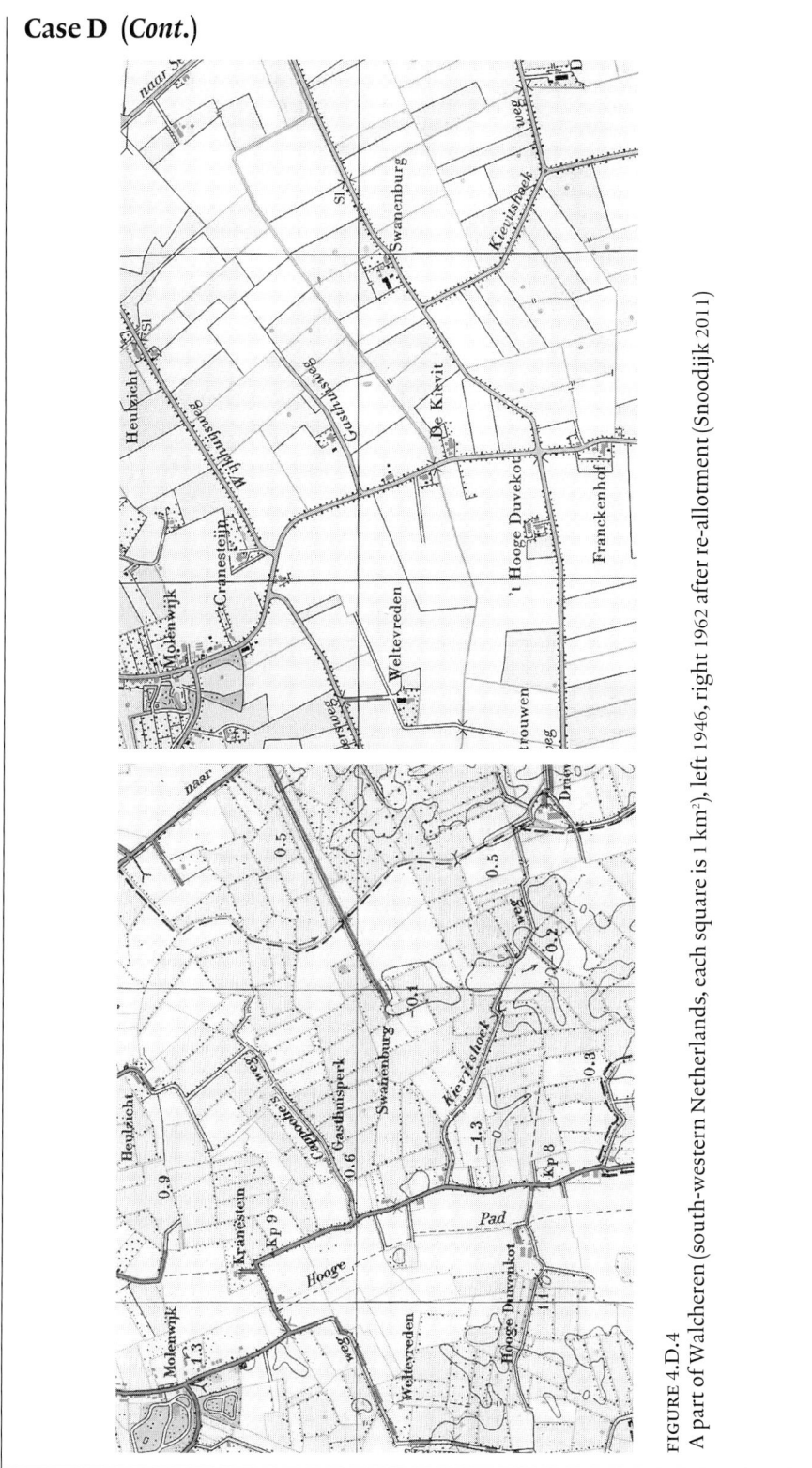

FIGURE 4.D.4
A part of Walcheren (south-western Netherlands, each square is 1 km^2), left 1946, right 1962 after re-allotment (Snoodijk 2011)

4.2 Explaining Rural Landscape Patterns

Even though the history of societal and farming developments is one of the main explanations of the great multiplicity in rural landscape patterns in Europe, this diversity is first and foremost related to the immense variation in the biophysical conditions for agricultural production. Today, the difference between the biogeographical regions in Europe is still a major factor in the differentiation of land-use systems that have developed over decades and of the resulting landscapes (Emanuelsson, Arding and Petersson 2009; Hazeu et al. 2011; Metzger et al. 2005). Through time, the natural processes have challenged society to organise itself in such a way as to survive and, at times, progressively increase to produce surplus products to sustain urban culture. Various coping strategies were employed to deal with constraints and to optimise the reliable production of food, feed and fibre (see Table 4.1). This led to the development of field patterns and field boundaries, which are typical of specific landscapes in certain locations. The field patterns were almost always associated with the use of natural or semi-natural landscape elements such as woodlots, forest patches, wetlands, tidal flats or mountain meadows. With time, and at an accelerated pace during the past century, many factors which shape farm systems have become increasingly homogenised in Europe and in the world, as leading technology, seeds and farm models have become standardised and widespread. Therefore, the underlying biophysical context has become less important and has less explanatory power with regards to understanding landscape patterns than 50 years ago. In any case, the combined effect of the biophysical context and farm system characteristics needs to be considered if we want to understand current rural landscapes.

The major physical limiting factors influencing farming and the associated traditional coping strategies have led to a multitude of landscape patterns across Europe, as summarised in Table 4.1.

The Main Types of Farming System in Europe

In many landscapes, the constraints have led to specific local or regional solutions and systems have developed which deal with these constraints and even exploit them, or at least become unique characteristics of these systems (Sanchez, Medina and Iglesias 2013). The reclaimed wetlands of the polder landscapes in the Netherlands are an example (see introduction to Chapter 1), while the Montados and Dehesas on the Iberian Peninsula (see introduction to Chapter 2) are examples of complex and multilayered systems which are adapted to dry climates and shallow soils and transhumance is an example of grazing systems adapted to the high annual variability in mountain climates and fodder availability, which would make a more sedentary land-use system impossible (Grove and Rackham 2003).

TABLE 4.1 *Global overview of limiting factors for agriculture and the associated coping strategies and landscape patterns*

limiting factor	region	coping strategy	landscape
summer drought	Mediterranean, Continental Europe, Alps	winter-spring cereal varieties; biannual fallow to accumulate water; perennial crops; community-organised water provision*; transhumance grazing*	small-scale irrigation; extensive cereal cropping with after-grazing
winter cold	Continental Europe, Northern Europe	winter-hard varieties; late sowing; (transhumance) grazing*; stables and barns	mosaic of horticulture, arable and grazing/forestry
spring precipitation	North-western Europe	soil drainage*; late sowing; grazing	small-scale parcel pattern separated by ditches; polder landscapes (see Text box I)
steep slopes	hilly and mountainous areas	Terracing*; perennial crops; permanent crops; grazing*	terraced agriculture all over Europe; alpine meadows; olive yards and vineyards
erosion and land degradation	unprotected and vulnerable soils everywhere in Europe	hedgerows; terracing*; perennial crops; shelter belts*;	*bocage* landscapes (England, north-western France, northern Germany and western Denmark); olive yards and vineyards
low fertility	sandy soils, peat bogs, acid soils, shallow soils, eroded soils	soil improvement by drainage*; application of litter, sods, marl, manure, nutrient-rich water*	wet-rice fields (Po plain); polder landscapes (Netherlands);
energy needs	Northern Europe Southern Europe	peat digging*; hedgerows*; energy crops;agri-forestry systems (fire wood, charcoal)*	Lake mosaics; bocages; coppice forest Montados and Dehesas

* *these activities were/are often community organised*

The conditions imposed by the context may be extremely limiting (Table 4.1). Severe limitations such as steep slopes, shallow soils, abundance of stone outcrops, extreme temperatures, scarcity of water or rainfall concentration in a short period, among others, limit land-use systems (Blume et al. 2016; Hazeu et al. 2010). Even today, when technological developments have made it possible to overcome many of the physical limitations, there are still costs and complexities to resolve which should not be underestimated. In some regions, more intensive and highly productive systems have developed, while in other regions, extensive systems still dominate. Lastly, there are regions with no agriculture where the dominant land use is timber production with different forestry systems. Therefore, to a large extent, the biophysical conditions explain the differentiation of land-use systems in Europe and, consequently, the resulting landscapes.

Several data sources exemplify this differentiation and these should be taken into consideration as an initial source of information when aiming to understand the complex mosaic of rural landscapes in Europe today. One example is the Environmental Stratification of Europe (EnS) (Metzger et al. 2005), which takes the form of a high-resolution stratification of the principal environmental gradients, combining variables of climate, elevation and oceanicity, as well as northing. It is a very informative contribution to understanding the differentiation of rural landscapes in Europe today and also to understanding the resilience of landscape structures over time. Other classifications, which also build on biophysical data, all support the assessment of the differentiation in land-use systems across Europe and particularly the understanding of the way these systems, due to their particular vulnerabilities and strengths, may be affected by different drivers of change (Hazeu et al. 2011; Van Vliet et al. 2015).

It is not only the biophysical conditions that contribute to the way land-use systems develop; farm property and tenant systems, which both correspond to a spatial economic and management unit, also affect land-use systems. In a classical economic perspective, agricultural structure is normally considered to depend on three main sets of production factors: land, labour and capital (Brouwer and Van Ittersum 2010; Hayami and Ruttan 1985). For many years, these have been used to describe and analyse the functioning of different farm systems, which then correspond to specific land-use systems and result in different landscape patterns and scales of organisation.

Successive classifications of land-use systems have been produced by combining the biophysical conditions with the farm structure and functioning. A classic example is the classification by Grigg (1974), which considers the whole world and identifies the main types of farming systems still found in Europe today: (a) Mediterranean agriculture; (b) mixed farming; (c) dairy production; (d) large-scale grain production.

(a) *Mediterranean agriculture* is characterised first and foremost by being adapted to particular climatic conditions: on the one hand, long summer droughts with rainfall confined to the winter months, which means that, unless crops are irrigated, they must be sown in the autumn, be harvested by early summer, or be drought resistant. On the other hand, there are the comparatively mild winters when temperatures are such that a variety of temperate crops may be grown. Next to the climatic conditions, there are the characteristics of the terrain, which is typically composed of summer-dry undulating coastal plains and mountains which have not only more rain in winter, but some in summer too. Thus, through the centuries, a typical Mediterranean *terroir* would grow a diversity of crops: wheat and barley in the plain, with sheep and goats grazing on the stubble, olives and grapes on the lower hills, patches of irrigated vegetables around settlements and pastures in the mountains used for grazing in the summer, as well as wood harvesting from the wooden and shrub areas in the hills and mountains. This has resulted in the well-known and admired Mediterranean landscape, which is composed of a mosaic of permanent and annual crops often combined in one plot or, at least, side-by-side plots with high multifunctionality (Pinto-Correia and Vos 2004). In the past century, Mediterranean agriculture has become more specialised, with vineyards expanding in the regions with particularly suitable conditions for wine production. The situation is similar for olive cultivation. Specialised, large-scale, irrigated crop production expanded as a result of dams and irrigation infrastructure. For vineyards, olive groves and vegetables, the specialisation of production has resulted in the homogenisation of the landscape and a clear reduction in the extent of the former Mediterranean mosaic. Large-scale extensive silvo-pastoral systems seem to have remained quite unchanged, but the cultivated annual crops in the under-cover have disappeared and the grazing density has increased with the support of fodder inputs from outside. Nevertheless, mixed and hybrid farming systems still prevail in many Mediterranean areas, and the same elements associated with a relative importance of permanent crops such as vineyards and olive groves remain a differentiating characteristic of Mediterranean agriculture and, thus, also of Mediterranean landscapes. The traditional 'heterodoxy' of Mediterranean agriculture with a high incidence of pluriactivity in farming and mixed family involvement in the farm units has made it less prone to follow agriculture development models based on productivism. For this reason, it is often classified as an agriculture which has been 'passed over' or marginalised (Marsden 2003). However, by maintaining this heterodoxy and particular mixed production, it is also an agriculture which may have more potential to adapt and take advantage of the opportunities offered by new multifunctional rural development approaches (Arnalte-Alegre and Ortiz-Miramda 2013).

(b) *Mixed farming* includes the combination of commercial crops and livestock and is found throughout Europe from Ireland in the west, to Central Europe and into Russia. It can be said to be the backbone of Central European farming, leading to the well-known farmed landscapes of Central and Western Europe, with some diversity of annual crops and pastures in the fields, hedgerows and small woodlots as diversifying landscape elements (Pitte 1983) and progressively, in the past 100 years, larger fields and more rational structures so that it is a landscape which is highly prone to simplification. Together with the enlargement of fields, hedge banks, hedgerows and stone walls have been removed or reduced in number and length, while some boundaries have been replaced by wire and moveable electric fences and, generally, there has been a net reduction in boundary length (Allen et al. 2011). Mixed farm systems today are highly commercialised and specialised and depend on external inputs of production factors. Furthermore, mixed farming sells most of its produce as unprocessed products. The intensification of farming and the progressive shift to a focus on livestock production took place in the nineteenth century, which was possible because the industrial strength of Western Europe and North America led not only to higher incomes and increased demand and the provision of fertilisers and machines but also to the export of manufactured goods, which has made it possible for these regions to import products which have been produced extensively from distant regions and, therefore, to increasingly specialise their production. Today, mixed farming is characteristic of densely populated, urbanised and industrialised societies, depending on industry for the provision of inputs. Even if we may still call it mixed farming, it has gradually evolved into specialised farming, which is focussed on livestock production, where annual crops are also produced as fodder for the animals so that a modern and specialised type of mixed farm has progressively been established (Andersen et al. 2007).

(c) Another substantially different land use system is *dairy production*. Although dairying has a long history, the modern dairy industry is a product of the past 100 years and, perhaps more than any other agricultural sector, it has been dependent on the growth of urban incomes, improvements in transport and advances in science and technology. At the turn of the nineteenth century, milk was produced, transformed and consumed only on the farm. The expansion of railways linked urban consumers to distant production areas and facilitated the transport and sale of fresh milk in cities. Further, technological developments transferred the production of butter and cheese to factories. Milk has always been produced in the mixed farms. However, the increasing demand for milk led to the emergence of regions specialised in dairying and, therefore, a specific farm type developed. Dairy farming is found mostly in Western and Northern Europe as well as the mountain regions, where grazing resources are naturally abundant and where dairy farming is often the only land use possible. There are also smaller regions specialised in dairying close to large urban areas. Dairy farms are highly capital-dependent, and while they were relatively small up until the mid-twentieth century, they are now becoming significantly larger. Due to the nature of its products, the economics of dairy farming is closely linked to the consumer and consumer demands. Furthermore, due to the type of products and the management of the processing industry, the cooperative organisation has been and still is strong. Dairy farms have become larger and more technology-driven, but their underlying characteristics remain the same. The associated landscape depends on the location of the dairy farms and the farm scale, but the presence of pastures, which can be combined with natural vegetation elements, and the presence of the dairy cattle in the fields creates a particular pastoral landscape which has a distinctive character. Pasture landscapes in mountainous and hilly areas are the clearest example of these pastoral landscapes. It is also within the dairy sector that we see the highest growth in the number of organic farms in different regions of Europe.

(d) Finally, *large-scale cereal production* is a product mainly of the nineteenth century when changes in farming methods, transport costs and urban demand accelerated its development mostly in North America, but also in Europe, in the countries of the former Soviet Union, central Spain, the Île-de-France in France and some regions of Germany, Denmark and the United Kingdom. In the course of history, large-scale grain belts have successively been displaced away from the major markets in urban centres as a consequence of rising land values. In the second half of the twentieth century, in areas located more centrally in relation to urban centres, grain production has been replaced by more intensive, mixed systems. Grain production has then in turn moved further away from the large urban centres replacing more extensive systems, which are characteristic of peripheral regions with more limiting biophysical conditions. This dynamic cannot be fully generalised, but was dominant for the most important areas of crop production during the twentieth century. Large-scale cereal production is dedicated mostly to wheat and rye, the only crops from which bread flour can be made, which can also be used for a variety of other products. Oats and barley have limited uses for human consumption, and are mainly cultivated as fodder crops and for brewing. These industrial cereal farms are, by nature, very large and specialised and are completely oriented towards selling the grain on the global market. Large-scale grain farms result in homogeneous, open and large-scale landscapes, where diversification is related only to the mosaic of different crops. Through a process of concentration, the number of farms decreases, while their size increases; the size of the fields becomes increasingly large, while the presence of diversifying elements, vegetation etc. is reduced to a minimum (Terwan et al. 2004). These are landscapes of low attractiveness for uses other than production.

Much large-scale grain production has developed in areas with poor soils and low rainfall in relation to areas of mixed farming and, thus, until some decades ago, large-scale grain production had low yields compared to those from mixed farming. However, yields have increased due to the rationalisation of farm structures, but mainly because of the increased use of inputs including heavy mechanisation.

This characterisation of farm systems, developed for the global scale, is very general. Since the middle of the twentieth century, farming systems have changed significantly in Europe, perhaps more than elsewhere in the world due to the extreme diversity and relatively small-scale organisation of European farmland.

The main trends have been intensification, often combined with extensification and even abandonment, and concentration and specialisation. *Intensification* means the pursuit of higher productivity through the capitalisation of agriculture, including investments in machinery and farm infrastructure and the increased use of biotechnology. Intensification has occurred all over Europe as much in the Mediterranean and the mixed systems as in large-scale grain and dairy production (Terwan et al. 2004). In contrast, extensification involves the opposite, i.e. a declining investment in production factors and an acceptance of lower productivity. While intensification takes place due to the pursuit of higher productivity, *extensification* often occurs without the specific aim to decrease productivity. Rather, it transpires due to, e.g. ageing farmers, difficulties in competing on more global markets, marginal location. Intensification and extensification operate at different spatial scales and in combination with other drivers (Wilson 2007) and act as a polarising trend as intensification and the surplus produced, which is sold on progressively globalised markets, has made production too costly and non-competitive in areas with more difficult conditions for agriculture. In Southern Europe in recent decades, trends of intensification in some areas and farm units have gone hand in hand with trends of extensification in others. Between 1995 and 2013, the total national Utilised Agricultural Area (UAA) in Portugal decreased from 3.6 to 2.3 million hectares, e.g., by more than 30 per cent. This reflects two trends: (i) the 40 per cent decrease in the number of small farms (less than 20 hectares), mainly in the northern part of the country, where the land of large parts of these farms has been abandoned or afforested, in some cases urbanised and; (ii) the replacement of low-productivity and state-supported grain-fed crop production (mainly cereals) with permanent pastures for extensive livestock production in the larger farm units mainly in the south of the country (Alentejo region), thereby resulting in a process of extensification at the farm and regional levels (Avillez and Carvalho 2015; Pinto-Correia and Breman 2009). This occurred alongside the implementation of large-scale irrigation projects, which established thousands of hectares of intensive irrigated agriculture in southern Portugal in a significant, state-driven intensification process as described in Case E later in this chapter.

Abandonment may mean that the agriculture use has been replaced by other, less demanding land use such as grazing in natural pastures or forestry, or even that the land is no longer subject to any form of management or direct human use (Pinto-Correia and Breman 2009). Abandonment is a well-known process in the most marginal land of Europe, be it in the southern or northern peripheries or in the mountainous areas. It results in the disappearance of many traditional farm structures and is often connected with depopulation and the ageing of the rural population, although this is not always the case. Along with the decline in the rural population and traditional farm systems, there is also a loss of many particular landscape character areas with primarily small-scale farms, which struggle to remain competitive in a globalised context and are located in remote locations, tending to be abandoned. The land gradually reverts to scrubland and later forest. In countries with a very high density of forest, the abandonment of former agricultural fields leads to a closing of the landscape and a severe loss of landscape diversity. In Mediterranean countries, shrub encroachment and the increased frequency of fires are most frequent. Nevertheless, non-competitive, small-scale farms, if located close to urban centres or the coast, or if they are accessible, tend to be maintained even if mainly for residential purposes. In such cases, land abandonment does not occur, but instead a process of countryside commodification takes place which is the result of consumption trends, as explained in Chapter 3.

The aim of *concentration* is to increase cost-effectiveness through large-scale operations, which is often achieved by increasing the size of farm units. Concentration has been and still is particularly prevalent in mixed and grain production in Central and Western Europe, although it also occurs in the commodity chain, through farms entering into contracts with a few or even just one purchaser by contracts of farms with few or a single purchaser (Bowler and Ilbery 1987).

Specialisation involves farms producing specific crops or products, answering requirements of optimised processing technology and of large and expensive standard equipment and machinery, defined in single-purchaser contracts, and stimulated by highly technology-driven consultancy. Specialisation also occurs at the regional level as a result of biophysical conditions which are better suited to specific production systems, but also the specialisation of the commodity chain and globalised competition. These trends have affected all farm systems, and the resulting structural changes in the last decades of the twentieth century in Europe have been profound, even if uneven in timing and extent. However, many of these described systems have maintained their underlying characteristics (Ilbery, Chiotti and Rickard 1997).

The patterns we see today are, thus, the result of changing processes which have affected the systems described. The differentiation in these farming systems is still characterising the European rural landscape. Nevertheless, to assess the effects that recent trends also have on the landscape, a review of the farm systems of today is required to shed new light on the current pattern and drivers of rural landscapes in Europe.

Geographical Distribution of Today's Farming Systems

The large types Grigg identified and described, which have been confirmed by the more European-centred literature, are connected to spatial locations as their descriptions also show. Nevertheless, they have not been spatialised. More recent classifications produced at the European level represent an improvement with regards to spatial detail and potential for monitoring with a more precise definition of the current main types of farm systems today. The European SEAMLESS (6th Framework Programme, 2005–2009) research project produced what is now a widely used classification of farm systems in Europe (Andersen et al. 2007; Andersen et al. 2006; Brouwer and Van Ittersum 2010). This classification is an improvement in relation to Grigg's or other similar classifications as it is based on the same farm characteristics (land-use distribution, main crop, livestock, size of the farm, labour intensity), but it is data driven, which, therefore, makes it possible to identify where the different types have higher incidence at the regional level and also where there are frequent combinations of types (hybrid farm systems). A detailed description can be found in Andersen and colleagues (2007). They identified 21 farm types according to different combinations of livestock production and six main farming systems: (1) field crops (annual crops of various kinds; what Grigg terms 'large-scale cereal'); (2) industrial crops (e.g. biomass production); (3) mixed farms (crops and livestock); (4) pastures and grassland (livestock on pastures); (5) permanent crops (olives, vineyards, fruit, mainly significant in the Mediterranean and a specialised version of Grigg's Mediterranean agriculture, as shown in the example in Case B) and; (6) horticulture. The farm types are, e.g. arable land mainly used with cereal, arable land with other specialised crops, dairy cattle on permanent grass, dairy cattle on temporary grass, land-independent dairy cattle. These can be further characterised by intensity (low, medium, high) and size (small, medium, large). The distribution of the main farming systems at the level of the NUTS-2 regions is shown in Figure 4.3. The map shows the dominant land-use system in each of the regions, even though there may be multiple combinations and detailed land-use mosaics which cannot be represented at this scale. Figure 4.4 represents the three intensity classes of these land-use systems in Europe, per NUTS-2 region.

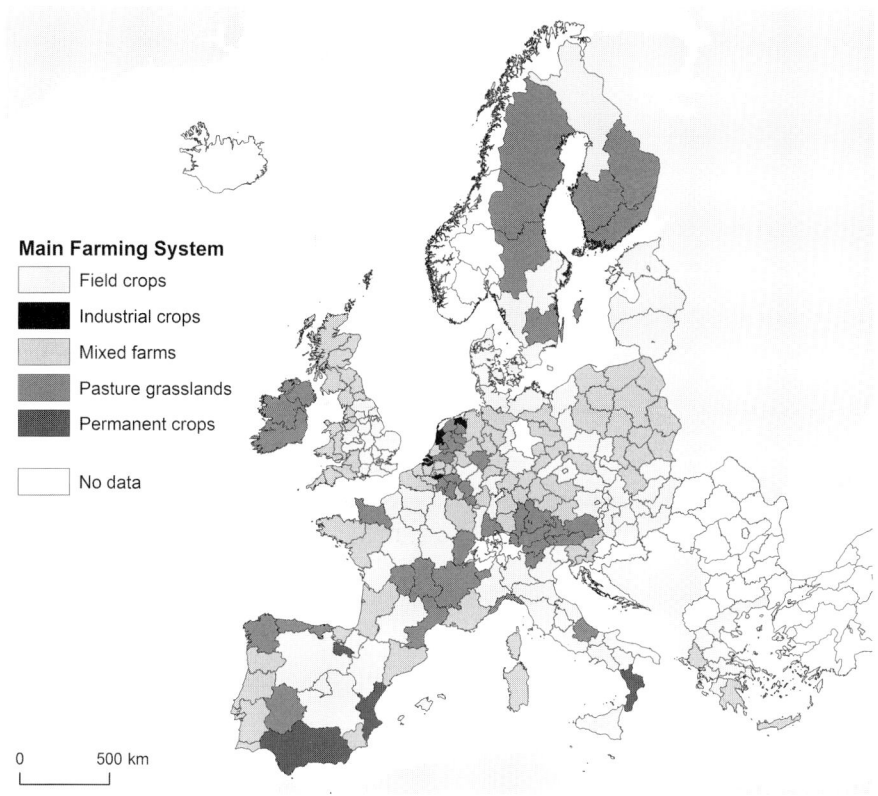

FIGURE 4.3
Main farming systems in Europe 2008
Source: Sanchez et al. (2013)

The combination of these two distributions is a very good proxy to understand the potential landscape pattern we can find in rural Europe today. Mixed farming, for example, which combines livestock, crops and pastures, is dominant in different regions of Europe from Portugal in the south-west to the northern UK in the north-west, to central France and Germany and Eastern Europe. However, the intensity is very different, being low in Portugal and the northern UK and high or medium otherwise. This means more extensive and pasture-based livestock systems in the two first cases and, thus, a relatively nature-friendly landscape (Oppermann, Beaufoy and Herzog 2012) and more specialised and crop-based livestock production in the other regions, which is reflected by more homogeneous and artificial farming landscapes. What is left out here is the forest area and the corresponding forest types, which for some regions (Scandinavia, mountain areas etc.) is highly important in terms of total land-use share and, therefore, also fundamental for assessing the landscape pattern and dynamics.

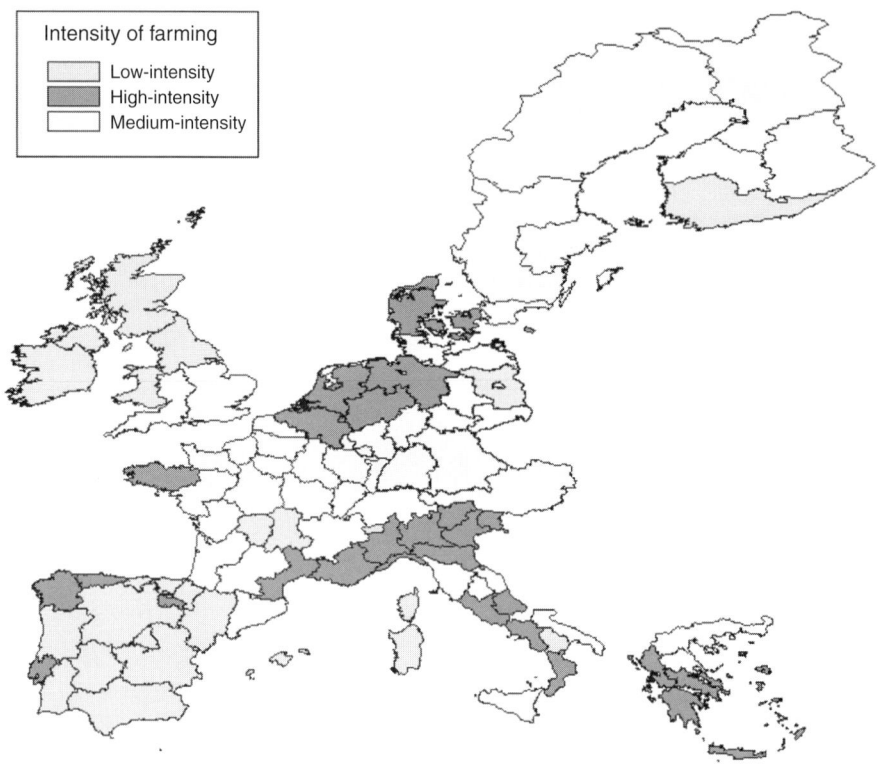

FIGURE 4.4
Intensity of farm systems in Europe, per NUTS-2 regions
Source: Andersen et al. (2007)

Case E Serpa, South Portugal – How Super-Intensive Olive Groves Are Drastically Changing the Landscape and Hampering Rural Life

Landscape Appearance

The region of Alentejo is characterised by gentle hills, poor soils, a Mediterranean climate and low-density population with highly concentrated settlements. There are a few larger towns and a few hard-rock reliefs originating in small mountain ranges, which stand out from the plain. The landscape in the region as a whole is characterised by extensive and large-scale land-use systems, mainly the silvo-pastoral system, Montado, but also extensive olive groves and open areas of natural pastures. It is, therefore, an empty, quiet, open and dryland landscape, which is characteristic of southern Iberia. In the past 15 years, this landscape has changed dramatically in some subregions due to the introduction of irrigated agriculture. The municipality of Serpa, located in south-east Alentejo and close to the border to Spain, is one of the areas where irrigation from the Alqueva dam has brought the most significant changes (Figure 4.E.1). Where there were crop fields and natural pastures and also traditional, low-density rain-fed olives, there are now intensive

Case E (Cont.)

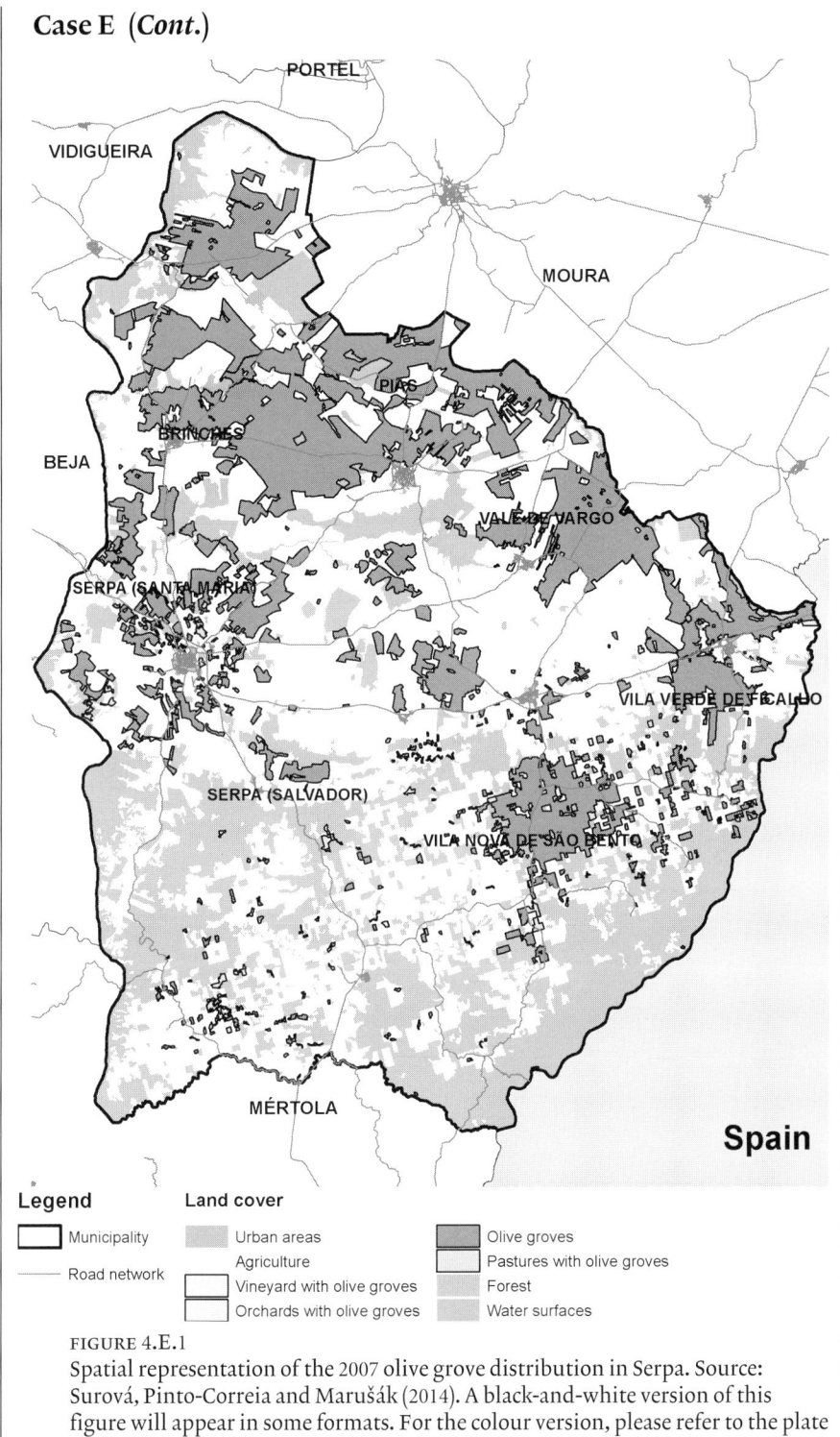

FIGURE 4.E.1
Spatial representation of the 2007 olive grove distribution in Serpa. Source: Surová, Pinto-Correia and Marušák (2014). A black-and-white version of this figure will appear in some formats. For the colour version, please refer to the plate section.

Case E (*Cont.*)

olive groves with regular and dense rows of small olive trees (Figure 4.E.2). These new olive groves already occupy more than 30 per cent of the municipality and this change has taken place in the space of no more than 10 years.

Landscape Processes

Since the beginning of the twentieth century, Portugal's southern region, Alentejo, has been considered the cereal provider for the whole country. Protectionist measures in the first half of the century resulted in an overall expansion and intensification of cereal production. However, soon the limitations of the soils, which are mainly shallow and with low organic matter, proved a strong limiting factor, while the rain-fed crop production has also proven extremely poor compared to European or worldwide average yields. Nevertheless, the region of Alentejo still has a mild Mediterranean climate with many hours of sunshine and dry summers and high temperatures throughout the year. Irrigated agriculture has high potential. Since the 1960s, the construction of small dams, which resulted in irrigation areas, has turned farmland of rain-fed agriculture into irrigated plots with high productivity for vegetables and fruit. In the beginning of the 2000s, the Alqueva dam was built on the Guadiana river, which is located in the east of the region, which resulted in the largest artificial lake in Europe. The dam was finished in 2002 when the filling with rainwater started. The main goal was to obtain a strategic reservoir of water and to make possible the intensification of large agricultural areas. There are 69 associated smaller dams, almost 2000 km of water tubes and irrigation channels and a total of 120,000 hectares of irrigated land, which is distributed in several irrigation perimeters. A highly technological and

FIGURE 4.E.2
Recently planted olive grove in Serpa where there was once a silvo-pastoral system

> **Case E** (*Cont.*)
>
> specialised farming is thriving in the region: maize, intensive olives, vegetables and even irrigated vineyards. This change is a true radical transition compared to the former extensive livestock-orientated farming as it is based on different products and demands totally different technologies and skills, new organisation facing the market and new types of farm enterprises. It has changed the rural fabric in a large part of southern Alentejo and it has had a profound impact on the local (and regional) landscape.
>
> The changes in farming systems were implemented with complete disregard for the landscape traits of the region, leading to a radical change in its character, the loss of remarkable elements and increasing fragmentation of the ecological corridors, particularly in the north of the municipality.
>
> **Conclusion**
>
> Irrigated agriculture is certainly leading to a new positioning of the Alentejo agriculture in the European farm production context. Similar processes have already occurred in Spain and it seems that it is now the turn of Portugal. A lot of money is at stake. The EDIA, the enterprise managing the Alqueva dam, estimates that 20,000 new jobs have been created in the region, some of which are located in Serpa.
>
> Nevertheless, the changes in land use have seldom been subject to an impact evaluation, and, particularly in regards to the landscape, possible impacts or mitigation measures have not been considered. The territorial dimension of this radical transition in land-use activities and the construction of the related infrastructure have not been considered and, thus, the change in the landscape has also been radical with no vision regarding what it should become in the future. The impacts on other landscape-related functions and economic activities such as the well-being of the inhabitants, tourism and recreation remain to be seen.

The intensity classification is a significant simplification of the farming systems across Europe. Nevertheless, this distribution of low, medium and high intensity also has profound impacts on the landscape. Where intensity is high are the more specialised landscapes with rational structures and not much room for nature or other nonproductive elements in the agrarian landscape. In contrast, where farming systems are low intensity, regardless of the farming system, there is a closer relation to the natural ecosystem and most often a more heterogeneous and not so production-driven landscape – historical remains and natural elements remain even in farm-based landscapes. In the middle are the medium-intensity systems, where the characteristics of the two extremes can be found and combined.

What is particular about agriculture and specifically relevant when we consider its impact on the transition processes registered in European landscapes is that agriculture is under continuous change. It is clear that some characteristics remain somehow stable: the biophysical context still plays a determinant role in the type of farming system that can be found in the different locations, and the production goal of different farming systems tends to remain the same, thus broadly maintaining the same land-cover characteristics or components, even if the scale of land-cover organisation may change. Therefore, the description of farming systems produced by Grigg (1974) is still relevant today in order to understand the large-scale differentiation of agricultural systems in Europe. Nevertheless, since the middle of the twentieth century, constant changes have taken place due to the restructuring processes in the agricultural sector itself. These include rationalisation in the use of production factors and structures, intensification in some areas and extensification or even abandonment in others, specialisation at the farm and regional levels and concentration of production on steady larger units (Woods 2005).

Most Recent Developments in Farming Systems

As we have seen in Chapter 3, many different drivers of change affect agriculture and the use of the rural space today, which are novel in nature in many cases and have not been seen before, such as the consumption of the countryside for amenity uses, or the quest to conserve nature and natural resources. In the farming systems themselves, the most considerable changes with impacts on the landscape are related to a quest for more productive and efficient systems within the productivist paradigm.

Novel farm systems have emerged in Europe in recent decades, mostly based on increased use of technology, with large landscape impacts. Technological development has enabled or reinforced new forms of intensification and specialisation experienced throughout Europe. The irrigation projects and advances in irrigation techniques in Southern Europe are a clear example of this. An increasing amount of farmland area is now being irrigated in Southern European countries such as Portugal, Spain, southern France, Italy and Greece (Sanchez et al. 2013). Where there formerly was an extensive and poor rain-fed agriculture, there are now intensive irrigated crops with yields of a level impossible to obtain before. These may be permanent crops such as olives and vineyards, but also vegetables and pulse crops. The disappearance of previous land-cover mosaics and the homogenisation of the landscape due to the intensification of production systems is the most visible impact, even if environmental consequences have been more frequently documented in the literature (García-Ruiz and Lana-Renault 2011; Lionello 2012). Private landowners install their own irrigation facilities, often using groundwater. However, most significant on a landscape scale are the large dams and related irrigation infrastructure, which have been constructed with state interventions. As shown in Case E, in such instances, the landscape change is radical. Solely with the construction of the Alqueva dam in southern Portugal, 120,000 hectares of land have become irrigated between 2005 and 2015. As the irrigation facilities are subject to compulsory payment for the land plots included in the irrigation perimeter, landowners are under economic pressure to use the water or otherwise to rent or sell the infrastructure land to others who will use the water in irrigated production. This compulsory payment mechanism has greatly accelerated the conversion to irrigated agriculture. Some owners choose to convert their production, while others lease the land or sell it, but in any case the conversion into irrigated production is guaranteed.

Another radical transformation in the production system, which leads to the replacement of former landscape patterns by new artificial landscapes, is the installation of **greenhouses** (and production under plastic more generally), e.g. the production of vegetable crops in controlled environments. Greenhouses are defined as all permanent structures with or without heating, which are covered by glass or plastic or other material that lets daylight through in which crops, transplants or ornamentals are cultivated (EC 2013a). In general, greenhouses are environments which can be controlled to a much higher degree than outdoor fields. Temperature, light, air humidity, water supply and carbon dioxide in the air can be regulated by the grower. In some modern greenhouses, even the access of pests and pathogens can be restricted or prevented. There is also soil-less production, either in substrates of organic or inorganic materials or as hydroponics, but the inorganic growing media and hydroponics are not allowed in organic cropping. Greenhouse production is related to high- and cutting-edge technology developments related to crop modelling, climate control and modelling, equipment, robotics and automation, energy, fertigation, water and growing medium management and plant protection. It implies, therefore, high investments and high running costs. Its expansion in different regions of Europe is related to the capacity to increase and control yields, expand the production periods and, therefore, increase the outputs and be more competitive on the global markets. The total area under greenhouse production is expanding in both Northern and Southern Europe and it was estimated to cover an area of 160,000 ha in 2010 (EC 2013a). Compared to other farming systems, the effect of greenhouses on the landscape is immense as the land surface becomes completely artificial with farmland landscapes transformed into built environments. Modern greenhouses tend to be significant in size and grouped in regions with particularly favourable biophysical conditions or access to large markets. They are, therefore, often concentrated, thereby resulting in the total transformation of the former agrarian landscape. South-western Portugal and southern Spain are two of the areas where greenhouses have been installed more recently, leading to this radical change in the former landscape.

Homogenisation and Diversification

Agriculture in Europe, as in the rest of the world, will face some serious challenges in the coming decades, including competition for water, resources, rising costs, decrease in agricultural productivity, competition for international markets, climate change and uncertainties regarding European policies' effectiveness with regards to adaptation strategies. Greenhouse production is seen as an alternative approach to overcoming some of the upcoming challenges, even though it raises substantial issues such as energy production and the transformation of land cover to form artificial landscapes.

Today, both irrigated and greenhouse farming are becoming increasingly attractive as investments and domains of economic activity for large-scale, corporate enterprises. Therefore, it is no longer solely an issue of scale increase for the family farm, which used to be the backbone of European agriculture. A fundamental shift in the structure of farming activities is taking place, particularly in some sectors and regions with large, often multinational companies taking over the production and other aspects of the value chain (Gereffi, Humphrey and Sturgeon 2005). The business strategy covering production, but also processing and marketing, will without a doubt replace the role of the farmers' organisations, mainly cooperatives, which have so far organised the value chain of some sectors such as dairy and wine. The capacity for scale increase in the production field, combined with the homogenisation of production factors and farm systems and the nested processing plants in the production units (as it happens with olive oil processors located in large-scale olive groves or with wine plants located in vineyards), will have an additional impact on the European agricultural landscape.

In recent decades, diversification, which has most often been aimed at increasing multifunctionality at the farm level, is increasingly taking place, though it remains a localised process. Such a development corresponds to the re-grounding and broadening trends at the farm level as presented in Chapter 3 (Van der Ploeg et al. 2002b), often combined with deepening. Diversification occurs as a reaction to the extreme trends of specialisation, concentration and market globalisation and is often driven by a search for increased sustainability through the regeneration of the internal resources to the farm and a reduction in external inputs at the same time as supporting higher landscape heterogeneity and biodiversity. Diversification is related to concerns for environmental quality and the balanced use of natural resources, new consumer demands and farmers' ideals of lifestyle and community attachment (Röling and Wagemakers 1998; Van der Ploeg 2009a). Thus, it is also related to the consumption of the countryside as such and corresponds to what has been described in Chapter 3 as the deepening, broadening and re-grounding of conventional processes at the farm level, which results in more multifunctional farms (Van der Ploeg and Marsden 2008). From a landscape perspective, a turn to a more multifunctional and diversified agriculture at the farm level is highly significant as it has a direct impact on landscape structure and function.

In recent decades, diversification and improved multifunctionality have been defined as a strategic goal in many local areas of Europe, often in those regions with more limiting biophysical conditions for intensive agriculture and where traditional forms of agriculture have been maintained until recently (Helming and Wiggering 2003). This strategy may have been led by local authorities or associations (Wiggering et al. 2010). All over the more favoured peripheries of Europe, there are many cases of success. The Goriška Brda region on the Slovenia–Italy border is one of these examples (Pintar et al. 2010). Facing an increasing and rapid change of former grassland and forest patches into vineyards in the local region, issues of water, soil and land management arise, which were addressed by innovative and inter-sectorial approaches and by the development of regional-specific strategies and technologies. These supported the emergence of some priority lead markets, also identified at the EU level: renewable energies, bio-based products, sustainable construction and recycling. The Brda consortium was created on the initiative of the local municipality and involving producers, but also regional stakeholders, policy makers and researchers in the fields of agriculture, spatial planning, environment and natural resource management. The strategy was based on a strong participatory approach so that all those concerned would feel involved and strong networks would be created. Existing social networks were used for a system of concentric circles in which organisations with a social character (voluntary firemen, rural women's groups, choirs, etc.) build the strongest links that other circles or networks, such as producers or local businesses, relate to. A trademark was created that secures the quality of the local products. Winegrowing was supported by an explicit orientation towards nature protection and the preservation of traditional cultural landscape elements (e.g. terraces), which was to provide a comparative advantage by increasing market recognition of the wines from the Goriška Brda. Innovations such as the use of specific local wooden posts were implemented to strengthen the uniqueness of local wine production. In combination with wine, fruit growing was also developed, mainly by full-time farmers, who used fruit as an important additional source of income. The local trademark supported the direct sale of fruit in local and nearby urban markets. Rural tourism was developed in connection with the wine and fruit production: quality farm tourism, wine shops, wine roads, annual and well-known events. With this, tourism contributes as a means of helping to preserve the cultural landscape, ensuring the cultivation of land and preventing the demographic decline of the region.

Tentative collective actions for diversification and the increased sustainability of agriculture on the local scale often result from obvious environmental problems such as those related to nitrate leaching and the resulting decrease in water quality (Vlahos and Schiller 2015). Even if initiatives are bottom-up, experience has shown that simply focussing on technological change in restricted bilateral negotiations between farmers and policy makers often leads to solutions of limited effectiveness in the long term. More integrated and visionary strategies, where farming is part of a more encompassing landscape and rural community goals, tend to be more successful in the long run (Wiggering et al. 2010).

At the farm level, on-farm renewable energy production is a classic example of a diversification strategy through the re-grounding of farm activities (Sutherland et al. 2015a), which may be based on multiple energy sources such as biogas production through anaerobic digestion, wind energy production, solar panels or mini-hydraulic plants. Re-grounding has an impact on the farming system as it creates an extra income from non-farm activities and allows the maintenance of less competitive farming systems, which means that it may have a double landscape impact as a result of the installation of the energy-producing mechanism and the preservation of a more extensive or diversified farming system. Re-grounding often corresponds to a socio-technical transition as described in Chapter 3 (Geels and Schot 2007, 2010) as this involves a socio-technical niche which remains largely unrecognised by both the agricultural and energy regimes until societal pressures open up windows of opportunity, leading to a greater recognition of the advantages of small-scale energy production at the farm level.

Diversification at the farm level as well as at the local landscape level can, thus, take many different forms, and even though it is not a mainstream development in European rural landscapes, it should not be underestimated as a powerful driver of change for many landscapes.

Persistence of the European Model: Family Farming

So far, the family farm is still representative of a substantial part of European farming, even in the most industrialised agricultural context, although the underlying societal basis has changed radically. This is due to an overall shift in society and the position of agriculture in society and the economy. Most farms are still based on family structures and, thus, there is no, or at least, a limited division between the owners and workers. Nevertheless, in recent decades they have become increasingly integrated into a modern market economy as they are inevitably integrated in the market patterns to sell the produce. Thus farms are subject to the conditions and demands of the global market (Primdahl and Swaffield 2010a; Woods 2005). Full-time farmers are intensifying production on the most suitable land and increasing the scale of operations. Such farmers may be referred to as full-time commercial farmers: they work full-time on the farm and primarily follow a market orientation so that their farm enterprise is highly, if not completely, integrated in both upstream and downstream markets (Van der Ploeg 2009b). The farm enterprise tends to be mostly specialised and oriented towards the most profitable activities with other activities being externalised. This type of farmer, who is assumed to be highly entrepreneurial, has typically been contrasted with the peasant farmer in the literature (Bernstein and Byres 2001). Peasant farmers are most often classified as old-fashioned farmers belonging to the underdeveloped and marginal parts of the world outside the modern economy. However, they may also opt for greater autonomy in the face of global markets and struggle to maintain a livelihood within farming, which is grounded on more control of their resource base and greater dependency on local rather than global networks (Van der Ploeg and Marsden 2008). Clearly, full-time farmers affect the rural landscape in different ways than small-scale, part-time and hobby farmers as they extensify the land use and plant hedges and thickets to a far less degree than small-scale, part-time farmers (Primdahl and Kristensen 2011).

The small-scale farm, which is often related to a peasant-like farming style, is still dominant in many countries and regions of Europe. To a certain degree, these small farms are becoming attractive for new entrants to farming and are, therefore, maintained or are reappearing, even in regions where they seemed to be going out of production. They include both traditional family farms and the activity space for new entrants, lifestyle farmers, who are often highly innovative and with good entrepreneurial skills (Sutherland et al. 2015b). In recent years, small farms have received increased attention in the political debate with greater recognition for the role they play in rural areas and the need to improve their economic and social conditions in times of structural change in the agricultural sector towards fewer and larger farms (EC 2011b). Despite the dominance of large-scale, specialised farming in markets, share of the economy, farm size and UAA occupation, small-scale farming does not seem to be disappearing yet (Ortiz-Miranda, Moragues-Faus and Arnalte-Alegre 2013; Sutherland et al. 2015b). Wide variation in farm structures across Europe and a lack of consistent data are amongst the main reasons why a commonly agreed-upon definition of small farms has been difficult to identify. To analyse farm structures and compare them across different countries, regions or over time, physical measures such as hectares of UAA or labour input per farm can be used. However, these measures are highly dependent on the type of farming and provide little information on the economic situation of a farm, with the economic farm size probably being the most appropriate criterion for understanding the resilience and needs of these farms (EC 2011b). Concerning the role in land use and other landscape changes, there are clear indications that lifestyle farmers are being underestimated. Even in an intensively farmed region with good agricultural conditions, in Danish and not particularly peri-urban conditions, lifestyle farmers are clearly the most active when it comes to landscape change (see Case F in Chapter 5). A reason why the role of lifestyle farmers in rural landscape change is underestimated may be that they (and their farmlands) are underrepresented in agricultural statistics due to their limited economic role (Primdahl et al. 2013b).

Thus, if technology, markets and networks have changed, the characteristics of the farmer and also of the family living on the farm have witnessed a similar degree of change since the middle of the twentieth century (Bruckmeier and Tovey 2009; Ortiz-Miranda et al. 2013a). Consequently we are now faced with an increasing diversity of farmers across Europe (Primdahl et al. 2013b; Sutherland 2010), and such diversity contributes to both the maintenance and the transformation of European rural landscapes. The recent trends in corporate farming discussed previously, which are rapidly spreading through Southern Europe, will certainly play an increasingly significant role in shaping farm structure and production in the future.

Giving up Agriculture: Land Abandonment

In contrast, the abandonment of agriculture and the abandonment of land are occurring on a large scale in the most marginal lands of Europe, although it may also take place on peripheral plots in intensively farmed regions. Agricultural abandonment occurs when farming ceases to generate a sufficiently high income and agricultural uses and activities are replaced by more extensive land uses, generally forest. In this case, the land is still managed, but the way it is used is normally less demanding than agricultural production. This results from an explicit decision by the landowner to change the land use. The drivers of such a change include natural constraints, land degradation, peripheral location, socio-economic factors, demographic structure and institutional framework (Pinto-Correia and Breman 2009; Terres et al. 2015). When land is abandoned, gradual shrub encroachment often occurs and, in time, the development of spontaneous forest, which is highly prone to fire in the case of the Mediterranean regions. The reasons for land abandonment are multiple and the process is complex, and not always the regions that were considered most at risk of abandonment at the end of the twentieth century are the ones where land abandonment is most observed today (Terres et al. 2015). In any case, the impact on the landscape pattern is significant. The financial crisis in 2008 and the related and recent development in the food security debate may lead to a reversal in the trend for land abandonment in the peripheral areas of Europe (Maye and Kirwan 2013). In Greece, Spain and Portugal, as well as in Eastern European countries, recent narratives include a return to the land and renewed use of former agricultural land (Dwiartama and Piatti 2016). In this respect, the fate of the land abandoned or at risk of abandonment is still an open issue (Terres et al. 2015). Also the decoupling of the CAP and the fact that a farmer may collect direct payments on unused land (it just has to be managed a little) have without doubt led to the abandonment of especially arable farmland. Figure 4.5a presents the change in cropland area in Europe between 1990 and 2006. As can be seen, the total area of cropland is shrinking, particularly in Eastern Europe and the Mediterranean.

Forest Land-Use Systems of Europe

As has been described in Section 1.3, it is not only farming that influences the condition and dynamics of rural landscapes; in many European rural landscapes, forest is the dominant land use, or at least it is as important as farming, which determines the character of the landscape alone or in a mosaic with farming (Figure 4.6). Forest landscapes have a history and character of their own (Stanners and Bordeaux 1995), which is strongly connected to the range of natural vegetation of the European continent (Bohn et al. 2003).

Forestry was the dominant land cover in Europe in prehistoric times. Through human intervention and activity, forests were progressively cultivated and former forest areas replaced by agricultural land. First forestry and farming were combined in what is today called agro-silvo-pastoral systems or only silvo-pastoral systems, which often led to open farming areas. This has been a continuous process for centuries. The lowest proportion of forest of the total land cover was reached during the early twentieth century, though with considerable differences between regions. Nevertheless, Europe's forest area has increased by more than 10 per cent since the early 1960s (Barbati et al. 2011; Stanners and Bordeaux 1995). A large part of this increase has occurred in Southern and Western Europe in spite of forest cover remaining stable or decreasing in many countries in Eastern Europe and the former USSR due to the over-exploitation of forest resources. This increase is partly due to the afforestation of poor soils unsuitable for agriculture and partly to afforestation or spontaneous regrowth on abandoned agricultural land. Many marginal areas, especially in the peripheral regions of Europe, as well as many Alpine and other mountainous areas have been the most affected by this forest increase, with the consequent disappearance

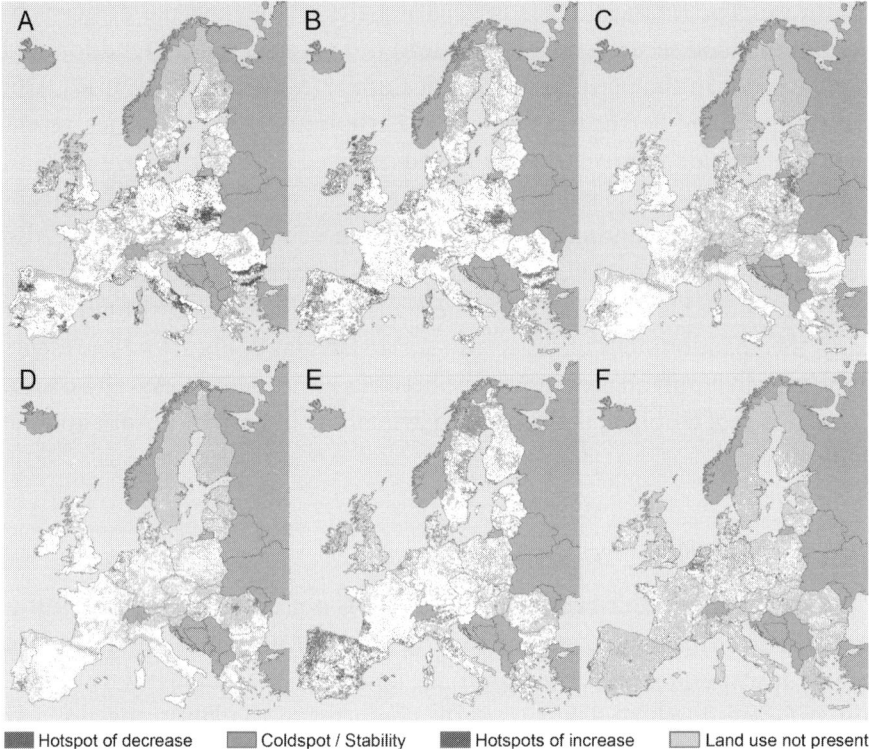

FIGURE 4.5

Hotspots of area changes among broad land-use categories between 1990 and 2006 (2000–2012 for (C) and (D)). Europe ((A): cropland extent; (B): pasture extent; (C): agricultural abandonment; (D): agricultural re-cultivation; (E): forestland extent; (F): urban extent). Hotspots include the 10 per cent largest change values (in positive and negative direction). Coldspots/stability areas entail the 10 per cent smallest change values (both positive and negative) as well as all unchanged areas. Areas outside hotspots and coldspots are in white. Source: Kuemmerle et al. (2016). A black-and-white version of this figure will appear in some formats. For the colour version, please refer to the plate section.

of the former farmed landscape. At the same time, the overall increase in forest cover has been followed by a drastic change in the composition and use of some of Europe's forest areas.

At present, some forests in Europe remain relatively undisturbed by human activity, which may be considered primary forests, where the forest dynamics follow natural processes. However, the majority are semi-natural forests, e.g. man-modified forest communities shaped by silviculture or agro-forestry, so that the forest structure and composition of species has been changed compared to the original natural situation and the potential vegetation. Tree species composition, notably, has been shaped and maintained by silviculture, which normally favours species with higher commercial interest; and, in this way, the dominant forest trees have a significant influence on the silvicultural systems applied in the forestry (or agro-forestry) tradition of each country such

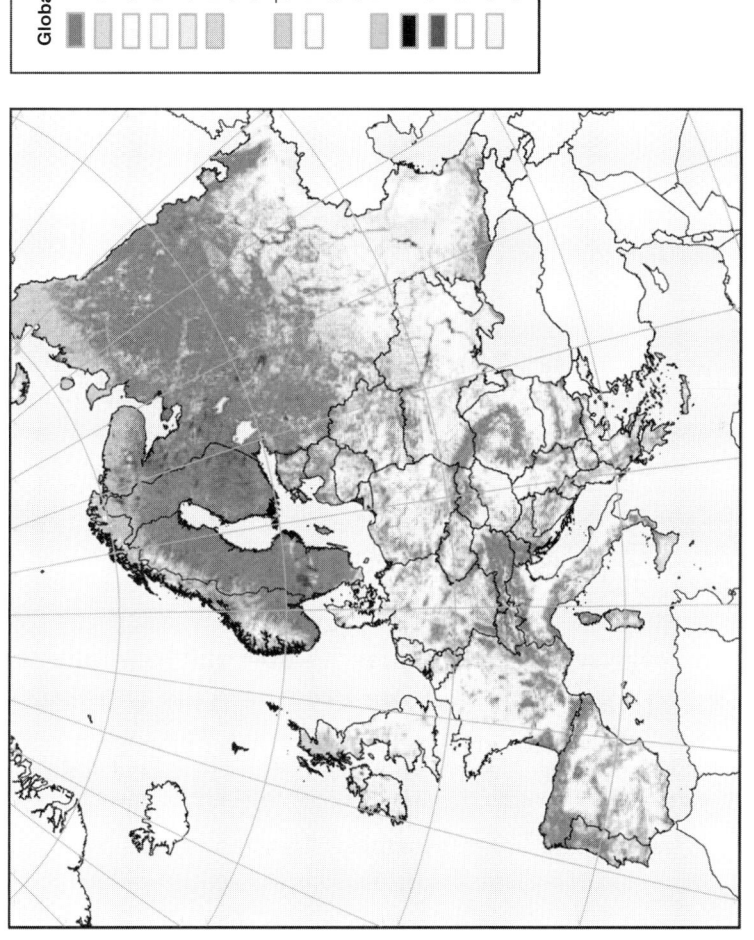

FIGURE 4.6
Forest distribution in Europe. Source: EEA (2007). A black-and-white version of this figure will appear in some formats. For the colour version, please refer to the plate section.

as even-uneven aged high forest, coppice, chestnut orchards, *dehesas* or *montados* (Barbati, Corona and Marchetti 2007). The plantation represents a particular type of forest as it is established by planting or seeding in the process of afforestation or reforestation (EEA 2007). The key factors in a forest that have a role in landscape appearance and dynamics can be classified as (Larsson et al. 2001): (a) structural (physical characteristics); (b) compositional (the biological component, i.e. tree species) and; c) functional (abiotic/biotic disturbance factors and management).

The most recent forest typologies produced at the European level consider 14 different types of forest (Barbati et al. 2007; EEA 2007). A forest type can be defined as 'a category of forest defined by its composition, and/or site factors (locality), as categorised by each country in a system suitable to its situation' (EEA 2007). These 14 types are: (1) boreal forest; (2) hemiboreal forest and nemoral coniferous and mixed broadleaved-coniferous forest; (3) Alpine coniferous forest; (4) acidophilous oak and oak-birch forest; (5) mesophytic deciduous forest; (6) beech forest; (7) mountainous beech forest; (8) thermophilous deciduous forest; (9) broadleaved evergreen forest; (10) coniferous forest of the Mediterranean, Anatolian and Macaronesian regions; (11) mire and swamp forests; (12) floodplain forest; (13) non-riverine alder, birch or aspen forest, and; (14) plantations and self-sown exotic forest. The typology in itself does not say much about the landscape, but it expresses the large variety of forest types in Europe, which needs to be considered if rural landscape transitions are to be understood. The map in Figure 4.6 presents the distribution of forest in Europe compared to the remaining land-cover types and illustrates how forest is extremely dominant mostly in the mountainous areas of Southern and Central Europe, as well as in the overall boreal domain of Europe. However, it also shows that in many regions of Central Europe forest is mixed with agricultural land and, therefore, has a significant role in the landscape pattern.

As well as farm systems and agriculture, forest systems and forest cover also change, although such changes are much slower and less visible than in agriculture due to the temporal scale of tree growth and thus of forest. Nevertheless, some very clear trends of change are apparent on the map in Figure 4.6, which shows the decrease and increase in forest land in Europe from 1990–2000. The Iberian Peninsula is highly dynamic, which is related to the relatively marginal conditions for farming. Through history, highly productive areas have tended to be maintained as specialised agricultural areas, and have directly registered the effects of intensification, specialisation and concentration. Marginal areas, in contrast, tend to be more unstable in relation to the impact of market and policy changes as their production systems are fragile in an increasingly globalised context (Pinto-Correia and Breman 2009; Pinto-Correia and Primdahl 2009). Thus, in the 1970s and 1980s, both Portugal and Spain witnessed a general trend in the growth of pine and

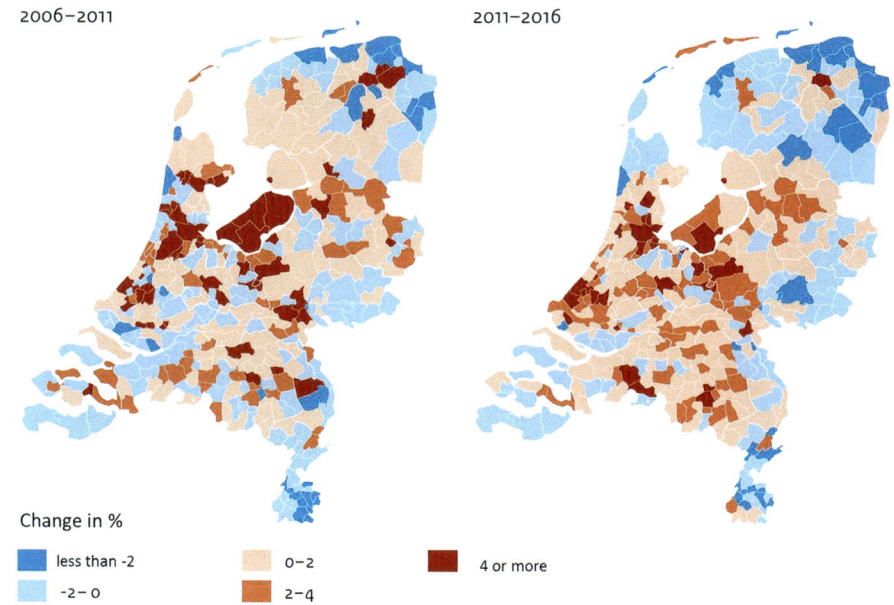

FIGURE 1.5
a. Population development per municipality in the Netherlands 2006–2016
Source CBS: www.clo.nl/nl210206

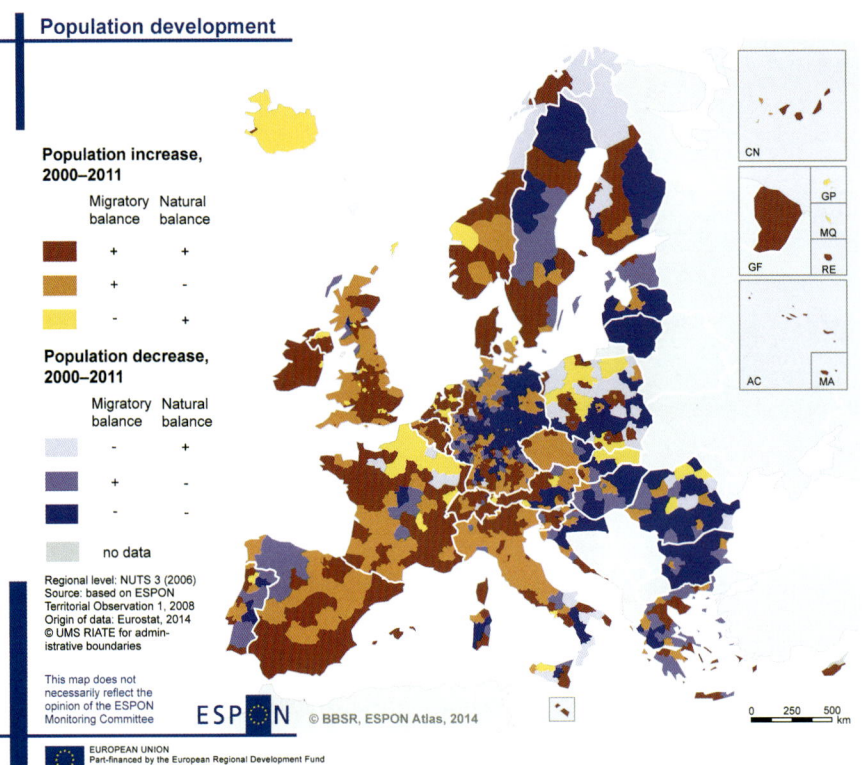

FIGURE 1.5
b. Population development EU 2000–2010
Source: www.Mapfinder.ESPON.eu

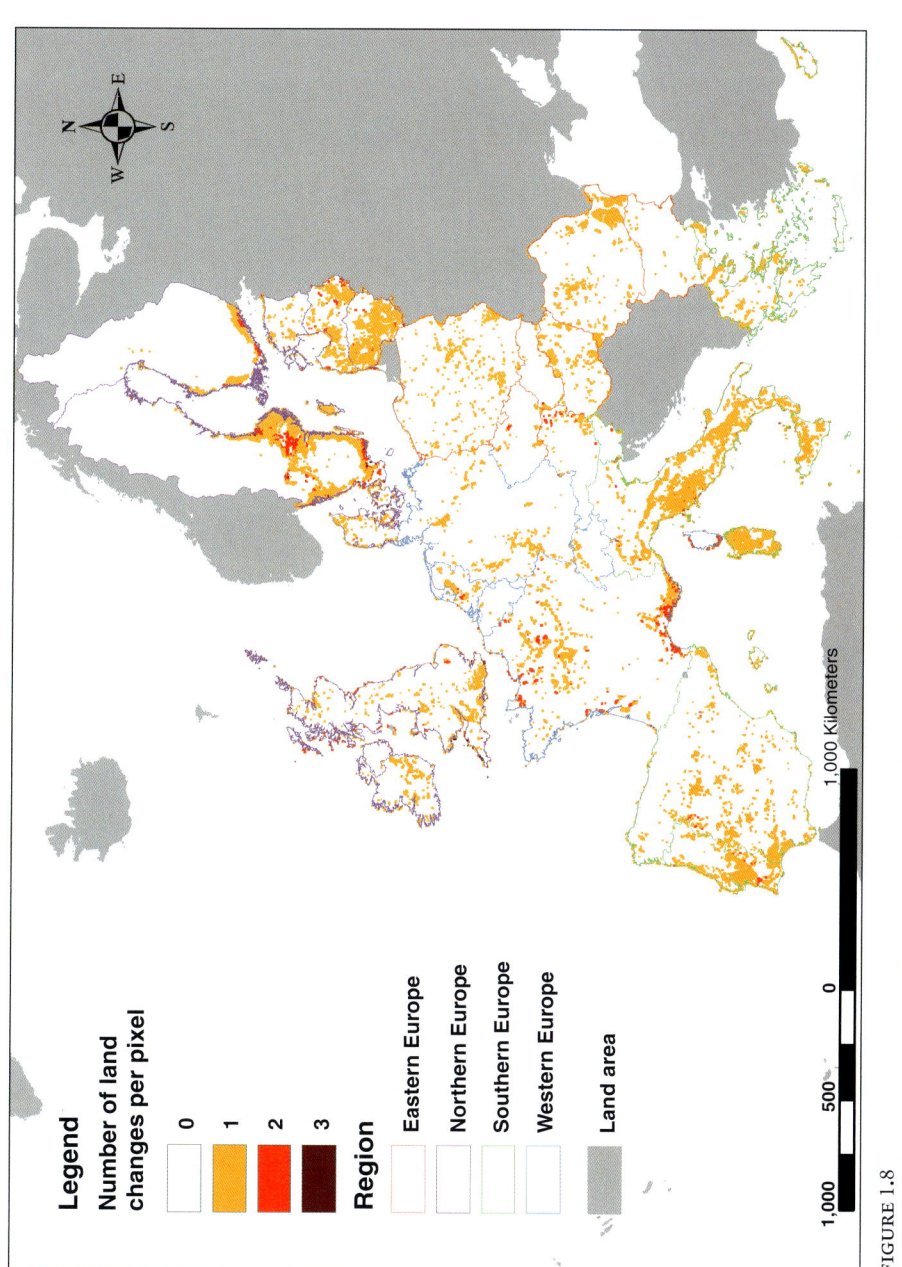

FIGURE 1.8
Generalised prime hotspots of Europe for the period 1950–2010, showing the spatial distribution of (multiple) land changes (Fuchs et al. 2013)

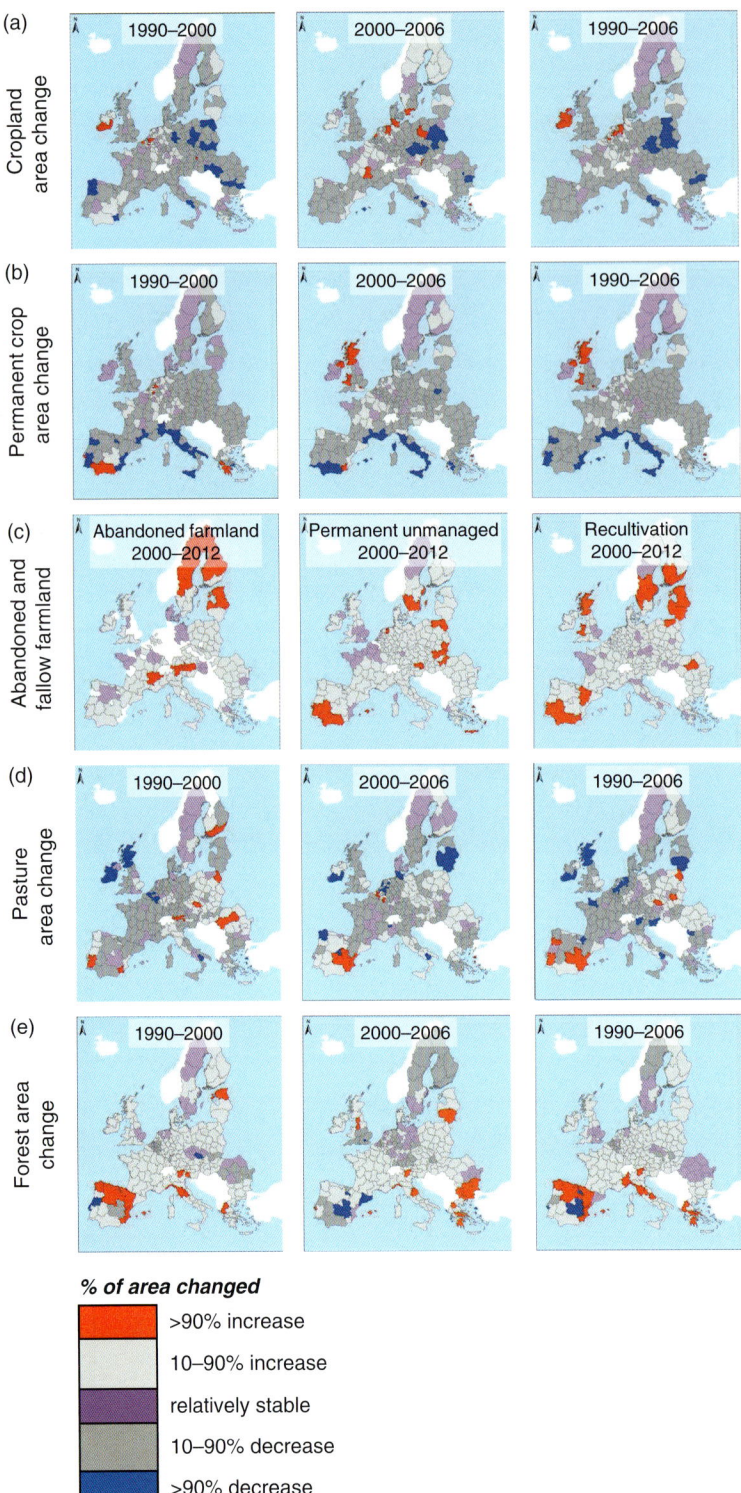

FIGURE 1.10
Hotspots of area changes in terms of broad land-use categories at the NUTS-2 level based on a range of data sources (Kuemmerle et al. 2013). Dark grey areas indicate NUTS-2 regions with area decrease, whereas light grey represents increasing area trends.

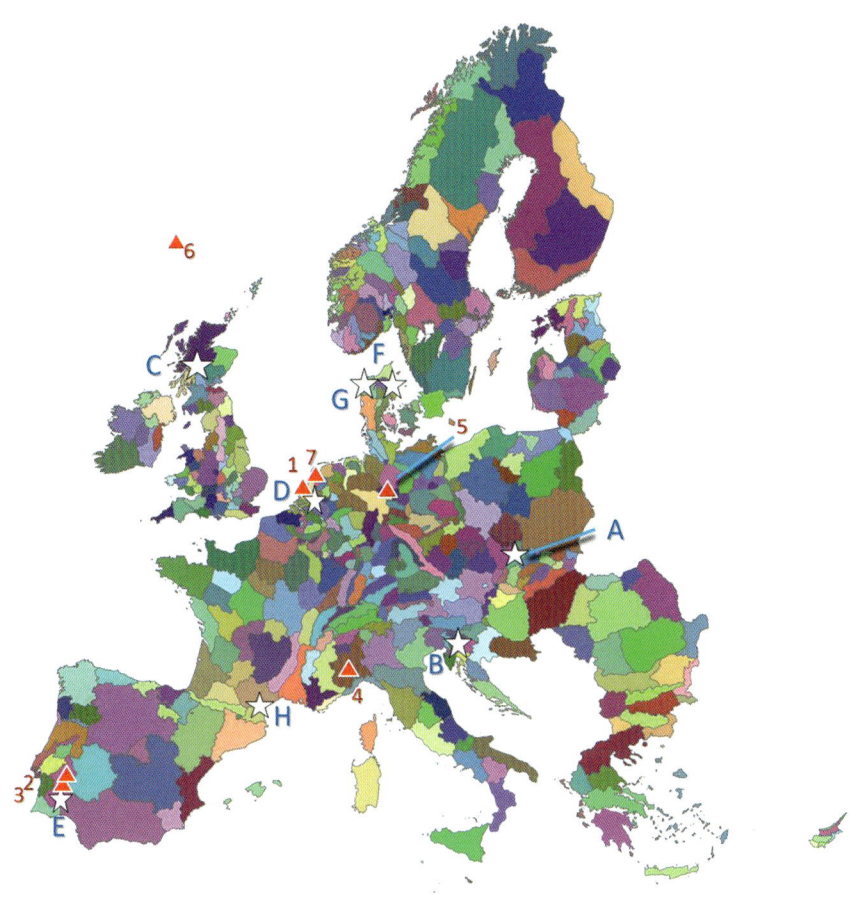

FIGURE 1.16
The locations referred to in the introductory stories to the chapters (red triangles and numbers) and in the cases (white stars and capital letters), projected on the map of the European Landscape Character Areas (Pedroli et al. 2017)

FIGURE 3.C.1
Location of the Great Trossachs Forest, Scotland (www.thegreattrossachsforest.co.uk)

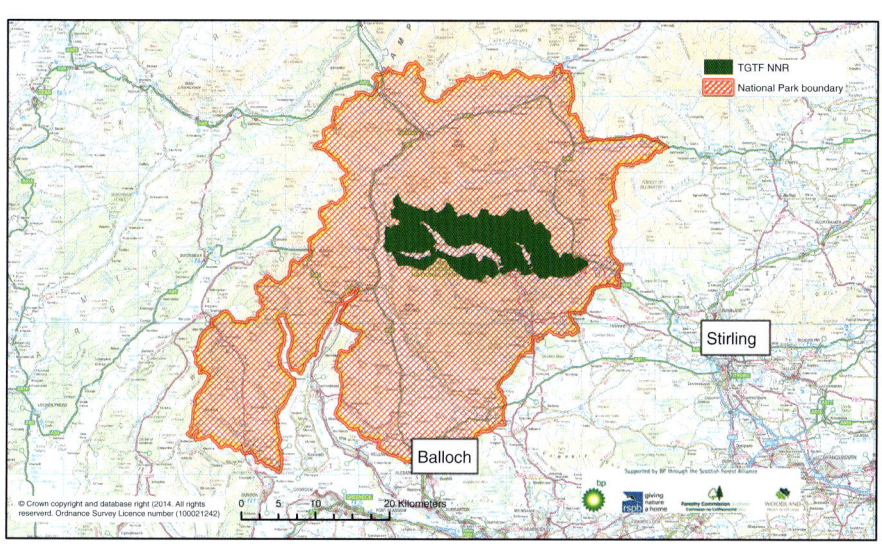

FIGURE 3.C.3A
Local area of the Great Trossachs Forest R
Source: www.thegreattrossachsforest.co.uk

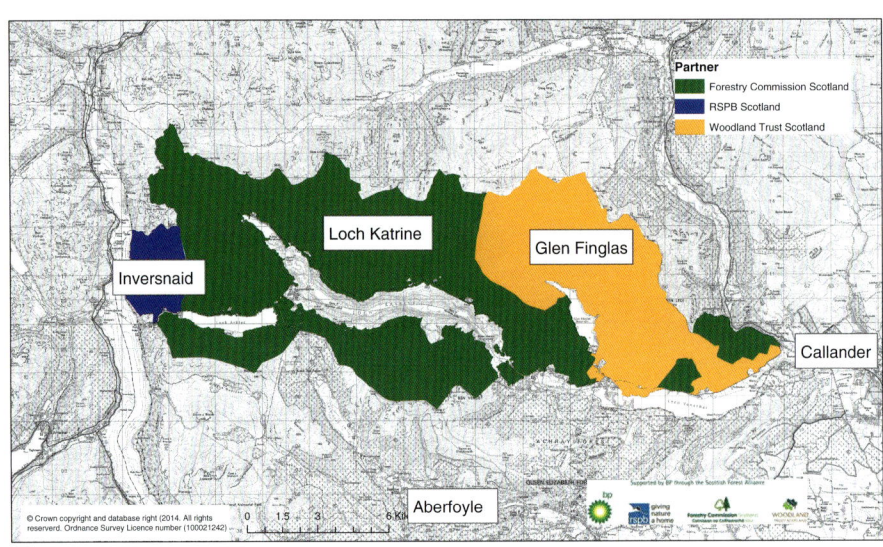

FIGURE 3.C.3B
Local area of the Great Trossachs Forest R
Source: www.thegreattrossachsforest.co.uk

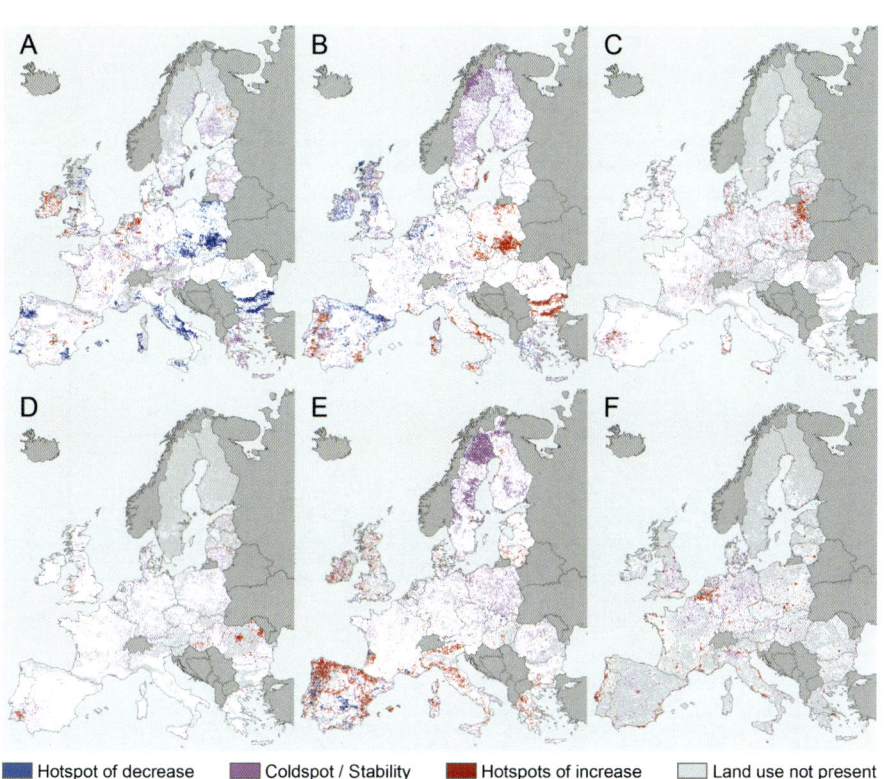

FIGURE 4.5

Hotspots of area changes among broad land-use categories between 1990 and 2006 (2000–2012 for (C) and (D)). Europe ((A): cropland extent; (B): pasture extent; (C): agricultural abandonment; (D): agricultural re-cultivation; (E): forestland extent; (F): urban extent). Hotspots include the 10 per cent largest change values (in positive and negative direction). Coldspots/stability areas entail the 10 per cent smallest change values (both positive and negative) as well as all unchanged areas. Areas outside hotspots and coldspots are in white.

Source: Kuemmerle et al. (2016)

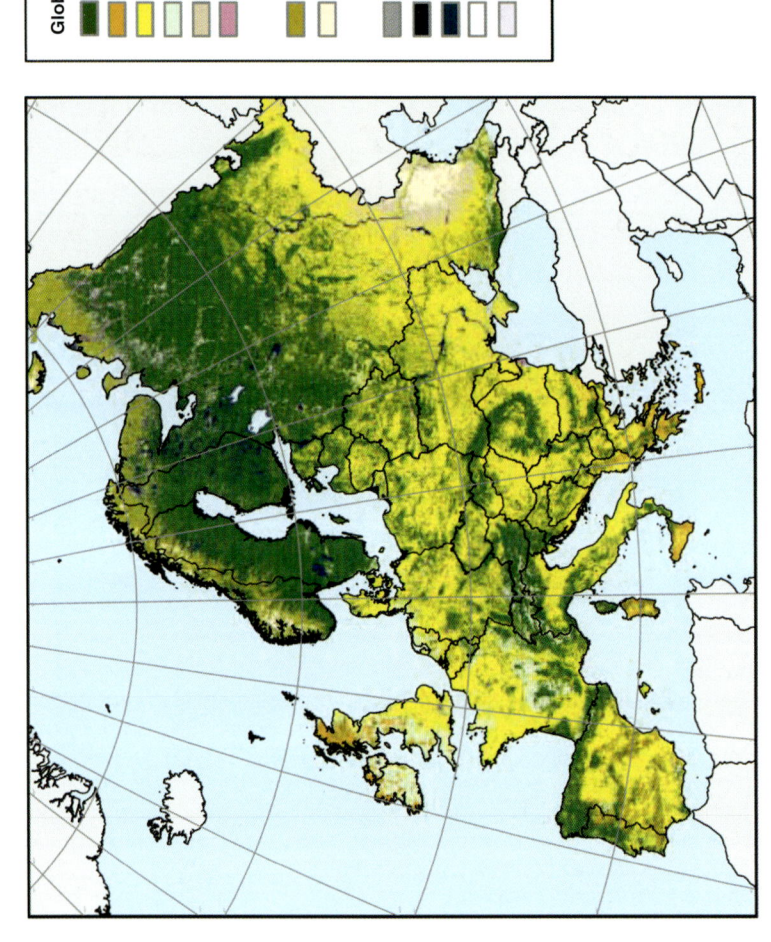

FIGURE 4.6
Forest distribution in Europe
Source: EEA (2007)

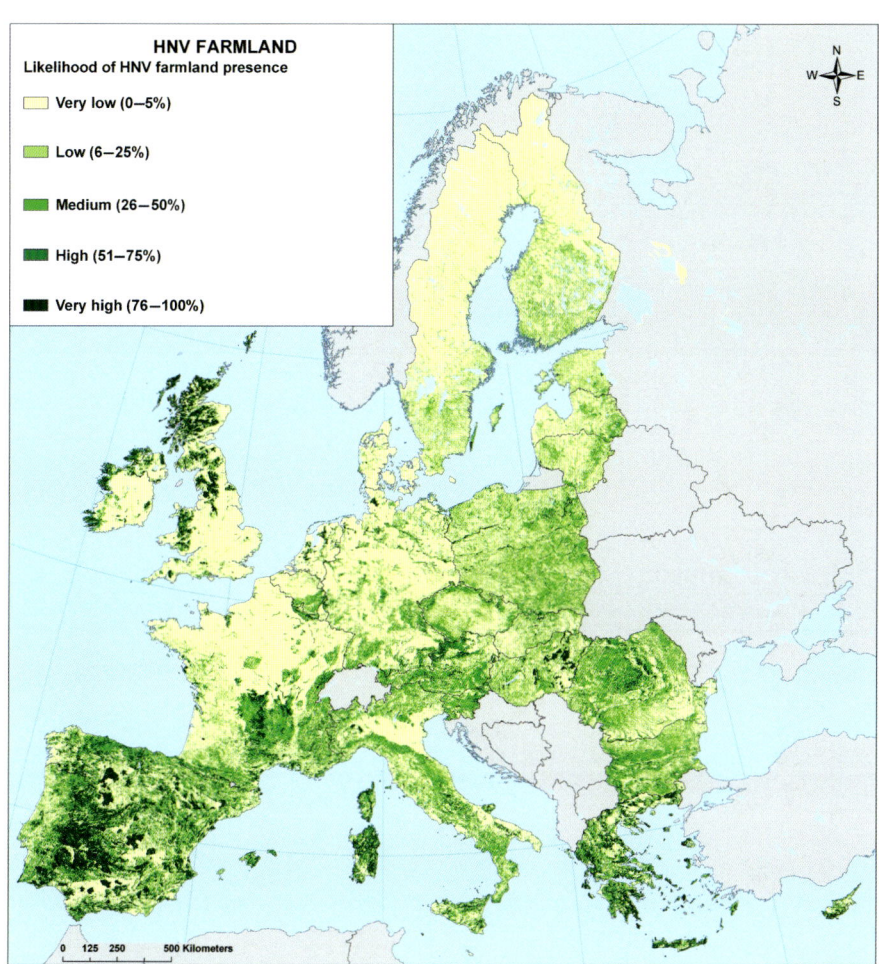

FIGURE 4.7
High-nature-value farmland in Europe: Southern and Eastern Europe, together with Scotland and the western UK, largely dominate regarding the presence and extent of this type of farmland, associated with extensive land-use systems as well as small-scale farm mosaics with heterogeneous land-use patterns.
Source: Paracchini et al. (2008)

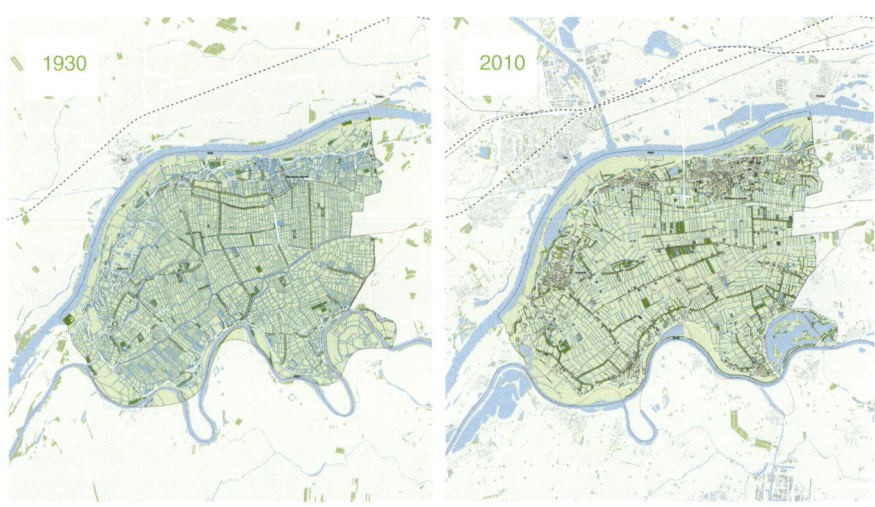

FIGURE 4.D.1
Maas en Waal area (Province Gelderland, the Netherlands) before and after land re-allotment
Source: RCE (2015)

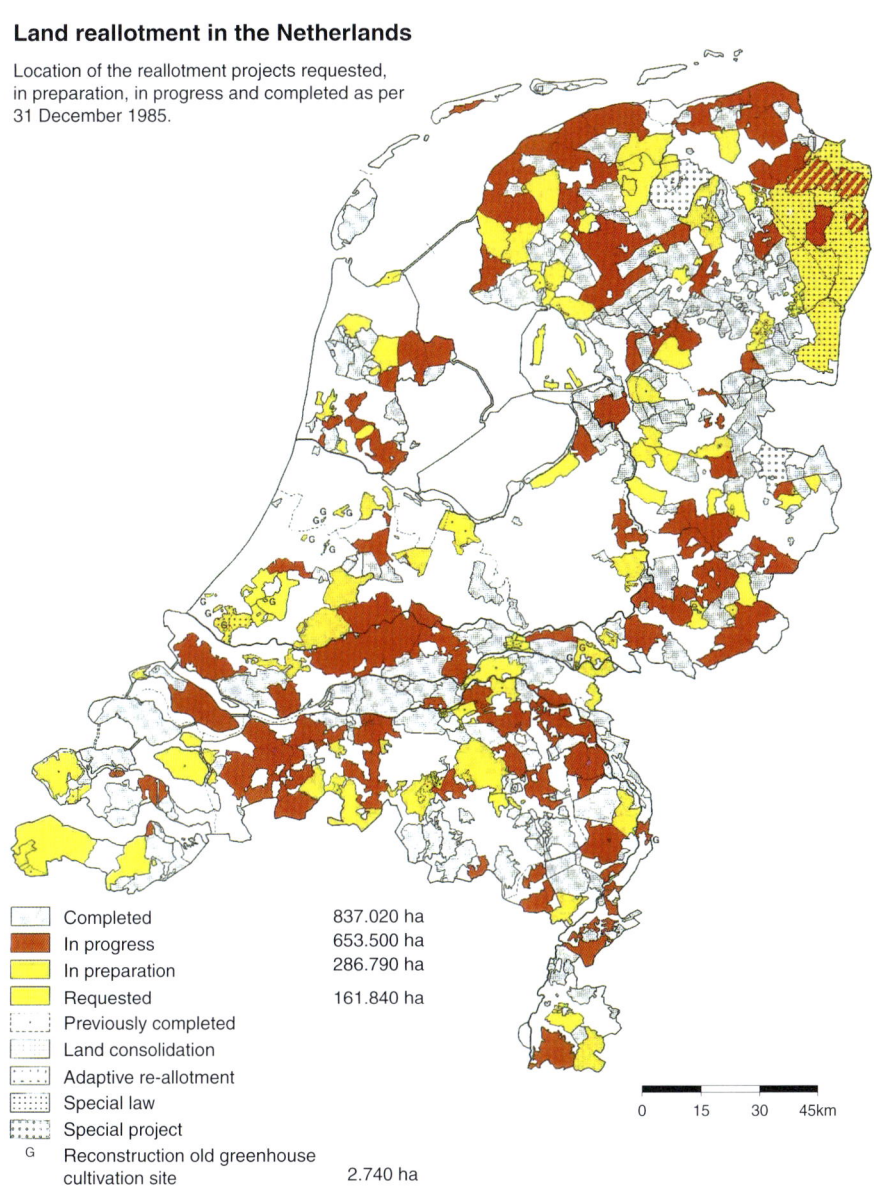

FIGURE 4.D.2
Land re-allotments realised in the Netherlands in the twentieth century (different colours indicate the status of the procedure as per 1985)
Source: Centrale Cultuurtechnische Commissie en Cultuurtechnische Dienst, Jaarverslag (1985)

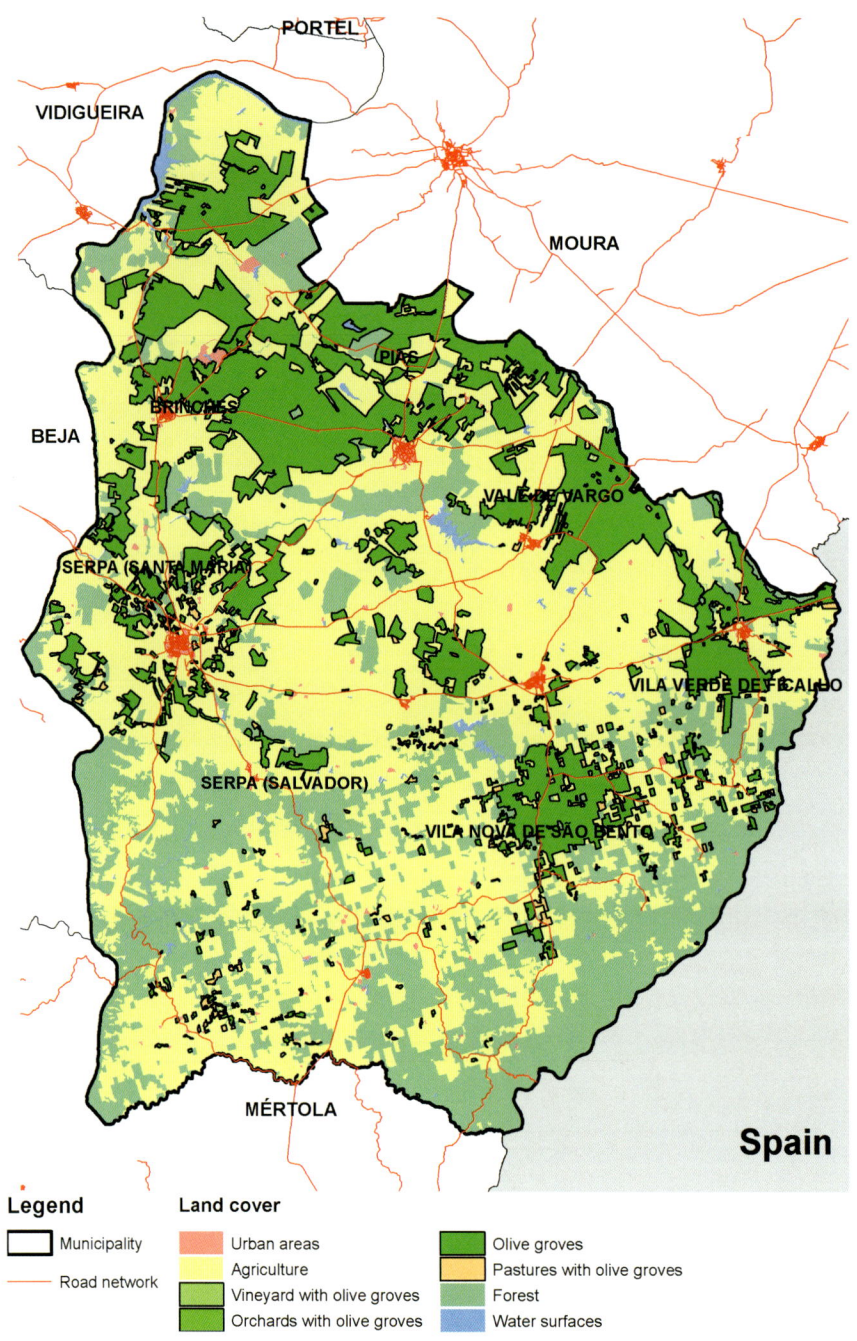

FIGURE 4.E.1
Spatial representation of the 2007 olive grove distribution in Serpa
Source: Surová, Pinto-Correia and Marušák (2014)

FIGURE 6.G.1
Topographical map of the village of Feuilla (left) and land use in the municipality (right)

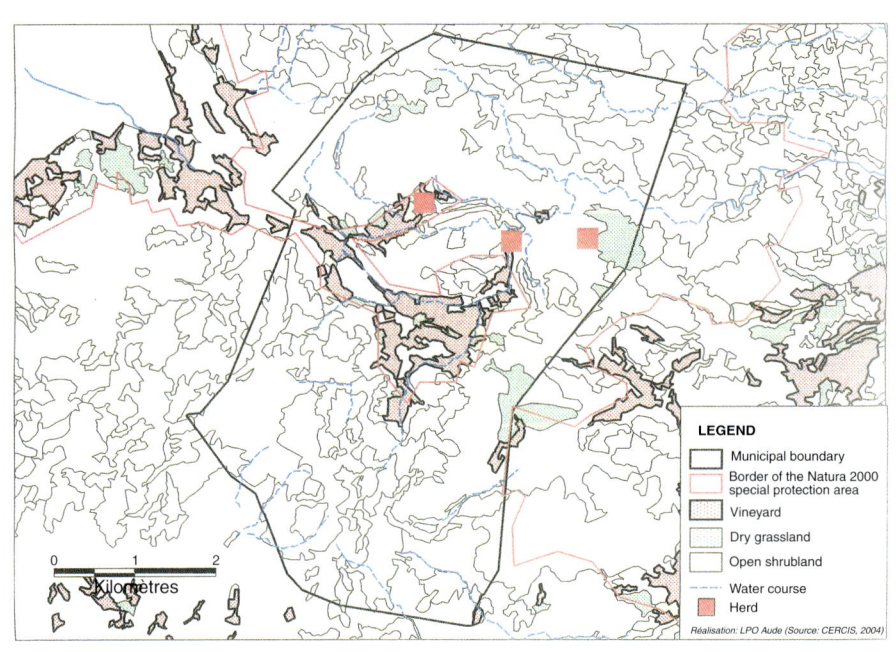

FIGURE 6.G.1 (*Cont.*)

mainly eucalyptus plantations in mountainous and hilly areas, focussed on the production of timber and replacing former crop fields and mosaic Mediterranean systems (EEA 2007; Pinto-Correia and Vos 2004). Later, in the 1990s, many of these mono-specific plantations, which were not adapted to the Mediterranean climate and were not properly managed due to insufficient means, became extremely prone to forest fires and, therefore, the forest area decreased between 1990 and 2000 (Figure 4.7). In the late 1990s, a new period of forest plantation began, this time supported by public policies, i.e. the CAP measures for the afforestation of agricultural land. Large areas of former crop and grazing land, with the most marginal conditions, were afforested once

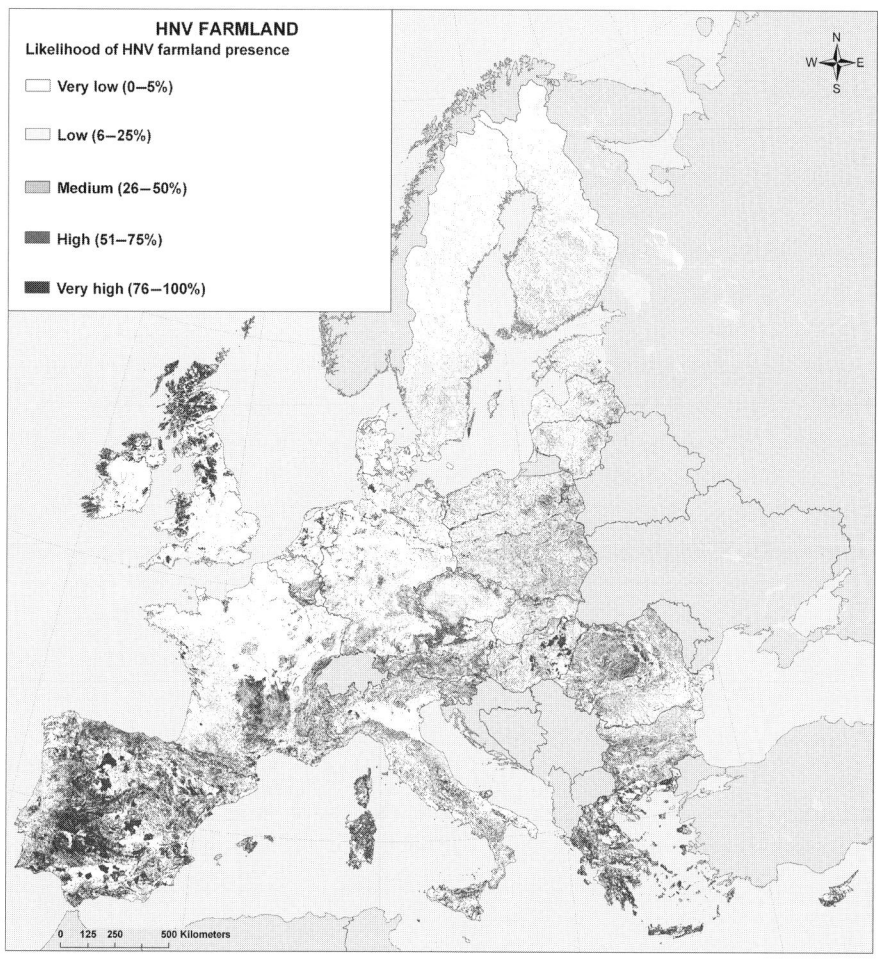

FIGURE 4.7
High-nature-value farmland in Europe: Southern and Eastern Europe, together with Scotland and the western UK, largely dominate regarding the presence and extent of this type of farmland, associated with extensive land-use systems as well as small-scale farm mosaics with heterogeneous land-use patterns. Source: Paracchini et al. (2008). A black-and-white version of this figure will appear in some formats. For the colour version, please refer to the plate section.

more (Barbati et al. 2007). This time there were guidelines with regards to which species should be planted in specific regions. For instance, in southern Portugal, the plantations were mainly pine (*Pinus pinea*), but were later combined with holm and cork oak (*Quercus rotundifolia* and *Quercus suber*). In northern Portugal, some of these afforested areas were also pine (*Pinus pinaster*) or chestnut trees. These plantations are now becoming mature forest stands, but they are also extremely prone to fire, at least in southern Portugal. It is unclear how the landscape in these marginal areas will be in the future, but certainly degradation due to diversity loss and forest fires and shrub encroachment are significant threats.

4.3 Global Challenges for Agriculture and Forestry in Europe

Coping with Climate Change, Water and Soil Conservation

At the start of the twenty-first century, agriculture and forestry in Europe are facing global challenges which are common to many other regions of the world. The extent of these challenges and the impact they will have on the structures and systems we have today, is still unknown. However, there is no doubt that there will be a change in the overall conditions (Costanza et al. 2014; Plieninger et al. 2016).

Over the next century, society will increasingly be confronted by the impact of global change including pollution, land-use change and climate change. These will have spatially differentiated impacts, but will certainly affect many regions of Europe (Metzger and Schröter 2006). Most relevant in Europe is climate change, which is predicted to result in a warmer and more variable climate. Affecting primarily Southern Europe, there are clear signs of deteriorating agro-climatic conditions, in terms of increased drought stress and shortening of the active growth season. In contrast, climate change may have beneficial impacts on Northern European agricultural systems, but the effect is still uncertain (Trnka et al. 2011).

The comprehensive study by Trnka and colleagues (2011) demonstrates how the observed warming trend throughout Europe (1.90°C from 1901 to 2005) is clear, although precipitation trends are more spatially variable: mean winter precipitation has increased in most of Western and Northern Europe, but has changed little in Central Europe. Furthermore, trends are negative in the eastern Mediterranean, while no significant change has been observed in the western Mediterranean. The review also shows how the effects of climate change and increased atmospheric CO_2 levels by 2050 are expected to lead to overall small increases in European crop productivity. However, several climate projections for 2050 indicate temperature increases greater than approximately 2.1°C, which would probably lead to a decline in the yields of many crops. There are different projections resulting from different models, but all agree on a consistent spatial distribution of the effects, leading to the need for the regionalisation of adaptation policy (EC 2009).

Further, the projected increase in extreme weather events (e.g. periods of high rainfall, temperature and droughts) over at least some parts of Europe is predicted to increase yield variability and, therefore, have a significant impact on the sustainability of farm systems. It is important to keep in mind that increases or decreases in temperature and/or precipitation are location-specific, and average temperatures and rainfall statistics can mask increases in the frequency, duration and intensity of these extreme events (such as droughts and floods), which appear to be on the rise in many areas (Dale, Efroymson and Kline 2011).

Agriculture is also itself a driver of climate change, releasing significant amounts of greenhouse gas to the atmosphere, as carbon dioxide (CO_2), methane (CH_4) and nitrous oxide (N_2O). Concerns about land-use change and its impacts on land-cover change and, thus, on global environmental change emerged in the research agenda several decades ago with the realisation that land-surface processes influence the climate. Agriculture is a major driver of land-use and land-cover change and, therefore, agricultural change also impacts climate change (Bessou et al. 2016). Land-cover change modifies surface albedo and, thus, surface–atmosphere energy exchanges, which have an impact on the regional climate. Terrestrial ecosystems act as sources and sinks of carbon and, thus, land-use change leads to global climate changes via the carbon cycle. Subsequently, local evapotranspiration plays an important role in the water cycle, and also depends on land cover and has an impact on the climate on the local-to-regional scale (Lambin and Geist 2006).

The fluxes of greenhouse gas are complex and heterogeneous, but in any case the active management of agricultural systems creates space for mitigation. For example, with existing technologies, agriculture has significant potential for cost-effective greenhouse gas mitigation through mechanisms which can be implemented in the production systems (Smith and Olesen 2010). These mechanisms can reduce emissions of methane and nitrous oxide and increase soil carbon storage. The approaches for increasing mitigation and reducing greenhouse emissions are intricately linked with other issues of agriculture sustainability and include: (1) reduction in soil erosion; (2) reduction in leaching of nitrogen and phosphorus; (3) conservation of soil moisture; (4) creation of microclimates to reduce temperature extremes; (5) diversity of crop rotation; (6) extensification or abandonment of land. A large proportion of the mitigation measures arise from soil carbon sequestration, which has strong synergies with sustainable agriculture practices and in turn reduces vulnerability to climate change.

There is, thus, a double challenge for agriculture with regards to climate change: to reduce emissions and to adapt to a changing and more variable climate. However, there appear to be synergies between the two challenges which still need to be exploited (Smith and Olesen 2010; Trnka et al. 2011).

In Mediterranean Europe, the effects of climate change can have a dramatic effect on the sustainability, quantity, quality and management of water resources, which are already often a limiting factor in production systems (García-Ruiz and Lana-Renault 2011; Lionello 2012). Precipitation is a key variable in the Mediterranean region as any future decrease would be critical for all types of human activities and, if further exacerbated by increased temperatures, it could lead to more frequent droughts. As a consequence, understanding the drivers of change which affect the land-use systems of Europe today means also taking the possible effects of these changes into consideration. Some of the processes of land-use intensification which took place in North-Western Europe would not be possible in Southern Europe due to limitations in water availability. Indeed, the recent conversion of large areas in Southern Europe into highly intensive and specialised crop areas has been possible only due to the irrigation facilities which have been built in recent decades (Case E). The issue of water quantity and availability is, thus, a limiting factor on land use in large parts of Europe and will continue to be so in the future (Sanchez et al. 2013). Infrastructure which allows the large-scale irrigation of formerly rain-fed fields is extremely costly to both construct and maintain, while irrigated agriculture brings with it other adverse impacts, mainly environmental, which need to be taken into consideration.

Also the soil, one of the determining factors regarding the sustainability of farm systems (Vanslembrouck and Van Huyenbroeck 2005), can be a crucial limiting factor for agricultural development. Despite technological progress, agriculture on farmland with good soils will continue to be favoured compared to farmland with shallow soils, a low level of nutrients and steep slopes. Constraints are related to the production capacity and the erosion risks. This is without a doubt a major issue in Mediterranean Europe, and it often leads to irreversible soil degradation (Grove and Rackham 2003; Van-Camp et al. 2004). Preventing soil degradation is, thus, frequently one of the main goals of public intervention, but still the exposure of fragile soils to erosion risks continues. By removing the most fertile topsoil, soil erosion reduces productivity and can lead to an irreversible loss of natural farmland and reduce the adaptive capacity of the soil (Van-Camp et al. 2004). Furthermore, erosion rates are expected to rise during the twenty-first century due to global warming, which involves a more vigorous hydrologic cycle, particularly in Mediterranean regions. Under these conditions, soil erosion by water is the main process of land degradation and, thus, is one of the key indicators for long-term assessments (Bosco et al. 2014). Besides soil erosion, soil degradation includes desertification, organic matter depletion and contamination (Van-Camp et al. 2004).

Acknowledging the role of soil in the sustainability of farm systems, measures to protect the soil have been or can be progressively introduced in many farm systems, which comprise zero or reduced tillage including ridge tillage (in which ridges are made in the field), shallow ploughing and rotation or scarification of the soil surface, all of which reduce soil disturbance compared to conventional deep tillage with a mouldboard plough. The mechanisms for greenhouse gas reduction are the same as those for zero tillage. The timing of tillage can also be taken into account as part of this measure. Other measures are catch crops and cover or intermediate crops, residue management, rotation and adding legumes etc. (Sanchez et al. 2013).

Nature Conservation

The long history of human use in Europe has resulted in primary uses and biodiversity having co-evolved through time to create cultural landscapes that are valued for their income generation as well as for their aesthetic, biodiversity and cultural values. Long-standing agricultural and forestry land-use systems are generally the source of the most valued landscapes and habitats (Hodge, Hauck and Bonn 2015). As farming and forestry systems became progressively more specialised and intensive, habitats and biodiversity came increasingly under pressure. The threats to nature resulting from changes in land-use systems have been acknowledged for decades now (Stanners and Bordeaux 1995). On the other hand, abandonment is also leading to the disappearance of former landscape patterns or a change in their components so that the associated nature value is declining (Renwick et al. 2013). Consequently, the concern for the state of the environment in agricultural and forestry landscapes and the societal pressure for more protection-oriented policies have been increasing. The main drivers of human occupation of rural space, as described by John Holmes (2006, 2012), presented in Chapter 3, show how protection has emerged as a main driver, together with the consumption of the rural space to contest the former dominance of production goals.

The main drivers of biodiversity change – habitat loss, habitat deterioration and eutrophication – led to the formulation of nature conservation strategies during the last decades of the twentieth century in most European countries (Lawton et al. 2010). In order to protect natural values and maintain a certain degree of biodiversity in the European rural landscapes, nature goals instead of production goals have gradually been introduced in marginal areas or small patches which had maintained a natural or semi-natural habitat. Later, spatial planning began to create a particular landscape structure by establishing connections between richer nature areas in the form of linear elements or corridors, linking habitat areas on different scales (Burel and Baudry 1999). Corridors favour the circulation and dispersion of species and have been widely used as they allow the connection between habitats, but they also provide the economy of space for nature conservation objectives in highly fragmented landscapes. From a debate focussed on high-quality patches and their preservation, there has been a progressive shift to a restorative approach which involves rebuilding nature through ecological networks to expand and link habitats (Selman 2012). Nevertheless, the size of the nature conservation areas, as well as the width and vegetation composition of the corridors, which is necessary to have an effect on the maintenance and circulation of different species, is very variable. Furthermore, if the patches and corridors become progressively more isolated, their effect is reduced: the patch and corridor approach has proven somewhat reductionist (Selman 2012). In most of Europe, the nature conservation paradigm has followed this approach of a segregated landscape, not being able to avoid, or even reinforcing, a high level of ecological fragmentation (Hodge et al. 2015). Therefore, there have not been any conflicts between uses, but rather the segregation of less-productive areas, which has resulted in low landscape diversity and nature condition. Nature conservation goals face a race against time to ensure that the creation of ecologically rich structures can match the pace of species extinction. However, so far, the increased resilience of the natural environment has not been achieved (Selman 2012). The integration of nature and productive land-use systems has often been a stated goal of public intervention, but it has not often been achieved (Van der Sluis et al. 2015).

In rural landscapes which are currently still dominated by extensive production systems (e.g. forest, silvo-pastoral systems and extensive grazing), or where small-scale agriculture is being maintained, high nature values can still be maintained. Focussing precisely on the extensive and small-scale land-use systems of Europe, the concept of high-nature-value (HNV) farming system, put forward in the beginning of the 1990s, exemplifies how nature values are dependent on certain farming practices, which adds a new societal value to production systems in addition to the production of food and fibre (Almeida, Guerra and Pinto-Correia 2013; Oppermann et al. 2012). The HNV concept was promoted to integrate biodiversity and environmental concerns in the agricultural sector in Europe, based on the assumption that low-intensity agricultural management could greatly improve the overall biological and landscape diversity of farmland (Andersen et al. 2003; IEEP 1994; Paracchini et al. 2008) and, from a more focussed biodiversity perspective, it has attained a more encompassing landscape-level relevance. The concept has been discussed for a while now in European fora, and the European Commission has produced three methodological guidelines for the identification of HNV farmland: the land-cover approach, the farming-system approach and the species-and-habitat approach (Andersen et al. 2003; IEEP 2007). The former is based on data from CORINE Land Cover and uses land-cover classes related to HNV farming to ascertain the location of HNV areas. It is the one used in the map in Figure 4.7 which was produced by the European Environmental Agency (Paracchini et al. 2008). However, the approach does not indicate whether a specific area is actually managed extensively or intensively. The second approach uses agronomic and economic statistical data (FADN) to analyse farming practices and it identifies the associated management pressures related to HNV farming systems to determine the occurrence and extent of HNV farmland. The species-and-habitat approach focusses on the distribution of species known to occur in association with specific types of farmland, which can help identify other types of farmland not discernible through the land-cover or the farming-system approaches.

Several EU Member States have already explored the application of EU guidelines to different European contexts (Pointereau et al. 2007; Ribeiro et al. 2014; Van Doorn and Elbersen 2012), which may support increased protection of these particular land-use systems, which are often fragile in a global context.

In many peripheral regions, the maintenance of traditional agricultural practices with a favourable impact on nature protection and nature conservation has so far been encouraged and supported due to the recognition of this role (Peneva et al. 2015). HNV farming has not only been an environmental solution; it also represents a broader concept for the economic and social sustainability of agriculture at the regional level as it has contributed to a greater understanding of the circumstances of peripheral agricultural areas in Europe which require: (i) new approaches and innovation which is not based on mainstream productivist farming; (ii) the acknowledgement of new roles for farming and; (iii) new farming identities. Therefore, the HNV concept has the potential to become a key component in the European model of agriculture for the future through its post-productivist and multifunctional characteristics, contribution to the diversity of rural areas and the conservation of biodiversity, all of which are combined with the productive use of the land. Normative institutions and funding opportunities such as cross-compliance and agri-environmental schemes are, nevertheless, crucial to maintain and enhance HNV farming systems, as well as strong networking and collaboration mechanisms between HNV farmers, but also between farmers and other stakeholders, so that farmers feel the added value of this type of farming is clearly recognised (Peneva et al. 2015).

Thus, there may now be a way of ensuring the continuance of farm systems and landscapes which are of particular value in Europe, but which are seriously threatened in a global market context, through implementing the ecosystem service or landscape service approaches (see Chapter 2), or through encompassing land-use classifications such as the HNV. These approaches make it possible not only to identify which systems deserve protection, but also to determine their location and the practices they depend on (Almeida et al. 2016; Almeida et al. 2013; Ribeiro et al. 2014). Based on this, specific management tools and support may then be targeted and implemented, which is discussed in Chapter 6.

Energy Production

The dominant resource extractions and land-use management activities involve energy. The use of fossil fuel is one of the key drivers behind increasing greenhouse gas emissions as well as land-use changes. Alternative energy sources (wind, solar, nuclear, bioenergy) are being explored to reduce greenhouse gas emissions. Bioenergy in particular needs to be analysed as it is directly linked to land use. Bioenergy is derived from biomass from: (a) agricultural energy crops; (b) forestry and wood-based industries, and; (c) farm, municipal and industrial organic waste (Smith and Olesen 2010). The spread of energy crops, in particular, in the first decade of the 2000s had a significant impact on the landscape patterns in some regions of Europe. These energy crops have been favoured by ageing or absent farmers, or in less-accessible plots in the case of fully active farmers, as they require less attention than many other crops. The expansion of energy crops depends on a combination of factors from the price of these energy crops to the volatility of the income generated by food crops.

As mentioned previously, biomass can be used in the generation of electricity, heat and biofuels. Yet energy production can have a wide range of effects on land productivity, land cover, albedo and other factors that affect carbon, water, energy fluxes and, in turn, the climate (Sutherland et al. 2015a).

In the near future, a substantial expansion in biofuel production will require an expansion in the range of available feedstocks and the introduction of advanced, second-generation conversion technologies. Such feedstock sources exist in agriculture and forestry residues and dedicated energy crops. These in turn can potentially be grown on land which is less suited for intensive production, thereby avoiding competition with food production, while the environmental impact on modern biofuel production should also be lower than the first-generation feedstocks (Pedroli et al. 2013; Smith and Olesen 2010).

Meanwhile, the climate influences the potential output, relative efficiency and sustainability of alternative energy sources. Thus, land use, climate change and energy choices are linked and any analysis should take this into consideration (Dale et al. 2011).

4.4 Conclusion

In this chapter, we have explored the use of rural land for production activities and natural processes, which have coexisted in the rural for many centuries: the production of food, fibre and energy, and the conservation of natural resources. All these processes were adapted to the natural conditions and shaped by technology and structural features. Until the middle of the twentieth century, agricultural and forestry were the dominant drivers of European landscape change. Different faming and forestry systems evolved, historically each of them intimately linked with the natural and cultural conditions. The production systems continue to change, but now in more regionalised ways and more linked to global drivers of various kinds. To understand current transition processes, these drivers must be included.

As in history, some of the drivers affecting agriculture and forestry and consequently the rural landscape are associated with technology, market development and policy – all of which becomes increasingly globalised. Others are linked to various forms of urbanisation which today intersect with changes in agriculture and forestry. In the next chapter, we deal with changing urban–rural relationships and how they have been and are influencing European landscapes.

5

Changing Relationships between the Rural and the City

Approaching the town of Wörlitz from the east – biking along the southern dikes of the Elb in south-eastern Germany – the landscape is slowly changing. The large fields of grain that dominate the picture gradually give way to old, tall deciduous woods. Suddenly, an eighteenth-century, tastefully designed embankment guardhouse appears. We are entering the Dessau-Wörlitzer Gartenreich, a regional landscape park, unique in a European context. Further west, historic buildings and tree-lined avenues come into view and around the corner, a picturesque ruin of a medieval castle appears next to a manor house in the Renaissance style. We have arrived at Wörlitz on the edge of the old park designed in the late eighteenth century in the romantic style. The traditional components of a romantic garden are certainly here; the sculptures, Roman temples, the (constructed) ruins, the stone bridges, the canals and lakes, the pastures and deciduous forest are all in place – laid out in addition to the castle, Wörlitz Palace, a park made for pleasure and for the joy of the eye.

However, there is more to it than this. Interwoven into this park matrix are grain fields and intensively cultivated vegetable patches. Combine harvesters and other agricultural machinery are found alongside lawn mowers, rose gardens and café gardens – not exactly an English-style garden. This park has definitely *not* been designed as a buffer between Wörlitz Palace and the productive countryside (Figure 5.1) – in fact crop cultivation has been included into the overall layout. Grain fields and rows of broad beans have become part of the aesthetic experience while also improving the farm economy. When it was introduced, this was indeed new thinking – first of all by Prince Leopold III Friedrich Franz of Sachsen-Anhalt, who got inspiration for the Gartenreich from his travels in Italy, France, the Netherlands and England (Küster and Hoppe 2010). Besides his interest in English romantic gardens and classicism, Prince Franz was also engaged in new European ideas about education and

FIGURE 5.1
From Wörlitz Garden, where grain fields are incorporated into the romantic garden design. Whereas the antique white temple in the centre is a classic element of a romantic garden, the productive grain field is not.

science. He was a child of the Enlightenment and the educational dimensions of agriculture and forestry were integrated into the overall development plans for the regions. In addition to the romantic garden of Wörlitz, five other gardens were laid out in the Dessau-Wörlitz region, all part of the so-called Gartenreich (Garden Empire), which has been included on UNESCO's list of world heritage sites since 2000. Each of these parks has a very different design; some have baroque sections, Chinese pagodas and exotic Mediterranean trees in special greenhouses – the orangeries – to survive the continental winter. Others are laid out as deer parks, which have the character of natural woodlands; still others were designed as peri-urban park lands for the city of Dessau. Between the parks, the regional landscape is unusually rich in cultural heritage of a high quality, including numerous historic buildings, old villages, historic dikes, ditches and other landscape elements.

The Dessau-Wörlitzer Gartenreich is a unique example of landscape design applied to a whole region, which was inspired by late eighteenth-century ideas regarding architecture and garden design, as well as ideas about the modernisation of agriculture and forestry. This was combined with the wise use of the natural conditions and the inherent limitations and opportunities of this fertile river landscape. Although it was designed more than 200 years ago, today, the Gartenreich is a highly sophisticated regional expression of a rural–urban relationship where rural and urban functions are woven together. The overall landscape is perfectly structured into a coherent whole with a strong geographical as well as historical identity, which serves a range of different functions, including agriculture, forestry, habitat management, housing, recreation and tourism (Andersson and Floryan 2005; Brandt 1989; Küster and Hoppe 2010; Ringkamp and Janssen 2000).

5.1 Introduction – between the Country and the City

As discussed in the previous chapter, European landscapes change when agriculture changes. However, structural change in the agricultural sector is by no means the only driving force as other factors also play a role in the overall processes which are transforming the landscape. In the previous chapter, we discussed how new concerns and demands regarding rural land are also affecting the landscape. These, among many others, are not solely emerging in the rural; they are linked to the urban sphere, population and processes. This chapter is about the changing urban–rural relationship and how it is affecting rural landscapes. As discussed in Chapter 1, a particular characteristic of the European rural space is that it has always been ruled from the villages or cities, which is not only because of market concentration but also because the public administration has traditionally been located in urban areas. Therefore, it is not possible to understand the rural landscape if we do not understand how it has been seen and organised from an urban centre perspective. Today, the vast majority of people in Europe and the rest of the world live in urban centres, and these citizens are the consumers of food and fibre, and use the rural space for its amenities. Thus the majority of voters are also located in urban areas, which means they have a substantial influence on the processes taking place in the rural. In this chapter, we discuss how to analyse the urban–rural relationship, as well as its consequences for rural landscapes seen from different positions. The ambition is to examine the relationships and rural landscapes from different perspectives, including historical background, as well as the current, highly urban influence on rural landscape change.

Many of the fundamental factors which are currently driving changes can be seen as part of urbanisation in a broad sense. A key issue here is the changing relationship between the urban and rural domains and the emergence of new functional linkages. This chapter starts with a discussion of common views of the urban and the rural – or 'images' as Williams (1973) terms them. We then present a historic outline of the changing urban–rural relations and how they have affected European landscapes. Current patterns are then described, followed by the presentation of a concrete case of changing landscapes in Denmark (Case H), which is an intensively farmed moraine area in Denmark where counter-urbanisation and hobby farmers' management practices must be included if one wishes to explain ongoing landscape changes. Subsequently, outdoor recreation, hunting and tourism are discussed as activities typically carried out by urban people.

5.2 Images of the Urban and of the Rural

On the front page of a Danish Agriculture and Food Council leaflet, which presents Danish agriculture for an international audience, is a bird's-eye view of a picturesque agricultural landscape (Figure 5.2a). It consists of a mosaic of (relatively small) arable fields, wood thickets, hedgerows, an inlet in the centre of the scene and in the distance the faint outline of a city. A few wind turbines and the yellow oil-seed rape fields suggest a modern rural landscape. However, there are no other indications of this. The photographer has managed to find a Danish agricultural landscape with an old farm in the foreground and no modern farm buildings in sight. Why has the umbrella organisation for Danish agriculture become so nostalgic when presenting today's agriculture? Why does Danish agriculture not stand up for its own modern buildings? or take an almost opposite position illustrated by Figure 5.2b, which is an image on the front page of a programme for an architectural competition on ideas for future agricultural buildings? Here we see new materials and forms and images of innovative agricultural structures, inspiring. But what about the landscape in which these building are located? At worst it is a green nightmare in the form of a mono-functional landscape; at best it is an undefined nowhere, deprived of any natural elements, cultural layers or visual character. Why do the architects who prepared the programme seem unable to imagine what a future rural landscape will look like – why do they prefer to depict the landscape as an abstract space?

Perhaps the difficulty they experience when dealing with both present agriculture and future rural landscapes is due to more fundamental problems when it comes to discussing and imagining rural futures more generally. This is strongly related to our concepts of the rural, as we discussed in Chapter 2. Since the differentiation between urban and rural started to make sense, the rural has been thought of as the space that provides natural resources and food (Mormont 1990; Woods 2011), as the territorial basis of agriculture and forestry, while trade, social events, meeting with others, and thus also the emergence of new ideas and trends, have been attributes of the urban. As the English literature professor Raymond Williams (1973) has convincingly shown through his readings of English literature over time, images related to the country are often images of the past, of childhood, safety and stability, whereas images of the city are about the future, change and risk. The present, in what concerns the rural, remains for the most part undefined because the past and future perspectives are separated. Why is it, asks Williams, that the optimistic utopias in English literature all take place in the country, whereas the dystopias have an urban setting as their

FIGURE 5.2
a. The front page of a promotion folder published by the Danish Agriculture and Food Council (2008). b. Front page of a programme for an architectural competition concerning farm buildings (Realdania 2006)

backdrop? Perhaps this explains why urban populations are increasingly looking at the rural space as a space of identity and leisure. These urbanites maintain the image of a space employed by traditional agriculture using local resources and shaping a cohesive rural community. However, this image seldom corresponds with the current reality in the countryside, let alone the Danish countryside, where modern, specialised, large-scale agriculture dominates.

These issues are further discussed in a short paper written about 10 years after 'The Country and the City', where Williams (1984) resumes his reflections on these contrasting images – this time with a critical view of modern English agriculture. He notes how farm units have been reduced in number, but increased in size; how urban citizens, including urban migrants, are criticising agriculture as environmentally destructive and cruel to animals as farmers constantly strive to increase efficiency. While Williams acknowledges that there is some truth to these critiques, he asserts that agriculture is still an important part of the rural economy and efficiency (and the necessary changes that go with it) should not be criticised per se in a hopeless debate between future-oriented farmers and nostalgic urbanites. However, Williams (1984, p. 215) argues that efficiency must 'never be reduced to a monetary criterion, or to a simple criterion by gross commodities. Efficiency is the production of a stable economy, an equitable society and a fertile world'. What Williams is arguing here is that it is imperative to see agriculture as an integrated part of the rural landscape, which implies that a decoupling of agricultural development and the rural economy is undesirable.

The terms 'country' and 'city' are both powerful and are full of meanings and images and are often closely related to how we see the environment around us and even how we shape our own identities and how we see ourselves (Woods 2011). However, in a European landscape context, not to mention the landscape transition processes we are facing in Europe today, it is essential to keep the rural and the urban together in a coherent analysis. This is what we try to do in the sections to follow using a combination of theoretical lenses and specific empirical examples.

5.3 Town and Country Relationships in History

Historically, there has been a strong mutual relationship between the rural economy and the development of European cities. As discussed in Section 4.1, throughout most of the second millennium, the rural landscape provided the material foundations for urban life. Food, fibre and energy were the main products traded, while the rural population exceeded the urban in number in all European countries until 1900, when the urban population in Great Britain increased to more than 50 per cent of the total population for the first time (Zeigler, Brunn and William 2003). Although rural settlement patterns of various kinds did occur in Neolithic European cities with high concentrations of residents, marketplaces and a number of central political, cultural and religious functions did not exist before antiquity. Consequently, urban–rural relationships in Europe did not exist before the Hellenic culture.

Both the ancient Greek and Roman states were essentially urbanised societies in which the city was the centre ruling over the surrounding landscape. In fact, the city state, the 'polis', was the principal characteristic of the classical world (Arnaud 1998). The city as the centre replaced former political systems characterised by people, tribes and clans. In the 'polis', the centre and the surrounding landscape formed a territorial unit in which the rural landscape was the primary supplier of food and other resources, while the city was the economic and political centre, which was usually rather small in terms of population and, according to Kitto (1996, p. 47), every citizen 'could see the fields which gave its sustenance – or did not, if the harvest failed; he could see how agriculture, trade and industry dovetailed into each other'. Only citizens, a privileged minority of males, were allowed to own land and, in the Greek kingdom, there was little exchange between other city states – slaves were moved, but citizens from one city state could not migrate or dwell in another. It is likely that the Greek colonisation and the associated migrations and creation of new city states were the consequence of overpopulation in existing city states. Trading within Mediterranean territories was widespread, but no urban networks as such existed.

During the Roman period, the urbanisation of Europe spread to Western Europe. Great parts of Europe became part of a Roman economy with one single market for luxury goods and imports from abroad. However, the densities of Roman cities varied enormously across the Roman Empire with the highest concentration of cities found in Italy and in southern Spain, while there were few cities in England or in the lowlands along the Scheldt, Meuse and Rhine. At the peak of the Roman Empire in the second century, a large number of cities had been established throughout Europe and simple urban–rural relationships evolved. However, on the whole, Europe was essentially a rural continent with a subsistence economy which dominated most people's lives. The urban influence over rural landscapes was limited – only a small proportion of the total population was part of or in contact with urban cultures. Limited mobility was a main reason for this – the Roman society was generally a pedestrian society, and cities had to get supplies from within a distance of 50 km and the vast part of it within 15 km (Arnaud 1998 with a reference to Frayn 1993). Rome was the exception to the rule: Rome and Portus (Rome's seaport) had huge warehouses to store enough cereals and olive oil from overseas to feed 1 million people for a year (Keay 2012).

Medieval Relationships

According to Mumford (1970, p. 14), the period between the collapse of the Roman Empire and the 'awakening' of European cities in the eleventh century was characterised by 'violence, paralysis and uncertainty, and a profound desire for security'. The fortified city became the answer to this, with new city walls – when possible built on the remnants of former Roman walls – as its expression. The provision of security was a key explanatory factor for the rapid growth of existing cities and the formation of new ones during the late Middle Ages. Other factors included political change (including political unifications) and the expansion of agriculture through reclamation and forest clearance (Emanuelsson, Arding and Petersson 2009; Mumford 1970, p. 18; see also Section 4.1). Compared to their rural hinterlands, cities soon developed into attractive places to live and the citizens who managed to acquire legal positions as townsmen or burghers (in continental Europe) could enjoy not only safety but also certain privileges such as exemption from military service. Citizens were free; they could move and take up crafts and trades. However, the price to be paid for this was rent, which citizens had to pay as well as tolls on products and taxes on profit to the lord (Pounds 1990, ch. 5).

In contrast to the ancient city, the medieval city did not rule the surrounding countryside. Rural landscapes were regulated by other rules. In great parts of rural Europe, citizens were 'bound to soil', their labour was partly controlled by the territorial lord and they could not move from the land to which they belonged. However, many did move in search of better living conditions and ended up in cities. In fact, cities depended on this urbanisation process. Without an inflow of rural people, cities could not have maintained their populations: the birth rate was lower and the death rate higher in the cities compared to rural areas. This was partly due to poorer environmental conditions in cities, and partly due to the fact that rural people who moved to cities tended to get married relatively late (Pounds 1990, ch. 6). More fundamentally, the medieval city depended on the surrounding countryside, especially when the former urban networks of the Roman Empire disappeared, which left the individual towns and cities isolated (Brunn, Williams and Zeigler 2003). Although the medieval city wall represented a clearly defined boundary between the city and the surrounding country, it was not really separating the urban and rural. Inside the wall plenty of rural functions were present, from extensive vegetable gardens to large stocks of pigs, poultry, horses, cattle and other husbandry.

European cities evolved and grew at great speed during the late Middle Ages. Twenty-five hundred German-speaking cities were founded during the course of four centuries (Mumford 1970, p. 23), and the urban population increased dramatically during this period, as did the population in general, and the walls of most cities expanded.

The Central City and the Colonised World – Renaissance Cities and Feudal Landscapes

From an organic point of view, seen as the direct connection between production of food, fibre and energy from natural resources in the rural and the consumption of these in the nearby towns or cities, the relationship between the cities and the surrounding countryside which characterised the late medieval cities gradually disappeared from the late fourteenth century. A major event, the Black Death in the fourteenth century, contributed to this development. Later, centralism, new military developments, colonialism and international trade contributed to the emergence of more uniform and authoritarian cities – the societal structure of the late Middle Ages was gradually replaced by new social orders in which the territorial state and centralised power located in main cities became the steering institution.

Of course, the cities remained dependent on food which was mainly supplied by the surrounding rural landscape, albeit to a lesser extent than their medieval forerunners. Capital cities and other urban centres increasingly became organised into networks of cities and trade and commerce expanded, which was partly driven by a new highly centralised ruling class and partly by colonialism and the expansion of overseas trade (De Vries 1984). Centralised feudal orders evolved and the local rural landscapes of Europe became ruled by centralised power – or by rural landlords closely linked to central power.

Although the total urban population increased by only a few per cent in the centuries between the Middle Ages and the emergence of industrial capitalism, there was a steep increase in the number of urban residents living off rent and tax revenues collected from urban as well as rural producers (De Vries 1984).

Military regimes developed on this basis fundamentally changed society in the Middle Ages. Gunpowder led to a revolution in warfare and, consequently, the old city walls no longer sufficed. New highly sophisticated fortifications were built around major cities, which could not be easily expanded. Cities became locked in and were forced to become more densely populated, and, as they could no longer grow by horizontal expansion, they had to grow vertically and, thus, more floors were added. Protection against aggressors became a state affair and big business.

> Protection gave way to ruthless exploitation: instead of security, men sought adventurous expansion and conquest. And the proletariat at home was subject to a form of government no less ruthless and autocratic than that which ground the barbaric civilisations of North and South America into pulp. (Mumford 1970, p. 87)

This proletariat was first and foremost a rural proletariat. The rural landscape still fuelled society and cities were centres of capital accumulation, knowledge production and political and religious power. However, in terms of overall population, Europe remained a rural continent and the better the regional conditions for agriculture, the more densely populated the region, although specialisation and differentiation of urban systems as well as rural landscapes affected migration patterns and urban growth (De Vries 1984, ch. 10).

From the late Middle Ages until the end of feudalism, agriculture went through what Emanuelsson and colleagues (2009) has termed the 'first agrarian revolutions', which basically consisted of improvements in cultivation systems, including the introduction of rotational systems, the production of livestock fodder from arable crops, the abandonment of the traditional fallow land system and the introduction of nitrogen-fixing crops. The development of these new systems started as early as the thirteenth century in Flanders and northern Italy, two European regions which were at that time already urbanised due to a flourishing textile industry. According to Emanuelsson and colleagues (2009, ch. 9), it was no coincidence that this intensification started where relatively large urban populations were demanding food. During the following centuries, new farming systems were introduced throughout most of Europe (Emanuelsson et al. 2009; Kjaergaard and Hohnen 1994). Furthermore, land reforms, which started with the eighteenth-century English enclosure and which subsequently spread to most of Western Europe, can be seen as part of this first agrarian revolution.

Overall, the urban–rural relationship changed during this period to become one of feudal dominance, which was increasingly controlled from capital cities, but this subsequently faded away to be replaced by land reforms and industrial capitalism (Figure 5.3). From separate domains, the political economies of the urban and rural gradually integrated during the transition from feudalism to industrialism. However, industrialism by no means evolved simultaneously throughout Europe. In great parts of the continent, including large parts of Eastern and Southern Europe, feudalism persisted well into the twentieth century.

The Industrialisation of the Urban and the Rural

Industrial capitalism, which emerged in the seventeenth century, dramatically changed the urban–rural relationship. Early industrialisation, which was associated with mining and the manufacture of textiles, actually began in the countryside, although such early stages depended on city-based communication and coordination (Vries 1984). The first mills in the Manchester area appeared in the countryside, which saw the construction of large factories and smoke chimneys as new components in the landscape image in the early stages of industrialisation. Later, towards the end of the eighteenth century, the industrial city developed polluting factories, densely populated working-class neighbourhoods, open sewer systems and miserable conditions in general. People moved into the city in great numbers, pushed by increasing rural populations combined with land reforms, which resulted in many peasants losing their land. These peasants were attracted by new job opportunities in urban industry. True industrial regions characterised by mining and heavy industry emerged throughout Europe (Brunn et al. 2003).

As discussed in Chapter 4, wheat and other bulk agricultural products started to flow into European markets from North America and Russia, which further stimulated the ongoing urban development and changing conditions for agriculture throughout Europe. This first stage of a globalised food market radically transformed rural landscapes in a large part of Europe. The effects occurred at different times in the various regions of Europe depending on how much agriculture was connected to the international markets, and were more pronounced in north-western Europe (see also Section 4.1). European agriculture could simply not compete with the large scale and relatively modern crop production of the North American Midwest. New transport technologies (railroads, propeller-driven steamboats) and the innovation of the silo made Chicago the new international centre for cereal trading (Cronon 1991), while, for example, in Denmark, where cereal was the most important export commodity, agriculture went into a deep crisis, which was overcome within one or two decades in the late nineteenth century through a fundamental restructuring of agricultural production from mainly crop production to mixed livestock production, which was organised into thousands of new local dairy and slaughterhouse cooperatives.

It was not only the physical landscapes that changed dramatically during the industrial period; cultural life and urban–rural relationships also changed. Wages, fixed working hours and time as a structuring factor were introduced and became part of an urbanised consciousness now penetrating society (Harvey 1985). High levels of subsistence economy and a cyclical view of time (day/night, winter/spring/summer/autumn), which characterised the rural way of life, slowly vanished. Technological innovations transformed European rural landscapes (Jepsen et al. 2015), which were sometimes driven from within the rural. A good example of this is the Danish dairies, which were owned and driven by local cooperatives, which introduced the industrial production of butter using new cream separators. Occasionally, new technologies were transferred from life science universities or urban industrial centres to agriculture in the form of new agricultural practices – with state-owned (or state-supported) advisory systems in which the farm advisor represented a new urban–rural relationship concerning the mechanisation of agriculture, e.g. the Netherlands. The rural landscapes – of which those in Northern Europe had already been transformed through the land reforms of the eighteenth century – went through major transformation from the late nineteenth century, which was mainly driven by technological innovation within the agricultural science – the second agrarian revolution (Emanuelsson et al. 2009) – including chemical fertiliser, pesticides, plant and livestock breeding and more efficient methods of feeding livestock. Also more general innovations link intimately with the Industrial Revolution, such as machinery based on fossil fuel and electricity. A good example is the mechanisation of field work (tractors, combine harvesters) and stable work (manure systems), which has transformed field patterns and increased the environmental impacts of agriculture in most of Europe during the twentieth century. Such processes of mechanisation freed labour, which was in turn in high demand from rapidly developing urban industries. The second agrarian revolution was, therefore, intimately linked with the urbanisation processes which changed the whole European geography from the nineteenth century onwards.

Together with industrial capitalism and an increasing urban consciousness, socialist movements emerged and a struggle for better conditions became part of political life throughout Europe. Workers organised themselves not only into socialist parties, but also into housing associations, sports clubs and other communities, which had a clear influence on the landscape, not only within the urban boundaries but also in rural areas such as parks and recreational facilities. To illustrate, this is how Pelle, a fictional young worker and socialist leader *in spe* who had recently moved from the country into a working-class neighbourhood in Copenhagen, experiences a sunny Sunday with his fiancée and her mother on the outskirts of Copenhagen around 1900:

> At the Triangle [border of Copenhagen City], they took an omnibus and bowled along the sea-front. The vehicle was full of cheerful folk. They sat there laughing at a couple of good-natured citizens who were perspiring and hurling silly witticisms at one another. Behind them, the dust rolled threateningly and hung in a lazy cloud about the great water butts, which stood on their high trestles along the edge of the road. Out in the Sound, the boats lay with sails outspread, but did not move; everyone was observing the Sabbath.
>
> In the deer park, it was fresh and cool. The beech leaves still retained their youthful brightness and looked wonderfully light and festive against the century-old trunks. "Heigh, how beautiful the forest is!" cried Pelle. "It is like an old giant who has taken a young bride!" He had never been in a real beech forest before. One could wander about as if in a church. There were lots of other people there as well; all of Copenhagen were out and about in the fine weather. (Nexö 1909, p. 30)

What we read in this short text is a fictional description of something which, at the time, was modern in Copenhagen: urban citizens leaving the city to enjoy a day in the countryside and to appreciate a holiday and the aesthetics of a beech forest.

New Roles of the Rural as Seen from the City

In Northern and Central Europe, recreation and a new (urban) appreciation of the countryside has become part of the urban–rural relationship in recent decades. Nowhere has this been more significant than in England where the *countryside* – the landscapes on the other side of the urban boundary –

has become of crucial importance to the national identity. It was the rural landscape, particularly that of the south-east, which had been made accessible by railroads and was full of scenery, monuments and 'honest people', which came to form the image of England. According to Macnaghten and Urry (1998), this image of the countryside was the result of English romanticism and the urban gentrification of rural areas combined with extensive outmigration of landless peasant workers, i.e. a new urban relationship. In other European countries, most of which were far less urbanised than England, rural landscapes also became popular recreational areas and significant symbols, although often with other values attached to them than the picturesque English rural idyll. The German Black Forest, the Swiss Alps, the Danish heath, the Dutch polders with their windmills, the Douro Valley with its terraces for the production of Porto wine and the Tuscan cypress-lined landscapes are examples of iconic European landscapes treasured for their aesthetic and symbolic values and linked to modernisation processes (of which urbanisation was always an important part) which began at the end of the nineteenth century (Macnaghten and Urry 1998; Olwig 1984; Schama 1995).

In great parts of Southern Europe, these relations have evolved in a different way. Industrialisation started much later and was much less developed so that Southern European countries remained deeply rural until relatively late – in Portugal and Spain, virtually until the end of the dictatorship regimes of the 1970s. The rural remained a place of production, which was distant from the city even if its products were exported to the cities, and was only appreciated for its recreational opportunities such as exclusive hunting or health resorts (e.g. thermal baths) by the elite.

As a reaction to the poor living conditions in industrial cities, which had become too large and dirty, ideas about clean and green cities developed throughout Europe. Although many European Renaissance and baroque cities were the subject of urban design on a grand scale, a widespread spatial planning tradition for European towns and cities, in general, did not emerge until industrialisation (Rasmussen and Olsen 1998). Ebenezer Howard's idea for future *garden cities* in the late nineteenth century was one of the first and most influential visions for a new approach to building cities, which involved connecting them with the surrounding countryside. In his 1898 work, 'To-morrow: A Peaceful Path to Real Reform', which was subsequently renamed 'Garden Cities of To-morrow', Howard outlines how new cities could be built to combine the best of the country and the city (Howard 1965). These garden cities should be built at a clear distance from existing urban areas so land could be purchased as cheaply as possible. A group of different people should form a company and initiate the process and new industries should be located in the garden city, which should then grow to a size of 30,000 inhabitants. Subsequently, a new garden city should be established further out and, in this way, a system of garden cities within a matrix of agricultural landscapes would be created which would be linked together by a railroad system. This integrated 'town-country' system encompassing the 'beauty of nature' and 'social opportunities', as well as 'fields and parks of easy access', would, according to Howard, be the third and most powerful magnet compared to the individual town and country magnets (see Figure 5.3) and would attract people. Howard asserted that the town, which symbolised society, and the country, which symbolised 'God's love and care for man' and the gifts and beauty of nature, should be unified. In a combination of lyrical prose and plain, practical writing, Howard argued that the garden city represented the best of both worlds, and that 'better opportunities of social intercourse may be enjoyed than are enjoyed in any crowded city, while yet the beauties of nature may encompass and enfold each dweller therein' (Howard 1965, p. 103).

Howard's idea, which also included detailed descriptions of how garden cities should be constructed from the bottom up without public support, was indeed an appealing one. In fact, it was so appealing that planners in many

FIGURE 5.3
Ebenezer Howard's famous three magnets – town, country and town-country (Howard 1965)

other European countries at the time actually claimed that it was their idea Howard was promoting (Hall 1996). In any case, a number of garden cities were established in England and in the rest of Europe during the early twentieth century, although none corresponded to Howard's idealised version (see e.g. Figure 5.4). Often industrial aristocrats were the initiators of garden city development and adopted Howard's ideas in various forms. For example, in Belgium, the garden cities were the result of direct contact between local people's cooperatives and their architects, who were engaged in social and environmental reform. In Vienna and Amsterdam, garden cities were built on the basis of high-density continuous public development rather than on the original cottage ideal, whereas in Paris, Brussels, Frankfurt am Main and Berlin, the garden suburb was considered a suitable tool for citywide planning and housing projects (Kafkoula 2013, p. 172). Even though these movements mainly affected the surroundings of the concerned cities, they created new images of a close relationship between urban and rural, and in particular, they represented new approaches to designing urban-rural areas.

The English new town movement and the idea of satellite towns attached to the main cities by efficient infrastructure developed out of the idea of a garden

FIGURE 5.4

From Hellerau Garten Stadt outside Dresden, Germany. This garden city, the first in Germany, was developed in 1908 and, from an early stage, performed residential, industrial, agricultural and a number of cultural functions (Kafkoula 2013). Today, Hellerau exists as an almost intact example of the garden city vision, although it is now a mainly residential suburb and a cultural centre.

city, which was green and urban at the same time. It is not an overstatement to say that the idea of a combined urban-rural place to live is still very much alive. In fact, it has had a significant influence on suburban and urban fringe landscapes throughout Europe. However, it is also fair to say that garden cities, when they emerged, were more town than country and, over time, they tended to merge with an expanding urbanised matrix so that they lost immediate contact with the rural environment as a consequence.

Interestingly, also in the countryside, movements began to promote the quality of rural life, especially those associated with the role of women. The Association of the Country-Women of the World (ACWW) was founded in 1929 (Van der Burg 2002, p. 268). In the Netherlands, the Association of Rural Women (Plattelandsvrouwen), founded in 1930 (Van der Burg 2002, p. 266), was a strong uniting and later also empowering force for women in rural areas. Theda Mansholt, aunt of later EU Commissioner of Agriculture Sicco Mansholt, was a fervent promoter of women's role in rural society in the first decades of the twentieth century (Van der Burg 2002, p. 1 seq.). The Association of Rural Women survives to the present day under the name Women of Today (Vrouwen van Nu).

If Howard was the father of the garden city, which attempted to combine the virtues of the city and the country, the British planner Patrick Geddes (1915) was one of the first to see the city as part of a wider region comprising both the country and city which, as a whole, relied on the natural resources available in the region, resources which had to be surveyed and assessed before the plan could made (Hall 1996). This view of the (regional) relationship between the country and city became the start of a regional planning movement. Geographers now joined the planning profession and the movement first spread to North America before returning to Europe, where the Greater London Plan from 1944 (Abercrombie 1945), the Copenhagen Fingerplan from 1948 (Egnsplankontoret 1948) and the Green Heart Plan (introduced in 1956, Werkcommissie Westen des Lands 1958; see also Van der Valk and Faludi 1997) for central Holland became iconic examples of how urban development should be seen as part of wider regional development in which rural landscapes played a role as places of food production, water supply and outdoor recreation. All these three regional plans may be seen as regional attempts to approach the urban–rural relationship in an integrated way, and they have all had a profound influence on how this relationship evolved in the three regions. Comparable examples can be found in many other post-war European regions (Hall and Tewdwr-Jones 2011; Vejre et al. 2007b).

The specific geography of the rural and the urban determined to a large extend the change patterns affecting European landscape – and still does. As the urbanisation of societies increased, so did people's understanding of the territory as naturally organised around the cities in people's. Thus, in a first phase, urban citizens' renewed interest in the rural resulted in a much stronger differentiation of the rural than had been seen previously. Close to the cities, where the population and commercial activities became progressively concentrated, the most densely populated and lively rural areas developed with multiple activities, while the more remote rural areas in marginal locations in relation to urban centres and population agglomerations remained emptier and less dynamic. This urban-centred organisation of the territory is surely linked to the dynamics of agriculture and the polarisation addressed in Chapter 4. However, the trends discussed in this chapter also show how the urban and rural were co-evolving and were bonded by functional relationships within regional boundaries.

The Post-industrial City and New Urban–Rural Relationships

Today's cities – here termed 'post-industrial cities' – represent changing urban–rural relationships with great implications for European landscapes, although the ongoing change patterns discussed in the next section are occurring in many different directions with open or at least unclear outcomes. However, what *is* clear is that the city is increasingly linked with other cities through economic, cultural and political networks on the regional as well as the global scale. The individual city is not – or only to a small degree – dependent on food production in surrounding rural areas. Globalised food markets develop and as cities become more connected in networks with other cities, the individual rural landscape becomes increasingly affected by (and vice versa) decisions and events in other parts of the world, cities and rural landscapes alike (Giddens 1990). When a Chinese food safety scandal concerning the contamination of milk powder became widely known among the Chinese public, European dairy farmers were affected, e.g. in Eastern Europe (Beldman, Daatselaar and Prins 2014, p. 67), firstly by a steep fall in prices on the milk powder market and a few years later by a boom in demand for 'safe' European milk powder and milk powder produced by other foreign countries from a rapidly growing Chinese middle class.

In this way, rural landscapes are becoming increasingly integrated in what Castells (2000a) has termed the 'spaces of flows' where interregional relationships are being superimposed on the urban–rural relationship. These spaces of flows are neither urban nor rural, but are definitely spaces in which the decision-makers are increasingly located in regional centres around the world – centres characterised by a high concentration of economic, cultural and political institutions surrounded by attractive landscapes, sometimes in highly urbanised settings, and sometimes in suburban and rural landscapes on the periphery of New York, London and other world centres. In such 'elite regions', large investments are being made to ensure that landscapes are well-functioning and attractive places to live in and visit. For other regions – covering much larger parts of the world – few or no such investments are given priority.

However, at the regional scale, the urban–rural relationship has been changing compared to the industrial city in that the dependencies are now recognised as being strongly integrated and mutual (CEC 1999). In polycentric regions, many functions such as housing, recreation, water supply, tourism and business should be understood within an overall regional context which is being increasingly affected by more or less globalised driving forces. Increased telecommunication – or 'teleconnection' (Seto et al. 2016) – is a central part of these processes, which overall means that the traditional urban–rural dichotomy is disintegrating (Ulied 2014).

Summing up, urban–rural relationships have changed dramatically since the simple organic relationships between the medieval town and its surrounding rural landscape to today's open and dynamic spaces of flows (Figure 5.5).

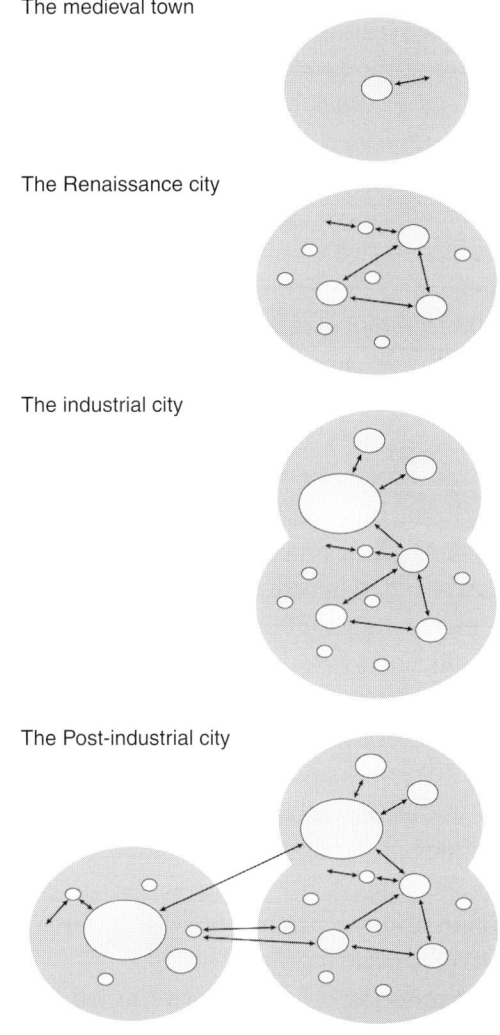

FIGURE 5.5
Urban–rural relationships during four major periods

5.4 Current Change Patterns in Urban–Rural Relationships

The Effect of Urbanisation on Rural Landscapes

Having shown in the previous section how the development of urban–rural relationships through time has contributed to the changing European landscapes as well as the change in views about these landscapes, we now turn to current change patterns which are being caused by urbanisation processes, and their connection with the agricultural change processes discussed in Chapter 4. First we outline the various forms of urbanisation processes and their impact on rural landscapes. Through a concrete case, we then present and discuss recent change patterns in rural landscapes affected by urbanisation.

Through a statistical analysis of population developments in more than 200 European metropolitan regions from 1951 to 1991, it is shown how four forms of urbanisation occur in various combinations depending on the regional context and stage of economic development (Champion 2001; Cheshire 1995). The four forms are: (1) *urbanisation*, which is movement from rural, often peripheral areas, to cities; (2) *suburbanisation*, which involves movements from city centres to the city fringe, most often as part of an urban expansion process; (3) *counter-urbanisation*, which entails movements from cities to rural areas and, finally; (4) *re-urbanisation*, which involves movements from suburban areas to city centres. There is also a more recent and more frequently occurring movement which could be called *re-ruralisation*, which entails people moving from the city to the rural, and subsequently moving from one rural location to another, thereby increasing the heterogeneity of the rural population and the dynamics of rural enterprises. This is intensified by increased mobility caused by flexible jobs, and stimulated still more by the emerging new professional networks of people and businesses across rural areas (Hedberg and do Carmo 2012; Woods 2007). All these processes, of course, have an impact on rural as well as urban areas and, when considered together in a global perspective, they reflect the increasingly mobile world described in Chapter 3, a world in which people, goods, energy, capital and information are flowing increasingly rapidly. Thus, as Holmes (2006) suggests, the dimensions of landscape protection and consumption are closely linked to such general urbanisation trends in society. As the share of the population who have an urban income and are living an urban lifestyle is increasing, demands linked with landscape services and public goods are changing, as are landscape management practices (Termorshuizen and Opdam 2009). With respect to rural landscapes, two of the forms of urbanisation mentioned previously – urbanisation and counter-urbanisation – have particular importance.

Traditional urbanisation processes have affected European landscapes for centuries and they are persisting as people continue to move from peripheral rural landscapes to cities, causing them to expand, while what were previously rural landscapes are becoming integrated into the urban matrix. In some parts of Europe, this process is associated with marginalisation in agriculture, leading to a decrease in rural jobs, while in others, it is part of a developing knowledge-based economy whereby universities and other, mainly city-based research centres are becoming more closely linked with new industries (IT, medical and biotech industries) (Antrop 2004a; Cheshire 1995). In Northern Europe (and North America) during the 1970s and 1980s, population growth became decentralised, which meant that the population of major cities declined, whereas that of smaller towns and rural villages increased as middle-class and upper-middle-class people moved out of the cities into the rural. From the 1990s onwards, this trend reversed and the population of many large cities is once again increasing (Antrop 2004a; Cheshire 1995). However, the full picture is more complex. In a study of 310 European cities, it was shown that a significant proportion of these cities (app. one-third) experienced continuous growth in population throughout the period 1960–2005, whereas others – mainly cities in Eastern Europe – have witnessed a decline in recent decades (Turok and Mykhnenko 2007).

Urban expansion is not only a matter of population increase. Changes in living standards, house size, industrial structure and urban infrastructure affect the amount of land consumed by urban expansion. Throughout the post-war period, European cities have continuously grown in size and these developments have had a significant adverse effect on, often very fertile, farmland and to a much smaller extent forest and nature areas with Tallinn, Helsinki and Istanbul being three outstanding examples (Figure 5.6) (see also Gardi

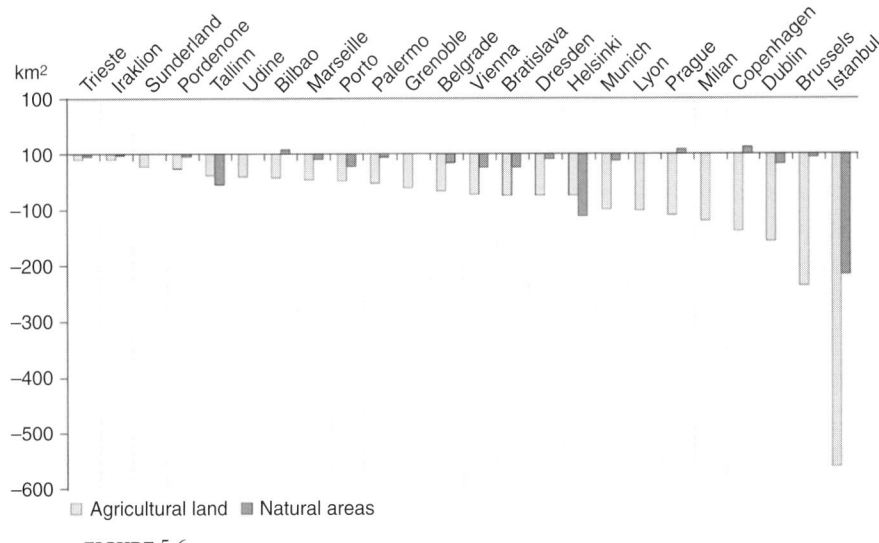

FIGURE 5.6
Rural land consumed by urban growth in a selection of European cities (EEA 2006, p. 34)

et al. 2015). The reasons for this are, first and foremost, that European cities mainly developed where conditions for agriculture were good, as described earlier, but also partly because natural areas and forests are protected against urban development in many countries. In a few cases, urban development has occurred in parallel with the expansion of forest and nature areas, such as in Copenhagen where a new, large forest was planted to the west of the city from the late 1960s to the late 1980s (Vejre et al. 2007b).

Due to urban expansion, urban fringe landscapes are under pressure and are often highly unstable, but are also characterised by a high degree of multifunctionality. For these reasons, urban fringe landscapes are often studied in relation to urban containment and planning control (Millward 2006); but it has also been argued that the urban fringe landscape is a landscape type in its own right and should be treated as such in research studies and planning practice (Antrop 1994; Qviström 2007; Scott et al. 2013). After all, the urban fringe is familiar to a large (although unknown) proportion of the European population who, like the three authors of this book, grew up in newly established suburban areas during the latter part of the twentieth century.

The way in which urbanisation is affecting different rural landscapes is described and discussed in the following section and the subsequent Case F.

Migrations from urban areas into the countryside are affecting European landscapes in a number of ways. First, counter-urbanisation may result in competition for land between traditional commercial farmers – both small-scale intensive vegetable farmers, who traditionally have been concentrated in peri-urban regions, and larger-scale crop and husbandry farmers – and incoming urban people, who move to the countryside or just purchase the land for various reasons, including simple land speculation. This means that commercial farming is being pushed out of peri-urban areas, by lifestyle farmers. The result is less intensive land use and reductions/changes in livestock numbers. In Northern Europe, horseback riding is a widespread phenomenon on such peri-urban farms (Bomans, Dewaelheyns, and Gulinck 2011), while in other parts of Europe, diverse types of livestock breeding are more frequent, as well as vegetable and fruit gardens (Pinto-Correia et al. 2015a). Despite the non-commercial profile of these small-scale productions, their role can be significant in terms of food provision. Still, this role remains often unseen, as the food produced is partly consumed in the farm household or transferred to the consumer through informal chains, and therefore not registered in any kind of official records (Dwiartama and Piatti 2016; Marin and Russo 2016). In a context of economic crisis and return to the land, as has been observed in Southern Europe in recent years, this informal and thus non-registered circulation of food in short supply chains is becoming increasingly relevant, as we discuss in the next section. The settling of urban newcomers not only leads to changed and less intensive agricultural land use, it often also implies changes

FIGURE 5.7
Montefioralle, near Greve in Tuscany, Italy. Although most of the former farmhouses are now used as second homes or as primary homes for urban incomers, the traditional architecture has been maintained fairly well and the picturesque landscape remains largely intact.

in landscape management, planning and design. New buildings, or the significant reconstruction of existing buildings, are common features of counter-urbanisation in many European landscapes. Sometimes the construction activity follows traditional rural architectural traditions (Figure 5.7), while at other times the traditions are reinterpreted or completely ignored. In some regions, construction has the character and scale of urban sprawl (EEA 2006) and blurs the urban–rural boundary, while often challenging the legitimacy of – or worse, the very justification for – the land-use legislation.

In summary, the new urban relationships can be illustrated by combining the two main dimensions of rural landscape (Figure 5.8). The conditions for agriculture and the degree of urbanisation determine, to a large extent, the dynamics of change in European landscape, as also illustrated in Case F (Primdahl et al. 2013a).

Outdoor Recreation

The recreational interests of urban incomers and visitors have a great impact on European landscapes. One example is horse riding, which affects many densely populated regions in Central and Northern Europe, including the construction of new buildings (including *manèges*), fences, racetracks,

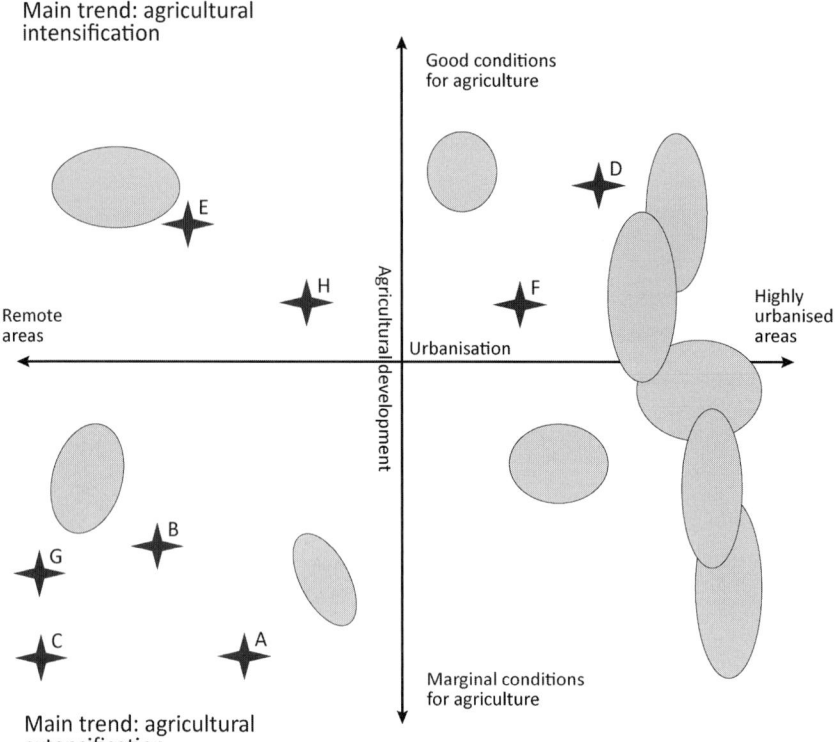

A Chapter 2: Land Abandonment and Tourism on the Súľov-Hradná, Slovak Carpathians
B Chapter 3: Pivka area, Slovenia. Can Biodiversity Conservation Be the Driver for the Control of Natural Forest Expansion in Former Agricultural Fields?
C Chapter 3: The Great Trossachs Forest, the Changing Character of a Moorland Landscape in Scotland
D Chapter 4: Land Re-allotment/Land Consolidation Schemes in the Netherlands
E Chapter 4: Serpa, South Portugal – How Super-Intensive Olive Groves Are Drastically Changing the Landscape and Hampering Rural Life
F Chapter 5: Counter-urbanisation in the Moraine Landscape in Hvorslev, Eastern Jutland, Denmark
G Chapter 6: Feuilla, France – Towards Revitalisation of a Village at Risk of Abandonment
H Chapter 6: Landscape Strategy for Karby Parish, Northern Jutland, Denmark

FIGURE 5.8
The two main factors affecting European agricultural landscapes: the natural conditions for agriculture and the degree of urbanisation. These factors should be seen in combination if future landscape changes are to be discussed. Also the local socio-ecological resources may be significant factors. The oval circles indicate local and regional landscapes where socio-ecological conditions and initiatives result in unique developments. Such developments are found especially in peri-urban regions. Based on Primdahl et al. (2013a, p. 801). For the reader's reference, in this figure we also placed the cases included in this book.

CASE F Counter-urbanisation in the Moraine Landscape in Hvorslev, Eastern Jutland, Denmark

Landscape Appearance

The landscape is composed of moraine plateaus intersected by a large river valley and a few water courses in a fertile, intensively farmed rural landscape approximately 35 km west of Aarhus. Most of the area is arable farmland, but some forests and semi-natural grasslands are also present in the area, especially along the river valley and in the gullies cutting through the moraine

Landscape Processes

Since World War II, changes in agriculture towards concentration, specialisation and intensification of production have been the major drivers of landscape transitions. Larger fields, new, large farm buildings and the disappearance of small, uncultivated landscape elements such as ponds and hedgerows are some of the major changes in landscape structure.

Since around 1990, however, significantly more new hedgerows and wood thickets have been planted and new ponds dug than have been removed. Landscape heterogeneity has gradually increased and the area of land not under crop rotation has expanded.

The key to understanding these recent developments is hobby farmers' landscape practices, which are more closely linked to the farm as a property than to agricultural production. When 350 owners of farms >2 ha in the Hvorslev area were interviewed, about two thirds of them chose 'good living place' as their primary motive for owning the farm, about 8 per cent mentioned primary production and about 30 per cent

FIGURE 5.F.1
Hvorslev landscape

CASE F (Cont.)

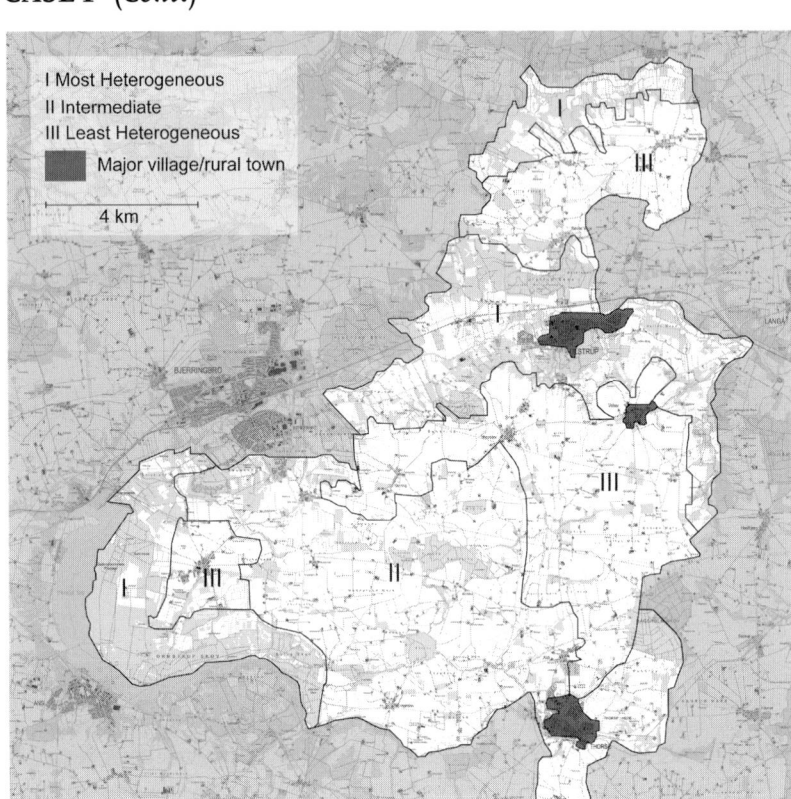

FIGURE 5.F.2
Situation map of the Hvorslev area

mentioned a combination of the two motives. From Table F.1, it appears that most of the hobby farmers and pensioners mainly see their farms as places to live, whereas full-time farmers tend to see their farms as places to live and produce (Table F.1). When these motives are combined with landscape changes (in land use and small landscape elements), it appears that farmers who primarily see their farms as places to live are actively changing their landscape to a much greater extent than the other groups. To understand landscape transformations in this landscape, farming aspects associated with the farm as a living place, especially residential functions, must be included.

Results: A Richer Landscape

This conclusion has many implications. A major implication is that the open landscape as a whole is slowly becoming more enclosed, more heterogeneous and richer in biodiversity. A second implication is that public policies should include the farmer as 'owner' when they define the policy target for specific policy measures and that they should communicate and include farmers in the policy process in different ways than they do currently when they see the farmer only as a producer. More

CASE F (*Cont.*)

TABLE F.1 *Farm owners' views of their farms and their occupational status, their 'landscape practice' and location of the farm (Based on personal interviews with farmers in Hvorslev-Bjerringbro, eastern Jutland, 2008; from Primdahl (2014, p. 217)*

How the farm is seen	The farm owner's main motivations for possessing the farm[1]:			
	A (good) place to live	A (good) place to produce	Equal share of both	Sum (Nr = 100%)
Occupational status ▼	———%———			
Full-time farmer,%	21	24	55	33
Part-time farmer,%	52	0	48	21
Hobby farmer,%	79	1	20	178
Pensioner,%	64	3	29	88
Others,%	-	-	-	3
All,%	67	4	29	323 farmers
Farm property, sum,%	44	9	47	7604 ha
Landscape practice[2] ▼				All
Hedgerows planted, metres/100 ha	681	225	303	469
Hedgerows removed, metres/100 ha	98	30	82	84
From land in rotation, ha converted/100 ha[3]	7.3	0	1.4	5.2
To land in rotation, ha intensified/100 ha	2.7	1.0	1.2	1.7
Share of farms with new buildings	38	38	35	37 (n = 119)
Share of farms with empty buildings	31	23	37	33 (n = 103)
Share of farm location within case area,% (Fig.5.F.2)3) ▼				Sum (Nr = 100%)
High level of landscape heterogeneity	82	0	18	50
Medium level	65	3	33	110
Low level of landscape heterogeneity	61	9	29	86
All,%	67	4	28	246

[1] The question was: Do you primarily own this farm property because it is: (1) a good place to live; (2) a good place to farm or; (3) an equal combination of the two?

[2] Landscape changes during the period 1996–2008 based on interviews with farm owner. All changes were recorded on maps during the interview.

[3] The whole area was divided into 19 homogeneous regions based on a standard topographic map at the scale 1:25,000. Subsequently, the regions were analysed for five variables (share of land with forest, slopes, wetland and metres of watercourse/ha, m/hedgerow/ha) and grouped into three types through a cluster analysis. Farms with less than 80 per cent of their property area within one region were excluded.

> **CASE F** (*Cont.*)
>
> fundamentally, the counter-urbanisation process which underlies these trends may result in more competition for farmland.
>
> **Conclusion**
>
> Although the area cannot be characterised as urban fringe, current landscape change is closely related to counter-urbanisation and hobby farmers' landscape practices. This landscape, therefore, provides an illustrative example of new urban–rural relationships and their role in landscape transformation. It is full-time farmers who for most of the nineteenth and twentieth centuries have formed most of the landscape structures we see today, and it is full-time farmers who work most of the present-day landscape. However, they are no longer the only key agents and not even the most influential ones if we look at the current change patterns.

riding trails, etc. Horse riding is a landscape practice which is mainly linked to counter-urbanisation and results in 'horsicultural landscapes', as they have been termed (Haigh 2008).

Outdoor recreation and tourism can also be seen as mainly 'urban' activities as the larger urban public could engage in these because of well-defined working hours and spare time which could be spent pursuing outdoor activities and travelling. Although recreation is an integrated part of any tourist activity, most outdoor recreational activities are not considered tourism if they do not include an overnight stay. Outdoor recreation affects landscape changes in several ways. The most common form, walking through the landscape, requires accessibility, including, first and foremost, walking trails. Places to rest and eat are also part of most landscapes used for recreation activities, and in many parts of Europe, extensive walking routes, including overnight stays, are important parts of the landscape. Other types of recreation, bicycling, horse riding and hunting (see next section), for example, usually also require specific features.

To our knowledge, there are no consistent statistical overviews of the extent and distribution of outdoor recreation on a European scale. For forest recreation, which in many countries represents a significant part of the overall recreational activity in rural landscapes, there is some information available regarding regulation and conflicts. From a survey based on experts from 26 European countries, it was found that public access and the gathering of non-timber products (berry picking, mushroom collecting) are regulated by legislation in almost all countries, while sports activities are regulated in more than 70 per cent of the countries (Pröbstl, Elands and Wirth 2009). In some countries, including the Nordic countries, Austria, Iceland, Germany, Switzerland and the United Kingdom, people have free access to all natural and forested areas. Among the most important conflicts associated with recreational use of the forest according to the experts were those between: (1) recreation and nature conservation, especially in Eastern Europe, where many public forests have recently been privatised; (2) recreation and forestry, especially in Norway, Sweden and Finland, where large-scale timber production dominates forestry; (3) different recreational user groups, especially in the densely populated central and Atlantic regions, and; (4) overcrowding in peri-urban and very touristy areas (Pröbstl et al. 2009).

In Denmark, extensive national studies of outdoor recreation have been repeatedly carried out, which have provided detailed information regarding the form and distribution of and change in outdoor recreation (Jensen and Koch 2004; Jensen and Tvedt 2012). These studies show that, in Denmark, forests are the most popular places to visit for outdoor recreation (70 million visits in 2008), followed by beaches and coasts (43 million visits) and agricultural fields (36 million visits). More interestingly, in a European context, the studies also show that between 1976 and 1998, forest visits increased in number by 15–20 per cent (depending on whether population increase is taken into account (15%) or not (20%)). Between 1998 and 2008, the numbers stagnated. The spatial distribution of the forest visits has also been analysed, which reveals that the closer the forests are to urban centres, the more used they are for outdoor recreation, and this seems to increase over time. If we look at the duration of the visits, it appears from Figure 5.9 that the shorter visits have increased in number, while the longer ones have declined. This development is consistent with the fact that more visits are being made to nearby forests and more visits are made on foot or by bicycle. The overall conclusion is, therefore, clear: the most-visited forests are located on the urban fringe and the location of rural open space in close proximity to the urban is a crucial issue from a recreational perspective. This may have meant that, for the urban European citizens of the 1960s and 1970s, when the private car was still a relatively new phenomenon, day trips and the feeling of recreation in the rural landscape started when they got into their cars. Today, half a century later, for the same type of urban citizen, recreation probably does not start mentally until they either leave the car at the destination or when they start walking or cycling from home.

The importance of nearby urban recreational facilities has been mostly studied in Northern and Central Europe. In Southern Europe, the use of the countryside and the rural landscape as a recreation area is not as widespread. Nevertheless, the demand for recreational opportunities is increasing in line with the increasing urbanisation (da Silva et al. 2016). Most walking and bicycle paths or other types of recreational infrastructure, which are built around cities of different sizes, are much used and are always considered as remarkable planning successes.

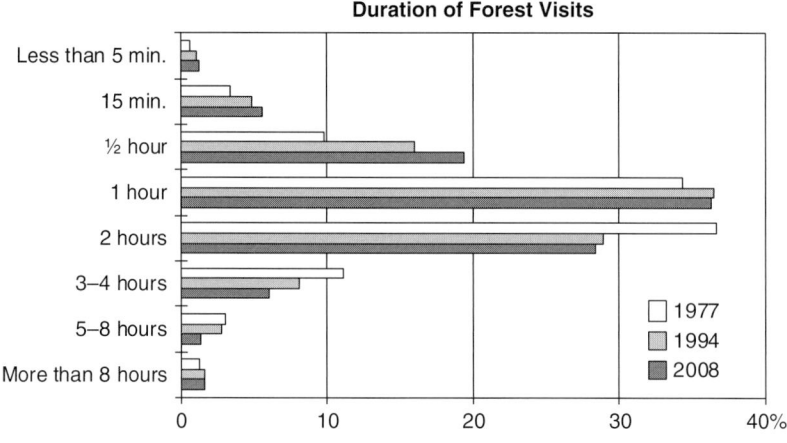

FIGURE 5.9
Changes in the duration of forest visits in Denmark from 1976 to 2008. Translated from Jensen and Tvedt (2012, p. 2)

The deer park north of Copenhagen visited by the fictional urban worker at the turn of the nineteenth century (see Section 5.3) is no less popular today. With 7 million annual visitors, the deer park is the most-visited forest area in Denmark.

Hunting

Hunting (including 'shooting'), which has spread across Europe, is an example of a recreational activity linked to counter-urbanisation and urban visitors. Whereas hunting historically has been an integral part of rural living (Emanuelsson et al. 2009, ch. 11), although highly contested in respect to rights, nowadays, hunting is mainly recreational and game and the right to hunt has, like many other aspects of rural landscapes, become increasingly commodified (Carolino et al. 2011; Woods 2007). Regions where there is significant interest in hunting are most often characterised by low-density farming or a predominance of forest. Under these conditions, when hunting becomes more significant economically and socially, it has a clear effect on landscape management and, therefore, it should be seen as a major influence in relation to the overall ecology of such areas and with respect to social conflicts. When farming is economically dominant and agricultural use is undoubtedly the most widespread use of the land, the importance of hunting as a management driver is considerably reduced. However, despite the direct impact that greater concern for hunting in land management would have on game abundance and distribution, the relationship may not be straightforward, while greater interest in hunting does not always result in more being done for game. The complex relationship between these processes is also related to the fact that the legal relation between hunting and property rights is not the same everywhere (Carolino et al. 2011). The right to hunt is directly linked to property rights in some countries (generally in Northern Europe, although hunting of elk and other big game often requires a special licence linked to a management plan), while in others, cooperation between landowners is required to manage hunting, and in others still, some forms of hunting or hunting rights are independent of property rights (more often in Southern Europe). These differences are related not only to the historical roots of the current land and use rights but also to too much hunting (and drastic reduction or extinction of game populations) and the mismatch between game habitats and property boundaries. The complexity of the forms of hunting management that are still found today, and the mismatch between production rights and hunting rights which exists in some regions, demonstrates how difficult it is to combine the enjoyment of public goods, i.e. the commodified use of the rural landscape, with production activities and other more-established land-use rights (Pelosi, Goulard and Balent 2010). Moreover, this complexity increases even more when hunting as a way to provide food and as an integrated part of land management (especially in forest management) is included.

In a Danish study on landscape changes in three rural landscapes, Primdahl and colleagues (2012) found that the farm owners who did not allow hunting on their land were, in fact, more active in establishing new, uncultivated landscape elements than those who did allow hunting. The farmers who did not allow hunting (representing 12–17 per cent of the land in the landscapes analysed) did not permit it because they had an interest in wildlife – they simply enjoyed watching game and, therefore, did not hunt. However, the group of farm owners who hunted (as opposed to owners who lent or rented out the hunting right) was the most active when it came to establishing new landscape elements, and this share has been growing over time, which indicates that hunting has become a major motivation for owning a farm.

The Development of New Urban Food Systems

In the meantime, the rural has also infiltrated the city, first in the form of compensation activities and spaces of enjoyment for the rural people who moved into industrial cities during the twentieth century, later as recreational spaces and green infrastructure and more recently as sources of safe and fresh food in the face of increasing struggles for food security (Cabannes and Ross 2015; Maye and Kirwan 2013). Urban farming has been growing in large European cities and new linkages between the urban and the rural are now being established.

An interesting development in the rural–urban relationship is the widespread appearance of urban agriculture initiatives, especially in Europe and

North America, where the phenomenon reconnects urban dwellers with the local production of food, ranging from brownfield requalification (Roth et al. 2015) to rooftop farming (Sanyé-Mengual et al. 2015). The topic of urban agriculture is discussed in several recent monographs, handbooks and proceedings (see e.g. Cinà and Dansero 2015; de Zeeuw and Drechsel 2015; Redwood 2012; Wascher et al. 2015). Urban agriculture is nothing new in itself. In fact, private, monastic and manorial kitchen gardens, and allotment gardens since the Industrial Revolution, have always played a crucial role in the food security of cities, except for a relatively short period at the start of the past century in the West. Still, the recent reappraisal of and innovations in urban farming may represent a changing attitude amongst modern citizens regarding food production and consumption, the benefits and constraints of which are still being studied (Mok et al. 2014). For instance, the claim that urban agriculture can guarantee food security has been criticised, especially regarding developing countries (Badami and Ramankutty 2015). Furthermore, convincing evidence of any benefit in terms of nutrition is lacking (Warren, Hawkesworth and Knai 2015). Nevertheless, Warren and colleagues assert that there is no reason to discourage urban farming: besides contributing to food security, it can contribute to the access of urban populations to fresher and more diversified food. Furthermore, cultural diversity is often mentioned as a key benefit of urban agriculture in that it promotes the integration of newcomers through social interaction around the agriculture (Crivits et al. 2016; Winter 2015), although the risk of reinforcing ethnic inequality has also been discussed (Reynolds 2015).

Increased awareness of food security, coupled with a desire for high-quality, fresh food amongst the elite, is increasingly leading to the emergence of closer links between urban citizens and agricultural production (Brunori, Malandrin and Rossi 2013; Hinrichs 2014). Besides urban agriculture, which is most relevant in larger cities and metropolitan areas, short supply chains, community markets and community-supported agriculture involving production in rural areas surrounding cities of all sizes are rapidly expanding. As discussed previously, these relatively recent movements have resulted in peri-urban areas and small-scale farms being occupied by newcomers, with the production of food becoming relevant once more after having seen a decline in former decades. This process has led to the revitalisation of the functional relationships between the country and the city through the establishment of new linkages based not only on direct producer–consumer relationships but also on new forms of community engagement, producer associations and cooperatives or multi-actor networks (Tregear and Cooper 2016). This process has been more pronounced in Southern Europe as a result of the economic crisis, with many young, qualified citizens searching for alternative economic systems (local systems) and sources of income (Dwiartama and Piatti 2016; Marin and Russo

2016). Local food systems, which are emerging in cities and are being promoted by local authorities and supported by local food or agri-food councils, are the clearest urban expression of these emerging processes. These new mechanisms are forming closer relations between the city and the rural as the connection to food production and food producers is re-established as personal, family and associative connections. As for the landscape, these new trends mean the reintroduction of farming-based patterns and elements in the land cover, in some cases similar to previous patterns, in other cases slightly changed due to new cultures or new production methods (Pinto-Correia, Almeida and Gonzalez 2016; Scott, Christie and Midmore 2004).

Tourism

When a 'visitor' stays overnight away from home, he or she is a 'tourist'. As the tourist may well be mainly using the landscape for outdoor recreation, the distinction between a tourist and an outdoor recreationist is not always easy to make.

An illustrative example of the economic and policy importance of tourism (and outdoor recreation) compared to agriculture for rural landscapes is the outbreak of foot-and-mouth disease in husbandry animals in the UK. The disease broke out in 2001 and spread rapidly to a great part of the UK, with high costs for the agricultural food sector. The economic cost for this sector was estimated to be more than £3.1 billion, with more than 6 million animals being slaughtered (Scott et al. 2004). However, the costs for the tourism/recreational sector were much higher, between £4.5 and £5.3 billion. The reasons for this were that, first of all, access to the countryside (including access via waterways) was denied to a great extent, including access to national parks and other recreational hot spots. Another important reason for the reduction in tourism and outdoor recreation was the burning of carcasses and the debate about the disaster, which made people change their recreational and travel plans (e.g. in the Peak District, UK, Figure 5.9). According to a national economic study, the loss in revenues from tourist and day visits was calculated to be £7.7 billion, with roughly one-third of the losses coming from a reduction in domestic tourist visits, one-third from day visits and one-third from overseas tourists (Blake, Sinclair and Sugiyarto 2003).

Such figures give some indication of the economic importance of rural landscapes as visiting places for (mainly urban) people, which of course has (or will have in the longer term) consequences for landscape management practices, landscape change and policy, including recreational access and facilities in general. Moreover, the case also illustrates very clearly the need to reflect carefully on the appropriate public policy interventions. For example, the strategy to combat the outbreak should include the interests of the tourism

FIGURE 5.10
The Peak District National Park in the UK, which is located in the middle of a number of large cities and receives 32 million visitors annually. Like the majority of national parks in the UK, the Peak District National Park was seriously hit by foot-and-mouth disease and the subsequent temporary closure of the park.

industry when making the choice between slaughtering or vaccinating. Thus, subsequent studies have shown that vaccinating would most likely have been more cost-efficient seen from a wider economic perspective instead of focussing on the agricultural sector alone. In a wider perspective, the case shows that the spatial relationship between farming and tourism needs to be further investigated (Scott et al. 2004; Sharpley and Graven 2001) in relation to changing urban–rural relationships in general.

Within a wider European context, there is no doubt that tourism plays a major role in landscape transformation, which has become increasingly influential over time. Tourism developed from a relatively limited activity among the upper class to widespread mass tourism for all segments of society (Urry 1990), and it is closely linked to urbanisation, to the division of time between work and leisure, to the emergence of vacation weeks (and weekends) and to reduced costs of transportation. During the nineteenth and twentieth centuries, seaside resorts developed throughout Europe. At first, they were connected by simple forms of transportation such as horse-drawn carriages; later by train connections between major cities and nearby tourist destinations (often coastal resorts), while today, air travel connects major cities to almost all parts of the world.

In Europe, mass tourism in the form of package holidays arranged by large tour operators started after World War II based mainly on bus trips. From the 1960s, this form of tourism exploded in popularity and the number of destinations increased. Gradually, the majority of bus trips were replaced by airplanes which transported mainly tourists from the northern countries to Mediterranean destinations in Spain, Italy, Portugal, Yugoslavia and Greece. The impact on the coastal landscapes in these countries has been immense. In some of the regions, small fishing towns have developed into huge coastal resorts, with sprawling high-rise hotels and apartment buildings intermixed with restaurants, amusement parks, roads, car parks, beaches and marinas (see Figure 5.12). Old cultural heritage sites were demolished or became dominated by new developments, and natural habitats were drained or destroyed. As well as the direct physical damage caused by such developments, overexploitation of the water resources, increased erosion and the fragmentation of larger habitats have also taken place. In many cases, the original rural economy associated with fishing and agriculture has been severely disrupted – or has even collapsed all together – due to a rapidly expanding tourism sector which depends on more continued growth (Hunter and Green 1995). Some regions have undergone such developments, but have managed to break the vicious circle of unsustainable development. Costa Brava in the Catalan region of Spain is an example of such an area where mass tourism led to rapid development followed by stagnation. Although redevelopment in the form of more sustainable tourist developments involving nature restoration projects and heritage policies has been on the agenda and proposals for rejuvenation have been discussed, no genuine renewal of the area and subsequent regrowth of tourism has taken place (Priestley and Mundet 1998; Sarda, Mora and Avila 2004).

This concentration of tourism activities in Southern European coastal areas has had a direct impact on nearby rural areas, speeding their progressive marginalisation: the active population has been attracted to tourism-related jobs, which are generally better paid than jobs in agriculture, and have left the rural areas – or at least have ceased to work in the rural (Pedroli et al. 2013; Van der Sluis et al. 2015). In the southern part of Europe, this was clearly related to the rapid abandonment of land in the last decades of the twentieth century as seen in, e.g. the inland mountainous regions of the Algarve in southern Portugal (Jones et al. 2011; Pinto-Correia and Breman 2009). Still, the villages were not totally abandoned as the active population who worked in coastal tourism would often return on a daily basis or at least at the weekend.

However, a form of counter-urbanisation in these inland rural areas has also taken place, subsequent to rural abandonment. The impact of the development of mass tourism in the coastal areas of Southern Europe has many faces and is surely changing over time. If we continue with the example of the Algarve and the relation between coastal tourism and the inland hills, what has taken place in the past two decades is a progressive occupation of former farm and village houses by incomers engaged in new activities, who have moved away from the coast due to increasing living costs and too much construction. New activities related to tourism, well-being and handicrafts are now widespread in the Algarve hinterland, which has created a new form of service-oriented rural community. These new people also partially maintain the landscape as they restart farm activities, which often take the form of specialised niche, organic or biodynamic production etc., which are viable due to the nearby tourism market and the constant flow of tourist consumers. There is, thus, a kind of 'Tuscanisation' of the Algarve hills, whereby the countryside is the location for multiple consumption activities as described by Holmes (2006, 2012; see Chapter 3), which is a relatively recent phenomenon surely linked to the mass tourism on the coast, and which is having a surprisingly positive effect on the landscape in terms of revitalisation.

With reference to Miossec's work on tourism development (Miossec 1976), Pearce (1989) presents a model of four stages of tourism development: (1) The area is 'discovered' by a small group of drifting visitors with vague or no attachment to this area, while local inhabitants have a diverse or even polarised view regarding the potential of tourism. (2) Once the first hotels and guesthouses have been successfully established, tourism facilities and transport infrastructure start to develop. (3) A hierarchical system of tourist facilities and infrastructure develops and local citizens either reject more tourism or demand increased regulation. (4) The number of tourists stagnates and the area may start a restructuring process involving nature and heritage reappraisal and landscape planning. New kinds of tourists may arrive and the traditional ones seek other areas.

This model overlaps significantly with Butler's now very widespread model of the 'tourist area life cycle' according to which, after having been discovered, a tourist area develops, consolidates and subsequently reaches its 'carrying capacity', after which it loses quality and new destinations emerge. Such development may create pressure to improve the overall quality of the tourist areas which developed during the early stages, but this may be too costly or not possible for other reasons (Butler 1980) (see Figure 5.11). Although models like this tend to oversimplify real life, it has certainly been highly influential and subsequent development has, according to the author (Butler 2004), to a large extent, confirmed its relevance. In relation to landscape development in the context of tourist areas, the model is useful in that it reminds us that rapid, mainly market-guided changes may well lead to unsustainable outcomes which are difficult or at least very costly to redirect.

The mass tourism concentrated in the coastal areas is far from the sole urban-driven tourist activity which is having an impact on rural areas. In particular regions with marginal conditions for agriculture (Scotland, Ireland,

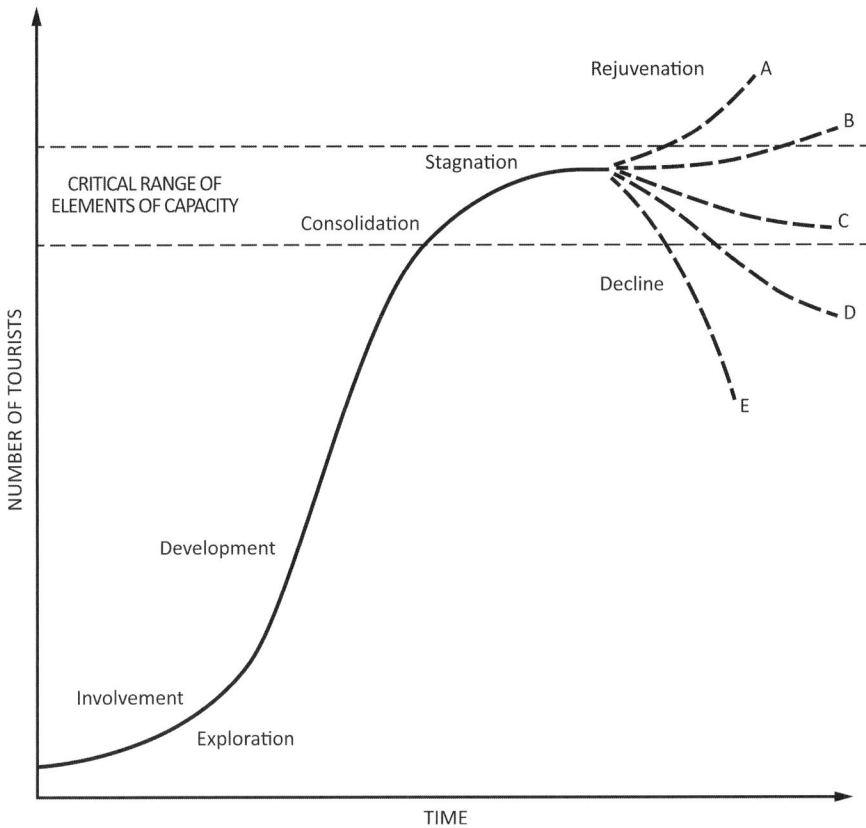

FIGURE 5.11
Tourism cycle of evolution after Butler (1980)

the Alentejo region in Portugal and the Andalusia in Spain, mountain areas all over Europe), rural tourism is a highly relevant driver of the rural economy. Rural tourism takes many different forms (Urry 1990). It may be closely linked to open-air activities and the enjoyment of nature, where the tourist is closer to nature and enjoys the landscape while, e.g. walking, cycling, horse riding or bird-watching, which is then classified as ecotourism.

In any case, tourism has developed in rural areas where farming has failed to be competitive on the world market, thereby providing an alternative source of income, which has been based on the farm or on other non-land-based business, and which has certainly created wealth and jobs. However, tourism may also be in conflict with agriculture concerning, for example competition of water resources – either for irrigation or for coastal tourist resorts (Swyngedouw 2015).

The main issue with rural tourism development is its relation to landscape maintenance. Rural tourism is strongly anchored in the landscape as a resource, as the tourist gaze is most often linked to an appreciation of the landscape. Nevertheless, tourism activities contribute to the management of

FIGURE 5.12
Playa de las Américas, Tenerife, Spain. Over the course of several decades, this resort developed on the basis of mass tourism from mainly Northern European countries.

the landscape only to a limited extent. When the farmer or farm owner organises tourism activities on his/her farm, the tourism income directly supports the maintenance of land management. However, when the tourism activity is related to tourism infrastructure outside the farm, more jobs and wealth will be created in the countryside, but land management activities are rarely supported. This is a significant countryside consumption dilemma, which we discussed in Chapter 3 and which remains to be solved with new ways of integrating activities in the rural and new forms of governance applied to land management and tourism alike.

Social Farming: Healthy Landscapes Make Healthy People

The beneficial effect on urban citizens of being in the landscape has long been acknowledged (Van Mansvelt and Pedroli 2003). Already in the nineteenth century, working the land(scape) and experiencing it were applied as therapy, for example in the garden of the Canada Lakeshore Psychiatric Hospital (Paine 1997), but also in various places in Europe. Interestingly, not only creating and working the landscape, but also experiencing and enjoying, it were considered therapeutic. Eating produce from that same landscape as another form of therapy was considered so self-evident that it was hardly ever mentioned. Several studies have shown that having a view of a park contributes substantially to the recovery of hospitalised patients, while patients who have a view of the wall of an industrial building recover less favourably (Larsen 1991; Larsen and Harlan 2006; Mooney and Hoover 1996).

Actively experiencing landscape in various ways, ranging from survival trips to school-farming, forestry weekends and long walking trips, is increasingly seen as an important tool to help people reconnect to the real-world qualities and, thus, to their basic sense of being. This is especially true for urban people, who have become increasingly disconnected from nature thanks to the large range of artificial elements that characterise modern city life (asphalt, concrete, neon lights, traffic, huge buildings obstructing the view of the sky, a high level of mechanical and electronic noise) (Van den Berg et al. 2007; Van den Bosch et al. 2015). Weekends in the countryside and holidays 'in the green' are appreciated by urban citizens as crucial for relaxing and recovering from urban stress (Groenewegen et al. 2006). Furthermore, the increasing appreciation of work-on-the-land as therapy for psychologically affected people and its social appreciation by mentally handicapped people point towards the importance of 'grounding' in the 'here and now of the place where you are' (Van Elsen, Günther and Pedroli 2006).

To enhance such grounding, it makes a substantial difference if farmers include the production of a diverse, locally specific and characteristic landscape in their farming style (Bohnet 2002). Such local specifics of landscape are often much more easily accommodated by small-scale farming such as organic farming rather than large-scale industrial farming (Di Iacovo et al. 2014; Hassink, Grin and Hulsink 2013; Pedroli et al. 2007b).

5.5 Conclusion: Substantial Change in the Functional Basis of the Landscape

Conflicting Urban-Oriented Interests in the Rural

Changes in land use and the composition of rural communities would not be possible without the management of conflicting interests. Conflicts are part of these processes. The different demands on the rural space today are emerging from a progressively more urbanised society and evolving needs and expectations. However, these are not always easily assimilated by rural communities. The new pressures on the rural space in the form of more specialised and industrialised agriculture, abandoned marginal areas, increasing demand for recreation and tourist uses and activities, energy production and the preservation of natural resources often create new tensions and conflicts.

The demand for housing in the countryside or the construction of tourist infrastructure not only occupies space and properties, which had formerly been strictly designated for agricultural use, it also makes land prices increase to a level that does not relate at all with the production capacity of the land. The increase in the price of land and other production factors significantly limits opportunities for maintaining agriculture, and often raises concern and tensions within the farming community. Even when farmers want to continue farming, or new farmers would like to take over, high prices and difficult access to land can make farming an impossible activity. Consequently the landscape, which is what attracted newcomers in the first place, may be severely affected by the disappearance of or change in agriculture in the long term.

Further, there are changes in the community composition, and thus the expectations about how the rural space should develop are not always compatible with the previous community dynamics. When the rural community transforms into a mixed community with many individuals with urban lifestyles and aspirations, these individuals find many of the attributes of the rural, which are related to farming, annoying, such as the typical noises and smells associated with farming activities. Conflicts and tensions arise then.

A balance between traditional and modern lifestyles in the countryside is hard to find as the newcomers are in many cases crucial for maintaining the liveability of rural communities. But so far, farming has clearly lost weight in the rural voices, in comparison to a rural where – still a few decades ago – agriculture was dominant. These emerging conflicts and tensions without doubt require attention and dedicated management solutions. In order to foster new, constructive community dynamics, creating opportunities rather than intensifying tensions and conflicts, new types of approaches are needed – that is what we will address in Chapters 6 and 7.

Synopsis: The Rural as Seen from the Urban

In summary, we cannot understand the rural landscapes of today's Europe if we do not understand the rural–urban relationship, which affects these landscapes and has done so over time, with numerous historic layers indicating these relationships. The relationships have changed dramatically since the Greek city state emerged in approximately 1000 BC (Table 5.1). During the first three millennia of European urban history, Europe was predominantly a rural continent, both economically and culturally. Cities and rural landscapes were mutually dependent, with cities relying on a rural economy until the Industrial Revolution. Gradually, agriculture has lost ground in terms of economic significance and in great parts of Europe, agriculture has not been able to compete on expanding markets and has become marginalised – a process

TABLE 5.1 *Six periods of urban–rural relationships in Europe*

Period	Urban–rural relationships
Antiquity	The city state as an economic and social entity with the same regulations governing the city and the rural landscape
The medieval	Organic mutual dependence
	The city provides security and specialised crafts. Represents civilisation.
	The rural provides food (although some food was produced inside the city walls), fibre and energy. Represents the uncivilised, barbaric.
	The rural landscape becomes increasingly farmed as populations increase and farming technologies develop. The city and the rural landscape are regulated by different systems.
The Renaissance and centralised city	More open and chaotic relationships
	The city provides a wide array of services (religious, educational, legal, commercial etc.) and commodities. The city is the centre for capital accumulation and consumption.
	The rural provides food, fibre, water, energy and men for the army. The rural as the centre for capital production. Appreciation of the picturesque in rural landscapes evolves. An educated rural elite slowly emerges.
	The rural landscape is dominated by the feudal order. Agricultural expansion continues.
The industrial city	The city provides an increasing array of services (including food processing), increasing the supply of jobs. The city is the centre of capital production.
	The rural provides food, fibre and energy (increasingly not local), recreational opportunities (including sites for second homes) and environmental services associated with water supply and waste management
	Railroads connect the rural to the city
	Land reforms replace feudal order. Barbed wire and later electric fences replace hedgerows, earth and stone walls.
	The rural landscape becomes mechanised – machines and chemicals replace manpower.
The garden city	Appealing ideas of combining the rural and urban in garden cities and the introduction of planning for metropolitan regions. The beginning of suburbia.

TABLE 5.1 (*Cont.*)

Period	Urban–rural relationships
The post-industrial city	The city functions as a cultural, economic and political centre linked to other centres through formal and informal networks. It provides an increasing array of services for the rural. Some European cities are shrinking due to ageing population and loss of job opportunities; others are growing fast.
	The rural provides residential and recreational opportunities and environmental services of various kinds. Local food production is gaining importance in some cities. Second homes are increasingly affecting rural landscapes in many regions.
	The rural landscape becomes (to highly varying degrees) urbanised and multifunctional.
The twenty-first-century city	Cities become increasingly ruralised through urban farming and conversion of parks to more nature near habitats.
	Highly qualified citizens move to the rural and use the city as a service provider.
	Food security and quality are drivers for much closer relations between citizens and agricultural production, inside or outside the city

which is ongoing. A consequence of this is a dramatic decline in the population of many rural regions and a substantial change in the functional basis of the landscapes.

From antiquity until today, cities have gradually become less dependent on their surrounding countryside for food, but other new dependencies have evolved, including the supply of clean drinking water, attractive rural settings for recreation and suburban residences. Today, the multifunctionality and attractiveness of surrounding landscape has an influence on the overall competitiveness of cities and the peri-urban landscape is a common issue in the spatial strategies of all cities. The Dessau-Wörlitzer Gartenreich introduced in the opening of this chapter represents a system of parks laid out with a regional matrix of agriculture and forestry. As a whole, the regional landscape represents a meeting point between romantic garden traditions mainly linked to the eighteenth-century English landscape architecture tradition and educational ambitions for modern agriculture and forestry linked to the

Enlightenment. In this way the Dessau-Wörlitzer Gartenreich can be seen as an example of a regional landscape deliberately designed to combine production and consumption needs.

Today, the rural landscape and the urban centre are much more closely linked functionally and over longer distances than ever before. Teleconnection is a significant factor (Seto et al. 2016) which is functioning within spaces of flows on a global scale (Castells 2000b). What is decided or what happens in urban centres is increasingly affecting rural landscapes in different regions and in multiple ways. Such developments not only challenge geosciences (Seto et al. 2016), they also challenge public policy and planning responses at all levels. The next two chapters address this challenge.

6

Landscape Policy and Planning – Managing Conflicts and Making Places

Far out in the Atlantic, between the coast of Norway and Iceland, the Faroe Islands rise from the sea. They have been inhabited for more than 1000 years by people who have historically made their living from agriculture – mainly sheep breeding. The sea and the coastline were simply too rough to rely on fishing as the major food supply until recent decades.

We visited Gjógv as tourists during a week-long fishing trip. The village had been recommended too often in guidebooks and was too close to our base to miss it. And we were surprised. Not so much by the well-preserved and still lively village, which is beautifully located in a small natural harbour – this was somehow to be expected. What came as a surprise was the overall landscape with the village located in the midst of a sea of grass and the timelessness of the landscape as a whole.

The open landscape of Gjógv is currently almost exclusively covered by permanent grasslands. Due to relatively thin turf and the hilly-to-steep terrain, overgrazing and, consequently, erosion have been a continuous threat to the Faroe landscape. As early as 1298 in the so-called Faroe Sheep Letter, the Norwegian king stipulated that the number of sheep on each individual pasture 'shall remain the same as it was in previous time unless men can see that it can accommodate more' (from Brandt 1987, p. 40). This 'unless men can see' statement is important as it does enable changes to be made to the livestock density over time, but it requires some sort of community agreement, which is difficult partly because the community's interest in avoiding erosion is stronger than the individual sheep owners' interest, and partly because more sheep would mean a change in the distribution of wealth within the community. A combination of community responsibility and hard-core legal infrastructure has proven a workable and sustainable solution from medieval times until the twenty-first century.

Today, the village of Gjógv is surrounded by permanent grassland grazed by sheep – even most of the former infield is now managed as permanent

FIGURE 6.1
Village of Gjógv on the Faroe Islands, July 2011. The closest houses on the right are new and have been built in a traditional style – some of them are being used as second homes and the village is composed of a mixture of houses of different ages.

grassland. However, agriculture is no longer of major economic importance. Fishing, although not from boats from the local harbour, and the developing tourism sector dominate the local economy, which is, therefore, becoming closely interlinked with the global markets for fish and tourism. The maximum stocking density for sheep is still regulated, but the sustainability of the local community and, thus, the landscape is much more dependent on changes in other forms of policy, most importantly resource management and market policies for fish products.

Public policy and planning and how these interventions affect European landscapes are the subjects of this chapter. We critically analyse and discuss the area both from a general institutional perspective and from a more concrete practical view with an emphasis on the specific domains of direct relevance for rural landscapes. The ambition is both to outline fundamental concepts and analytical frameworks concerning landscape policies and spatial planning and to discuss recent developments in the most important policy domains and in approaches to landscape governance. We start with the developments from traditional regulations to more integrating forms of governance. Then we discuss the relationship between competences linked with landscape management and change on the one side and public regulations of these as they appear in policy and planning on the other. In Section 6.2, we discuss the policy-making process, including the steering instruments available to guide landscape management and the factors influencing implementation. In Section 6.3, we outline recent developments in the most important policy domains affecting rural landscape: land-use policy and spatial planning, agricultural policy and environmental policy. Finally we close the chapter by a brief discussion of two promising approaches to more integrated and more involved policy and practices.

6.1 Landscape Policy – Between Different Notions of Governance, Competence, Agenda and Instruments

Public policy intervention in landscape management is one of the oldest forms of public regulation. Currently, such policy and planning measures constitute a diverse and dynamic body of regulation in all European countries. Public policies (including spatial planning) which are designed and implemented to affect the protection, management and change of rural land use and landscape features and the acquisition of property are considered 'landscape policies' in this context, although 'landscape' as such does not represent any established policy domain (Scott 2011). Over time, such policies appear, disappear and reappear in different forms and with different aims. They are most commonly sectoral policies, which are issued from individual policy domains such as agriculture, water resource management, nature conservation, land use etc. Some of the policies, including the oldest, were introduced to protect crucial resources such as the grass turf of the Faroe Islands. Others are meant to stimulate development in agriculture and forestry or to contribute to multifunctional landscape development mainly from a local perspective, which includes many of the current policies. Common issues include the regulation of land-use rights (including building rights), access to grazing, firewood, timber and other resources, hunting and fishing rights and, in more recent times, recreational access, nature and cultural heritage conservation, environmental protection and landscape management. In this context, spatial planning is seen as a special category of policy, which is future orientated and traditionally involves extensive public participation.

Recently, territorial dimensions have been included in many policy domains such as nature conservation, rural development, water management, agriculture etc. However, overall, landscape policies remain, to a large extent, fragmented and uncoordinated. They deal with conflict management from a range of perspectives as well as rural place making.

On a practical level, rural landscapes are managed and changed by primary agents, first and foremost farmers and local inhabitants (see e.g. Case G). As described in Chapter 3, farmers manage the landscape from different perspectives and while being influenced to various degrees by external drivers such as the market, technology and public policy. Traditionally, changes in the market and new technologies have had an immediate influence on landscape decisions and practices, whereas public policy has been seen as a (reactive) frame for such changes, although this is overly simplistic. First, the different types of drivers influence each other. Large open markets such as the EU, for example, would not exist without strong and highly centralised market policies. So-called market-driven development would not occur without support from market policy regulation (Harvey 2005). Second, public policy has become more proactive and often functions as the instigator of change. For these reasons, analysing the various roles of public policy on landscape management is no easy task, never mind the overall impacts of policy compared to other drivers and the key agent's own values and goal.

CASE G Feuilla, France – Towards Revitalisation of a Village at Risk of Abandonment

Landscape Appearance

Feuilla is a 2400 ha municipality of about 100 inhabitants in the *département* Aude, southern France. It is part of the Languedoc-Roussillon, situated 20 km from the Mediterranean coast on the foot slopes of the Pyrenees, the land of Catharism. The village is located in a valley covered by Mediterranean *garrigue* (shrub land), some woodland and vineyards amidst sparsely vegetated limestone hills up to 700 m (Figure 6.G.1). Ruins of an old windmill where the access road crosses the hilltop indicate the dominance of strong winds in the area (Figure 6.G.2). Feuilla is situated within the Regional Nature Park (PNR) Narbonaise en Méditerranée, and one of the 22 municipalities participating in the 2010–2021 Park Charter.

Landscape Processes

Feuilla has a rich history of cereal and grape cultivation, horticulture and grazing, local glass industry, lime, ochre and charcoal production, as well as of garrigue products (honey, fragrant oils, lavender). Although one would expect flocks of sheep grazing the Mediterranean vegetation, shepherds are hardly roaming the area anymore. Since the beginning of the twentieth century, the village has undergone a demographic and socio-economic decline, even if a certain increase in population has been observed since the 1990s. At the end of the nineteenth century, the main road was diverted to pass by the village; today it does not have a school nor a café anymore, and the church is in use only on rare occasions. Today only some 100 ha of vineyards remain, worked by young winegrowers for the cooperative winery in a nearby town. The historic nucleus of the village itself – a densely built and fortified ensemble – is partly in ruin (Figure 6.G.3). A number of houses have been converted into second homes and in summer the population can amount to more than 250 people. Few original inhabitants remain in the village, but many owners of second homes are former inhabitants or their relatives, who still have strong bonds with the village. Fifteen per cent of the newcomers are foreign, often well integrated into village life as well.

CASE G (Cont.)

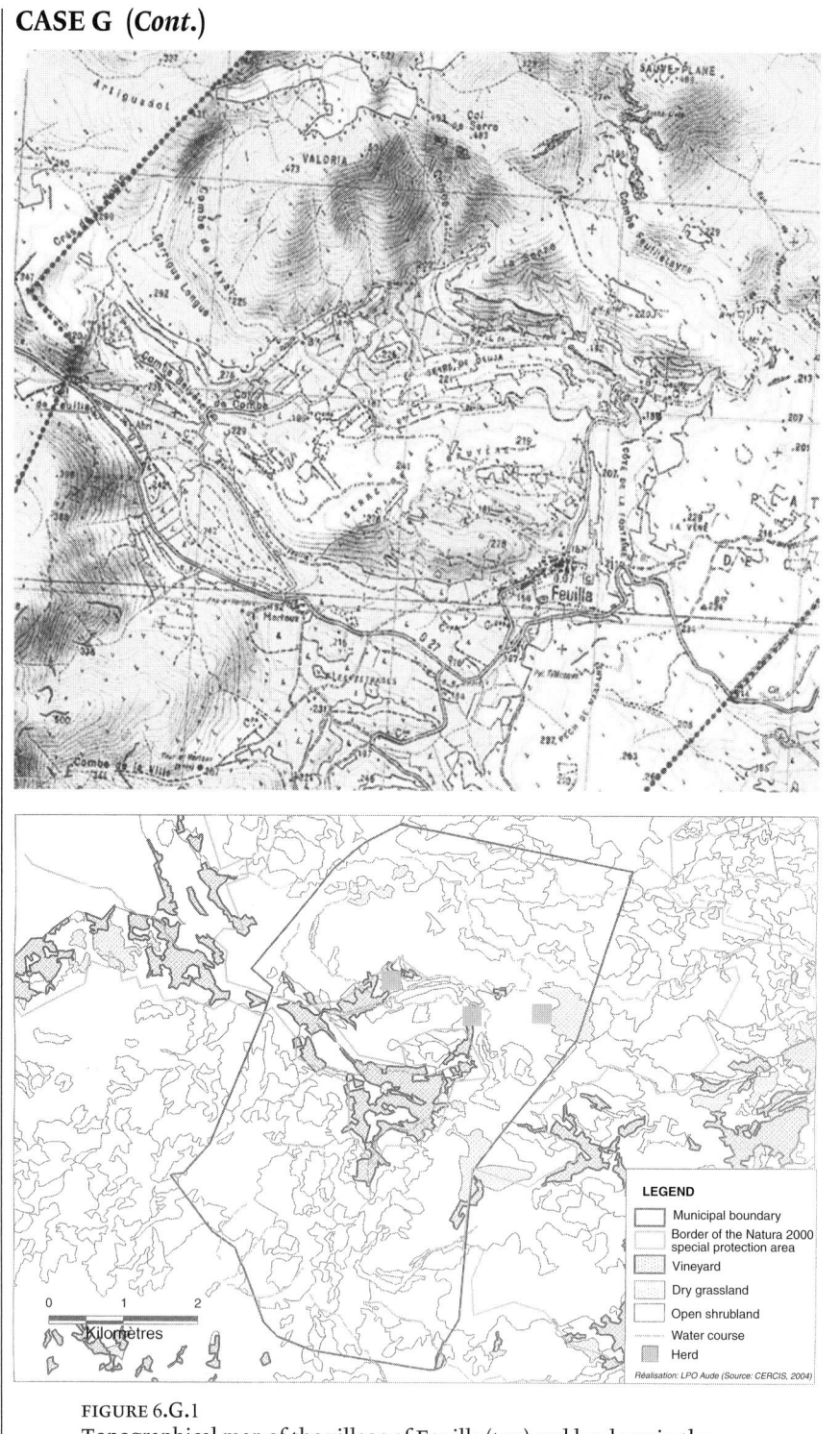

FIGURE 6.G.1
Topographical map of the village of Feuilla (top) and land use in the municipality (bottom). For the colour version, please refer to the plate section.

CASE G (*Cont.*)

FIGURE 6.G.2
The access road from the east with the remnants of the windmill to the left

The municipal council is committed to the flourishing and revival of the village rather than to acquiesce in a gradual degradation. Although the level of public services and infrastructure is limited, hope has been placed on an increasing number of newcomers. The installation of wind turbines on the surrounding hilltops would bring appreciable economic benefits, but the village has deliberately refrained from this development to not have its horizon spoiled by these structures and rather to reinforce the natural character in accordance with the PNR.

Indeed the natural and cultural heritage of Feuilla is a rich resource. The uninhabited Mediterranean garrigue and macquis harbour a wealth of plant and animal species, inviting discovery during pleasant hikes. The historical nature of the village includes the fortified village centre, remnants of a watermill in the main stream and a small hamlet developed as a zero-energy initiative. Moreover, a vivid tradition exists of village feasts renowned in the region, e.g. *la Fête de l'Ancienne Frontière* (the Feast of the Old Frontier, i.e. the former border between Occitan and Catalonia, France and Aragon, that ran through Feuilla) in June each year.

FIGURE 6.G.3
Feuilla in its surrounding macquis landscape

CASE G (*Cont.*)

Although Feuilla is a bit off the routes, the major opportunity for the future is considered to be rural tourism and perhaps the installation of a photovoltaic energy project. A hikers' accommodation is currently being projected in a joint undertaking with the neighbouring communities to be able to host walkers and cyclists. Also riding tourism is being welcomed.

The village council has embarked on designing a zoning plan (which did not yet exist) to guide the land-use changes and to prevent uncontrolled second housing development. It is the intention to develop a spatial plan based on the local identity, natural and cultural heritage, and giving room for new initiatives.

Conclusion

Feuilla may serve as an example of many small villages throughout Europe. In this case, new inhabitants, partly owners of second homes, bring strong intentions to build on the heritage values that were gradually considered by the original inhabitants as no longer feasible for further maintenance. New functions of recreation and tourism, and associated small business, have started to play a role in village life. These had been viewed with scepticism by the former inhabitants (most of them now living elsewhere), but since such initiatives bring new life and merge well into the character of the place, these are now more easily welcomed with appreciation. This is the basis for stronger interest from regional authorities as well, supporting the improvement of tourist networks, e.g. related to the precious Mediterranean ecosystems, the Cathar history or local products.

Government and Governance

The relationship between top-down, command-and-control regulations and local, more or less, collective decision-making changes over time and between countries. In recent decades, there has been a clear trend in the form of a move away from central government focussed on the pursuit of the common public interest based on authority and majority rule, towards a more network-based governance approach to dealing with change which is more proactive, involving coordination and negotiation among multiple stakeholders (Louglin 2014; Rhodes 2007). However, traditional government and more recent governance approaches coexist. Today, market policies represent a highly centralised and *government* style of regulation, which is controlled by the detailed, top-down execution of decisions and a legal system that refers to the European Court. The Common Agricultural Policy (CAP) is the single most influential market policy with regards to European rural landscapes and will be discussed further in Section 6.4. Over time, the political discourse concerning the CAP has developed from a policy mainly administered by a single agency (DG Agriculture in Brussels and national ministries of agriculture) to a policy implemented more broadly by a number of agencies which are increasingly interconnected through networks. However, this has not fully transpired in practice. The core of the CAP still represents a mainly centralised and traditional way of steering. The rural development programme is supposed to be more decentralised, but in many countries the decentralised agencies, e.g. local action groups can act only within a strict framework which is set centrally by the state administration, which means that local autonomy is limited. In other countries, the rural development part of the CAP, as well as other policies, is governed through complex networks involving independent public bodies, private organisations and voluntary institutions of various kinds, which have some degree of autonomy from the state (Rhodes 2007). Such a *governance* style of policy has been introduced (or reintroduced to some extent) in Europe, partly in response to a highly fragmented public domain, and partly as a way forward in dealing with increasingly complex policy issues. There are numerous examples of these new governance approaches to landscape policy, including some of the local action groups under the rural development programme of the CAP, the implementation of the NATURA 2000 network and the action plans for the individual habitats.

The Florence Convention (European Landscape Convention, ELC), which by 2016 had been signed and ratified by 38 Member States of the Council of Europe, represents a recent and novel approach to landscape governance (Jones et al. 2007). In Article 5 (Council of Europe 2000), the ELC states that the individual party (the individual state) must:

- Recognise landscapes in law as an essential component of people's surroundings, an expression of the diversity of their shared cultural and natural heritage, and a foundation of their identity (5a).
- Establish and implement landscape policies aimed at landscape protection, management and planning (5b).
- Establish procedures for the participation of the general public, local and regional authorities and other parties with an interest in the definition and implementation of the landscape policies (5c).
- Integrate landscape into its regional and town-planning policies and in its cultural, environmental, agricultural, social and economic policies, as well as in other policies with possible direct impact on landscape (5d).

This is understood and followed in quite different ways in the various countries which have ratified the Convention. While combining different governance levels may well be in the spirit of the Convention, it does not mean that the ELC in practice is implemented through initiatives taken at different levels. In Iberia, for example, the ELC has mainly taken the form of a landscape classification, expressed in a landscape character map and the corresponding descriptive characterisation of the landscape, all published in the early 2000s (Cancela d'Abreu, Pinto-Correia and Oliveira 2004). These classifications recognise differentiated patterns and functions in the landscape character areas identified, and have most often been produced in an articulated approach across scales, allowing for downscaling and even upscaling. The landscape character areas identified have been used as a basis for the organisation of different sectoral plans, as the forestry plans or the regional spatial plans. Nevertheless, the classifications produced are not recognised in the same way by those who work with day-to-day decisions regarding the landscapes. In the daily land management there has been no increase in landscape awareness due to these classifications being produced, though they show significant potential as a tool for communication between stakeholders with different perspectives and on different scales (Loupa-Ramos and Pinto-Correia 2016).

Territorial and Spatial Competences

From an overall rural landscape point of view, the current policy systems that guide landscape change in Europe are in crisis, as we outline in more detail in Section 6.3. One of the roots of this crisis is the fundamental relationship between landscape management and change, which we term 'landscape practices', on one side, and the public regulation of management and change, i.e. 'landscape policy', on the other.

In his theoretical analysis of the landscape dimension of environmental management, Hägerstrand (2001) states that the division of land and water into spatial domains reflects the links between society and the physical landscape. Two competences are of interest here; the practical landscape actions and the regulating landscape practices. In European landscapes, the practical actions in the landscape – the daily management practices and changes – generally take place within the boundaries of properties, which may be owned by the state, or they may be common land which belongs to the local community, or private properties belonging to individuals (including families) and firms. Practical landscape management and change take place on such properties, and the power to make these decisions and actions is termed 'territorial competence' (Hägerstrand 2001). However, this competence is by no means unlimited or restricted solely by economy or technology as it is also constrained and guided by various forms of legislation, as well as other policy interventions. Furthermore, the property patterns themselves (including property rights) are in fact established, maintained and changed through policy and politics. Properties and property rights have historically been areas of conflict and struggles between those who maintain the rights and those who demand alternative regulation of the rights, or more fundamentally, an alternative distribution of properties and rights (Egoz, Makhzoumi and Pungetti 2011). The competence to regulate landscape practices has been termed 'spatial competence' (Hägerstrand 2001), acting within clear boundaries such as the borders of nation states, regions and municipalities. In each of these the spatial competence is attributed to sectoral bodies which always see themselves as responsible for the territory. The relationship between the two competences is as crucial for the rural landscape as it is complex.

> An invisible landscape of legal rural, human intentions and plans is hidden in the visible physical landscape. All rules and regulations imposed from above are ultimately filtered through this invisible landscape with results which might differ considerably from those originally intended.
> (Hägerstrand 2001, p. 54)

Hägerstrand's 'invisible landscape' refers to the complex mix of landscape agents and how they act individually, usually on the scale of the property, and how they interact through networks. An alternative term for the filtering process mentioned by Hägerstrand could be 'landscape governance'.

Two Policy Agendas

If we look at the spatial competences which specifically and directly affect European farmers and other primary landscape managers, two agendas should be emphasised: the *market policy agenda*, which from the very start has been *the* key agenda of the European Union (Fearne 1997), and the *sustainability agenda*, which emerged in Europe in the 1970s and which became a core part of the European Union in the mid-1980s. Both are international policy agendas with the WTO as a key institution for market policy, and the UN's programme for sustainable development as the main international institution for dealing with environmental and social policy (Dwyer and Hodge 2001; Primdahl and Swaffield 2010a). The sustainability agenda has clearly evolved in response to the market agenda – both internationally and within the EU (Commission 1987; Lowe and Baldock 2000). Over the years, this has become an important policy agenda involving a high number of public officers and a high profile in the public debate. One characteristic of the agenda is that it has steadily grown in terms of the issues covered and the significance of the competence so that it is now located on all levels from the local municipality to the UN. Objectives and rules often filter down through the various levels, becoming more and more landscape specific at the lower levels.

The market policy agenda, on the other hand, is highly centralised and most policy decisions are now usually made at the EU level or the national level for countries outside the EU, both of which are increasingly in compliance with WTO objectives (Sturgess and Dalton 2000). Overall, the agenda has also been characterised by deregulation and expanding markets, which means that there is (almost) no competence remaining for local and regional bodies. The agricultural policy and land market policy are now decided in the EU for most of Europe and it is, of course, not possible to include any potential landscape impacts of the policy decisions except for very general concerns. Consequently, the integration of environmental and social concerns into the market policy agenda can take place only at the high levels. In effect, this means that the two agendas function quite separately until they meet the agents in the local landscape, that is at the farm level. For this reason, but also because of the underlying increasing mobility described in Chapter 5 and the fact that territorial competences increasingly are held by absentee owners (including companies) who may not have any attachment to the landscape, this is developing into an asymmetrical meeting in which highly globalised technologies and market forces are becoming increasingly influential and the local farmer and the local community are losing autonomy. Consequently, it is difficult to integrate the market policy agenda with the sustainability agenda, which also means that a happy marriage between the spatial competences and the territorial ones is a challenge (Figure 6.2).

We return to the concepts of landscape governance, competencies and agendas several times in the remaining sections of this book. First, however, we take a closer look at concrete policy developments.

6.2 The Policy Process

Policies Affecting Land Use

Public policy and spatial planning (termed 'policy' in the following) function in different contexts and in different ways. The focus here is on policies which affect natural resource management, landscape management, rural land use and primary functions such as agriculture, forestry, recreation and tourism. Policies including plans are usually designed, implemented and redesigned in processes involving: (1) a perceived problem – occasionally combined with deliberate visions – which initiates a need for policy intervention; (2) a design phase in which goals and objectives are clarified and a choice of instrument(s) is made; (3) implementation of the policy in question, and; (4) evaluation of the effects and outcomes and possible recommendation for policy redesign (Hogwood and Gunn 1984; OECD 1997). The four stages are, of course, closely connected. The objectives must be directly

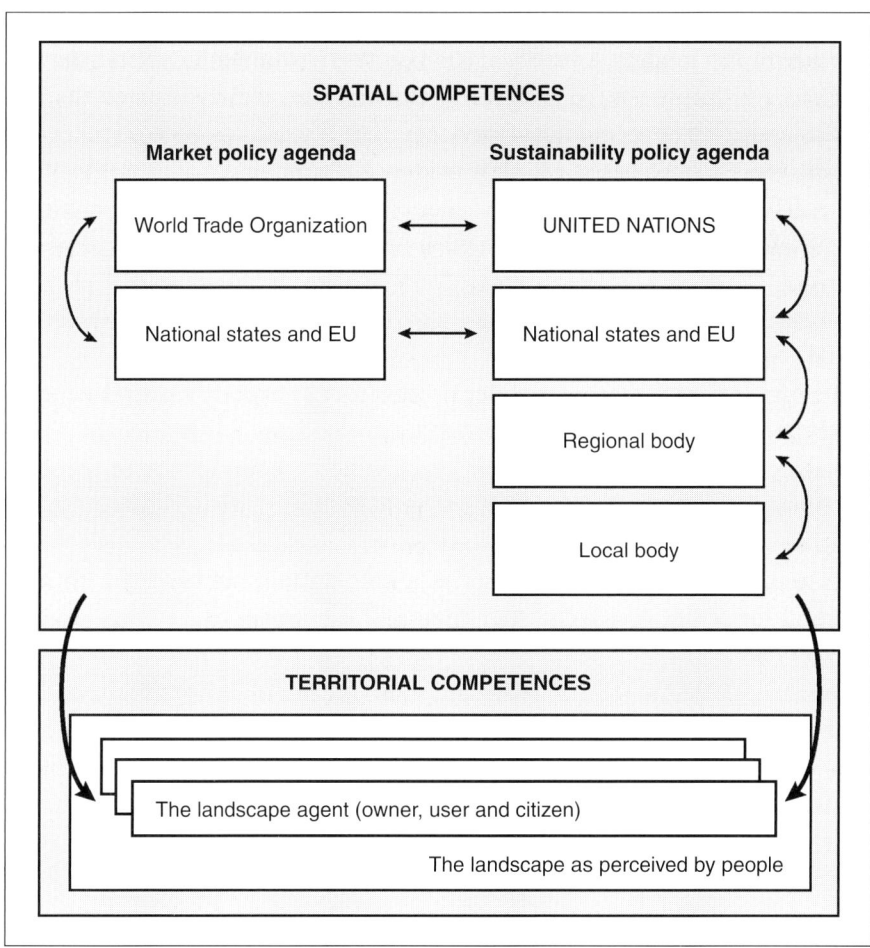

FIGURE 6.2
Relationships between territorial competences and spatial competences. Adapted from (Primdahl and Swaffield 2010b, p. 10) and inspired by (Dwyer and Hodge 2001).

linked to the problem or vision to be pursued, while the instruments must function well, meaning that the policy target – the farmer or the habitat manager in a nature reserve – must behave as assumed. And this behaviour in turn must lead to the intended outcome. All landscape policies are implemented through agents, and for landscape policies to work there must be clear and causal relationships between the effects of the policy instrument in question on the target's behaviour and the environmental outcome. One way to ensure that such relationships exist is to establish so-called impact models and evaluate the model before (*ex ante*) and after (*ex post*) the policy in question has been implemented (Rossi and Freeman 1993).

Agri-environmental policies can serve as an example to illustrate the problem with impact models. A survey of 60 EU agri-environmental policy schemes designed and implemented in seven Member States clearly showed that, in most cases, such impact models were not explicitly used, while those that were implicitly used were based on poor evidence (Primdahl et al. 2010b). Other studies have confirmed the poor evidence base that supports EU agri-environmental schemes, which is one of the most important types of landscape policy in Europe (Carey et al. 2003; European Court of Auditors 2011; Kleijn et al. 2001; Kleijn and Sutherland 2003; Primdahl et al. 2003). We return to agri-environmental policies in Section 6.23.

The problem of initiating a policy process occurs usually within a context where there is a policy history, i.e. some current policies to relate to, and a policy hierarchy, i.e. policies at a higher level, which are implemented through the policy in question or vice versa, and policies which are designed at lower levels. This means that a clear-cut linear process as outlined previously rarely exists and Figure 6.3 be seen rather as a schematic framework which may be useful for policy analysis rather than a linear model of a typical process (DeLeon 1999).

The formulation of overall goals and more precise objectives also does not occur in an isolated phase. The perception of the problem often has built-in goals – water eutrophication is a problem because the goal is to achieve aquatic habitats that are not severely polluted by high levels of nitrogen and

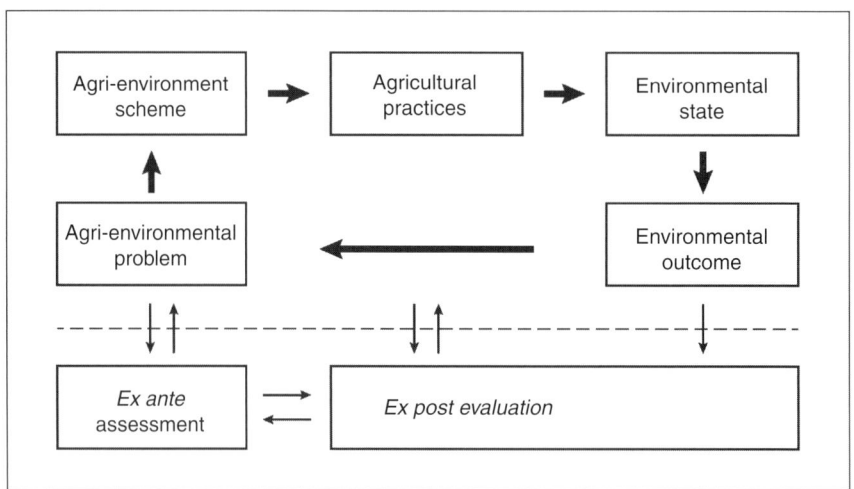

FIGURE 6.3
An impact model of agri-environmental policies and the two principal types of policy evaluation. Assessment of policy before implementation and evaluation of how the policy-affected agricultural practices are having an impact on the environmental status and with what outcomes (Primdahl et al. 2010b, p. 1246).

phosphorus. Goal setting is, nevertheless, an important part of the policy process, which involves policy makers and other agents (public and private organisations, NGOs, owners and users) agreeing on the nature of the problem. Where do the agents want to go? Who is expected to do what? How should multiple objectives be handled? What should be considered success? And what should be done if the objectives are not achieved (a plan B)? Hogwood and Gunn (1984, ch. 9) provide a checklist of such questions to address in relation to objective setting, which is useful both for policy making and for analysing policies which have been implemented.

Policy Instruments

The other main dimension of policy design concerns the instruments or policy tools. Schneider and Ingram (1990) have developed a conceptual framework based on assumptions regarding the behaviour of the policy target group. The framework is highly relevant for landscape policy because a wide array of policy instruments is applied in this field. For example, if changes in the environmental law introduce a restriction on fertiliser application, it is generally assumed that farmers will comply and reduce their application to the level prescribed because it is likely that they will obey the law.

Five types of instruments are identified:

(1) *Authority tools* are statements granting permission, prohibiting or requiring action more or less precisely described. Statements may be targeted at other public agencies or they may be targeted at citizens and they assume the policy target is responsive to statements, although they are often accompanied by other measures such as various kinds of sanctions.
(2) *Incentive tools* rely on concrete payoffs, either positive in the form of payments, for example, or negative (disincentives) in the form of an environmental tax, for example. In general, incentives assume that the target group will react rationally as 'utility maximisers' and probably will not take action without the incentive.
(3) *Capacity tools* provide the policy target with information, training and resources to promote certain types of decisions and actions. It is assumed that the barriers to these are related to a lack of information, skills, specific machinery or other resources.
(4) *Symbolic and hortatory tools* include persuasion and labelling aimed at changing the policy target's perception without altering the tangible payoffs. Arguments appeal to the target group's values and preferences such as planting native tree species instead of exotic species because the former increase biodiversity.

(5) *Learning tools* are used in situations where the problem is well described, but the necessary policy actions are either unknown or uncertain. The assumption is that the policy targets can learn from experience and can identify the tools needed. Experimental programmes where the policy objectives (reduced nitrogen leaching, for example) are well defined, but it is up to local policy makers to identify solutions in cooperation with the policy targets, for example. Adaptive governance (Folke et al. 2005) may utilise learning tools as part of the institutional set-up.

Although the five types of tool described represent a useful framework for addressing the diversity of motivations and preferences among policy targets, the different tools overlap considerably when it comes to how they function in practice. A specific information policy which offers farmers free advice on habitat management may be characterised as an incentive, capacity-building or symbolic tool depending on how three different types of landowners would respond to the advisory service offered. The same is the case for many law regulations such as the protection of specific habitats. Do such regulations mainly function by authority, or are they effective primarily because of the associated sanctions? In other words: is it an authority tool or an incentive tool? The answer probably depends on the individual landowner as well as the social culture of the country or region in which the regulation is applied. For these reasons, we prefer to distinguish between three types of policy tool: regulatory measures, incentives and information – both advisory and education. Selman (1988) and Gilg (1996) have used similar terminology, although they distinguish between disincentives and incentives, while they also include public ownership/long-term management as a policy option. To illustrate the diversity of instruments applied in European landscape policy and to show some of the potential strengths and weaknesses associated with these, we have listed a number of examples of three types of instruments in Table 6.1.

The different instruments can often be successfully combined and support each other, but they may counteract each other if incorrectly combined, thereby reducing overall effectiveness. When EU agri-environmental policies emerged in the late 1980s, they were generally designed as single incentives which were, more or less, focussed on specific issues such as semi-natural grassland management or reducing fertiliser application. They were not considered in an overall environmental context such as afforestation schemes to promote the planting of forest on valuable semi-natural grasslands, for example (EC 1999). In other cases, farmers received incentives to refrain from polluting, which violated the polluter pays principle. The introduction of cross-compliance measures in the early 2000s has without doubt improved coordination and the integration of different EU agri-environmental instruments (Benneth et al. 2006).

TABLE 6.1 *Policy instruments to guide landscape management and change. Please note the content is generalised and principal – exceptions may be found in most examples shown and transactions costs are included in assessments (+ = strength, ÷ = weakness)*

Type	Examples	Efficiency – strength and weaknesses
Regulatory measures	Property regulation	+ contribute to farmland and forestry protection + protect farm structures through restrictive rules to avoid subdivisions ÷ may be bypassed by speculators or may be in conflict with overall market regulation
	Land use	+ may avoid undesirable land uses (e.g. urban sprawl, afforestation, deforestation) ÷ reactive instrument, cannot (usually) be used to ensure desirable land-use change
	Nature conservation	+ prevent habitat removal or undesirable disturbances ÷ cannot prevent abandonment or ensure desirable habitat management
	Access	+ ensure recreational access rights on a general level within certain parts of the landscape (usually uncultivated areas such as beaches, forests, permanent grasslands) ÷ may cause continuous conflicts between owners and users
Incentives/ disincentives	Agricultural subsidies	+ maintain living agricultural landscapes through supporting agricultural investments, products or income + contribute to food security + ensure reasonable income for small-scale farm families ÷ increased pressure on the public budget and may increase food prices ÷ disturb land market to the benefit of current owner and the disadvantage of future farmers ÷ may unintentionally redistribute funds from producers to owners ÷ disturb food markets and reduce innovations
	Agri-environmental payments (e.g. five-year management agreement)	+ facilitate desired changes and the maintenance of landscape management practices which would otherwise cease + rewarding farmers who have contributed to landscape enhancements ÷ short-term agreements – may not ensure long conservation of biodiversity, drinking water, recreational access etc. ÷ contributing to commodification of the landscape

TABLE 6.1 (*cont.*)

Type	Examples	Efficiency – strength and weaknesses
	Tax (and tax reduction or exception)	+ reduce environmental impact or resource exploitation (e.g. pesticide tax, water tax) + generate funds for proactive initiatives (e.g. habitat restoration, new walking paths) ÷ may lead to undesirable actions by actors hit by the tax in question
Information, advisory, education	Information	+ easily applied and relatively low-cost option which can be efficient when lack of information is a barrier to solving the problem. Web-based manuals for wild life improvements (creation of new ponds and removal of water course barriers, for example) ÷ not effective if the policy target is unmotivated or directly against the recommended action
	Advisory	+ can be highly effective when given in the context of change. Advise on selection of species given in connection with farmers' establishment of hedgerows or small woodlands, for example
	Education	+ courses, workshops, series of lectures etc. can be highly effective in the longer term if target groups change awareness on the issue being taught

Implementation

Once a policy has been formally approved, it is put to work, it is being implemented. Some policies and plans, e.g. a water course restoration plan, may be implemented more or less instantly following approval, while it may take others several years or more to have an effect on the landscape. In any case, they will have to be implemented through peoples' decisions and actions as indicated in Figure 3.4 in Chapter 3. No plans or policies affect the landscape directly. Even the bulldozer driver reshaping a water course as part of implementing a restoration plan is making decisions during the work.

Besides these concrete objectives and the choice of instrument(s), several factors influence policies and plans when implemented. For example, agri-environmental measures are included in the rural development programme as it is framed by a central regulation, and once implemented, written agreements, usually for five years, are signed by farmers. The central regulation prescribes, in some detail, the conditions under which measures must be designed and implemented. Usually, the EU funds 50 per cent or more of the costs. Besides pursuing environmental goals, the EU regulation is concerned with issues such as risk of overcompensation, violations of the polluter pays principle, monitoring environmental change and evaluating policy effectiveness. The Member States (or in some cases, regions) then design the specific measures they want to implement, which are then approved in Brussels. The measures are then offered to farmers – sometimes in targeted areas and sometimes generally to all, sometimes with specific sets of requirements, sometimes with options enabling a tailor-made agreement. The individual measures may be offered in many different ways, e.g. through a website, and farmers are then expected to fill in an application form and start implementing the agreement. Alternatively, the farmer may be contacted in person by a public agency, a private advisory service or an NGO, who may try to convince the farmer to join and sign an agreement or may simply offer advice about how to apply and what to apply for. The way in which contact is made, how the farmer is advised and by whom is likely to affect the implementation of the specific measure and the EU agri-environmental measure more generally.

Winter (1990; Winter and Nielsen 2008) distinguishes three factors that affect policy implementation. First, there is the way *organisations* – individually and together in networks – influence decisions and the behaviour of the policy target, the farmer, for example. An organisation often has a different view of the goals behind the policy measure in question than the policy body above it had when it designed the policy. Therefore, the organisation will attempt to influence the policy while implementing it. Also the way organisations, including private cooperations, influence the implementation affects what happens to policy implements. Are they in line pursuing the same goals or are they obstructing, passive or simply not cooperating well? Not all organisations given a role in implementation necessarily support the policy in question (Sabatier 1986).

Second, *the field worker* – or the 'street-level bureaucrat' as this agent has been termed in a famous study of social policy implementation in the United States (Lipsky 1980) – is the one who is in direct contact with the policy target. This individual's resources (including time, knowledge, funds etc.) and discretionary power will affect the policy. Historically, in many European countries, the farm advisor has been a key agent who acts as a field worker, although the privatisation of advisory services (or the removal of subsidies supporting such services) in many countries has reduced or changed the role of farm advisors (Curry and Winter 2000).

The local community and NGOs are becoming increasingly involved in policy implementation and often play the role of the field worker. In particular programmes of the CAP, local action groups from specific NGOs are given responsibility for establishing the link between the authorities and the policy target.

Third are the values, decisions and behaviour of the policy target, which in the context of this book is *the primary landscape manager*. Curry and Winter argue that the issue of environmental concern in European agriculture is well established and continue:

> A cornerstone of the successful implementation of agri-environment measures is the extent to which farmers are both willing and able to carry them out. In turn, this willingness and ability are dependent, to a considerable extent, upon the skills and attitudes of farmers, centrally influenced by the research, education, training, information and extension services that have, for a long time, supported European agriculture.
>
> (2000, pp. 108–109)

As discussed in Chapter 3, policy tools and policy implementation affect individuals, who then make decisions regarding their day-to-day actions, which in turn affect the landscape. These individuals are influenced by different sets of drivers and their own value set. Sometimes they have the landscape in mind when they make decisions, but often they do not. The decision process and the way policy tools affect this process is, thus, more complex than a simple driver-state-response linear conceptual model. The primary agent manages the landscape from different positions – as producer, as landowner and as citizen – and it is crucial for successful implementation of landscape policy that these positions are included in the design as well as institutional arrangements surrounding implementation.

In summary, with regards to the landscape policy process, we note that several processes and stages are involved in the design and implementation of landscape policy, although in practice, they often overlap and are combined. It is useful to distinguish between the design and the implementation of the policy in policy-making practice and when analysing and evaluating the effects and outcome of policy if one wants to learn from the policy processes. Furthermore, it is important to separate the effects (or outputs) policies have on human practices and their landscape outcomes if one wants to understand how policies affect the rural landscape. If a habitat conservation scheme does not work so that the habitat does not develop in the right way, or biodiversity is declining, there may be several reasons. For example, the scheme may not have been designed correctly, the management prescriptions may simply be inadequate or the incentives offered to farmers for managing them may be too small. Alternatively, it may be the implementation which has failed; farmers may not have been adequately informed, or they may not have been listened to when they expressed their views concerning the contract. The framework presented here can help to analyse the policy process and thus improve it.

6.3 European Landscape Policies – Developments and Current Challenges

As stated in Section 1.5 (the second tension mentioned), no European landscape policy, as such, exists – landscape is simply not an explicit part of the European Commission's competences. However, many European policies have significant effects on the landscape, and some even contain regulations that explicitly refer to landscape (e.g. some of the agri-environmental measures). At the same time, the development of the EU in combination with the development of internationalised market policy and sustainability agendas, as described in Section 6.1, has resulted in radical changes and a high degree of convergence between national landscape policies. In this section, we briefly outline the major trends regarding these developments and close by reflecting on the major challenges facing European policies in relation to the landscape today. Although Europe is more than the EU, we do focus on the EU simply because the vast majority of countries are EU Members and therefore must comply with EU legislation. While briefly discussing evolving food policies, we have focussed on three major policy domains affecting European rural landscapes: land use and spatial planning, agricultural policy (including agri-environmental policy) and environmental policy. Energy and climate policies are also affecting rural landscapes, but they are for the most part a more indirect influence in the landscape, and are therefore left out to limit the extent of the chapter.

Land-Use Policy and Spatial Planning

Rules concerning user and owner rights over the resources of the rural landscape belong to the oldest form of legislation in Europe. In medieval Scandinavia, the laws were actually called 'landscape laws' (Jones 2006; Olwig 1996). Before laws were written down and given formal status, oral laws existed as did customary rights, e.g. grazing rights for specific places derived as a result of many years of (accepted) practice. Such customary rights still represent important rules for European landscapes. In Denmark, an example of a right linked to customary law which is still very much alive today is the right to gain access to a piece of land as a result of continuous use. Thus, if a person has frequently walked on a specific route – a path on the land – for 20 years, he or she can claim the right to walk on the land, even if it is private with no public access (Ramhøj 2009). This right takes different forms in various European countries, but is found in many different regions. In Portugal, it is termed '*uso capeão*' and is still often used in court decisions concerning conflicts between neighbours. Other uses which are not linked to specific features of the land in question such as the right to collect firewood in a forest may also be granted if the oldest people living in the area can confirm that this has been taking place during their lifetime (40–50 years). Although such rights in most European landscapes may be of limited importance concerning current territorial competences, customary law has historically been important in defining user rights in the rural landscape.

In modern times, land-use regulations have been split into a number of domains. Today, land development such as urban expansion, rural housing and infrastructure construction is usually regulated through land-use and planning legislation, i.e. through regulatory measures. Although the EU has had a significant effect on urban and regional policy, the land-use and planning domains do not belong to the EU competence except for funds (including the European Investment Bank) that support regional development projects (Williams 1996). A so-called European Spatial Development Perspective was formulated during the 1990s (CEC 1999), but it ended up being a rather vague policy document which had very little influence on European spatial policies (Faludi 2004). In the words of Hall and Tewdwr-Jones:

> the ESDP was stripped of the maps that alone could have given it some operational force, so remaining a set of metaphors. And the same, at least for now, seems to be the fate of territorial cohesion. But this is perhaps the inevitable result of trying to develop coherent policies for one of the largest, and certainly the most diverse, unit of territorial governance in the world. (2011)

Nonetheless, land use and planning play a significant role in guiding European landscape changes, although the policies vary considerably across the Member States. Urban expansion and development in rural areas (outside 'urban zones') has traditionally been mainly regulated (at least since World War II) through regulatory instruments, which have often been linked to zoning systems and designations (Department of the Environment 1989). Examples include the Danish rural and urban zones (Primdahl 2014), the Dutch spatial development plans (Faludi and Van der Valk 1994; Van der Valk and Van Dijk 2009), the Portuguese municipal directive plans, which have structured the whole territory seen from the urban centres (Gaspar 2006), and the English 'green belts' (Elson, Walker and Macdonald 1993). The British green belts together with national parks, areas of outstanding natural beauty and other protected designations exclude more than 40 per cent of the British total land area from large-scale developments of any kinds (Hall and Tewdwr-Jones 2011, p. 11).

The degree to which the planning systems regulate land use in rural landscapes varies and is probably the result of contingencies, but there is no doubt that it is often related to how a mainly urban population views the rural landscape. In England, for example, hedgerows are protected in the planning system mainly for their scenic, historical and environmental value (Hodge 2016), while the Danish coastal protection zone in a similar way protects scenic and environmental values. However, at a general level, rural areas are covered by planning regulations to only a limited degree, while agricultural structural developments, as a rule, are not regulated (Hodge 2016; Primdahl 2014). Thus, properties that are defined as agricultural may often also have less restrictive building regulations than other properties. In the Netherlands, Denmark and England, for instance, farmers who want to build a new 'farm building' – a house for machinery, for example – do not need permission to construct a 'rural building'. Furthermore, rules for subdividing and the opposite, merging, rural properties are affecting rural landscapes. Also, other types of rural properties may be linked to special regulatory measures such as forest properties, second homes etc.

As urban and rural land uses are often regulated through different legislation, the very definition of 'rural' and 'urban' is of course important for how the spatial competences affect landscape change everywhere. This is potentially a problem for two reasons. Firstly, agricultural buildings and machinery are becoming increasingly larger with an equivalent impact on scenery, terrain conditions, traffic etc. The need for regulation is simply growing. Secondly, the agricultural landscape is becoming increasingly urbanised. The fact that more than half of all farm owners in a typical Danish agricultural landscape say that their primary motive for owning their farm is that it is a good place to live (see Chapter 5) may undermine the justification for having different regulations for farms compared to other rural properties. More generally, with increased mobility and much more dispersed settlement patterns than previously, the functional distinctions between urban and rural land uses are becoming blurred, as shown in Chapter 5, Table H.1. Different land-use regulations for urban and rural land may therefore lose legitimacy.

Although land-use regulations are usually used to define what may not be done (without consent), they may also on rare occasions be used to ensure specific management practices. However, as the use of regulatory measures to ensure management practices may have a role to play in some contexts, we briefly mention three examples of such measures which are currently in use. First, owners of forest and silvo-pastoral land in Portugal are obliged to maintain 10 metre-wide belts of ploughed land along all national and municipal roads to prevent or contain forest fires, while they are also forbidden to cut the most valued and slow-growing trees in their forests such as cork and holm oaks, which form the tree cover/canopy in the Montado silvo-pastoral system (Pinto-Correia and Fonseca 2009). Second, Dutch farmers are obliged to maintain canals and ditches (including dredging) according to the specifications of the water board that comprises their farmland (for which they democratically elected the governors for ages) (Van de Ven 2004). Finally, the Danish Forest Act, which was reformed in 1804, requires forest owners to maintain their forests as production systems. Historically, this was to ensure the future supply of timber for the navy. Rules which are still in force include the obligation to replant after cuts and to maintain a coherent canopy (Fritzbøger, 1998).

Rural Landscape Planning – Between Conflict Management and Place Making

The general aim of landscape planning, just like any other spatial planning, is to ensure well-functioning spatial development which produces more attractive environments that would not emerge in the absence of planning. This is ensured through a number of processes and plan solutions where different stakeholders participate and cooperate with professionals and public authorities. Overall, landscape planning has two purposes: conflict management and place making (Healey 1998).

Conflict management is a traditional function of spatial planning with land-use regulations as the key instrument as described previously. First, conflict management is about the public interest versus private rights, which in a rural landscape context means establishing the concrete framework for development rights in a broad sense, including housing, agriculture, industry, mining, tourism etc. Recreational access rights are usually also a central issue here. What areas should be protected, what developments should be promoted and where and which management practices are needed for conservation of significant habitats, cultural heritage feature and scenery. Second, it concerns conflicts between private landowners and users. For example, attempts are made to prevent potential conflicting land uses and constructions. Third, conflicts between short-term, local gains and long-term, sustainable development on a regional scale represent a third type of conflict management for landscape planning. Key instruments with respect to conflict management include land-use legislation (planning law often in combination with sectoral law), different types of development plans, zoning ordinances and designations.

Place making is different to conflict management as the focus here is on change and enhancement rather than regulation. Place is about developing landscapes to become (or to contain) better places from a range of perspectives. In urban planning, place making has a long-established role in focussing on the quality of public urban spaces, functionally, symbolically and aesthetically (Healey 2009). For rural landscapes and regions, this tradition is by no means as pronounced as it is in urban and regional planning (Primdahl et al. 2013a), which is in fact a problem closely linked to the lack of academic and professional discourses about rural landscape futures, as discussed in Section 5.2. On a general, abstract level, rural place making is about the same fundamental conceptual issues concerning 'place' as in urban and regional planning. According to Healey (2004, pp. 48–50), it is about *scale* (either in nested hierarchies or as spatial-temporal 'reach'); *position* (in respect to borders versus relational networks); *internal organisation* (spatial differentiation/place characteristics); *identity* (materiality, imagination and meanings); *development* (conceptualisation of change and pathways); and *the representation of ideas* (of spatiality and place qualities). On a more concrete and specific level, rural landscape place making is fundamentally about collaborative processes concerning the future structure and character of the landscape in relation to the functional needs emerging from the new urban–rural relationships (see Chapter 5).

Healey (1998) argues that for planning to be effective, it must incorporate both the conflict and the place-making dimension. The legal infrastructure must be brought together with the right combination of policy instruments, spatial frameworks, place designs and collaborative processes to guide landscape change. We return to this in the final parts of both this and the next chapter.

An Example of the Regional Level: The Tuscan Landscape Plan

Against the background of a certain reluctance towards the development of national and regional landscape policies in Europe, an interesting example of a regional landscape policy – one of the few in Europe – is the Tuscan Landscape Plan approved in April 2015 (Regione Toscana 2015). It may serve as an exemplar of an approach that could profitably be followed in many other regions as well.

This Landscape Plan – of which the earlier (2007) approved Piano di indirizzo territoriale (PIT), a territorial development plan, forms an integral part – represents a clear attempt to propose a visionary way forward to balance the various public and private interests in landscape on a regional level (Pedroli 2016). The plan reflects a polycentric vision of the urban–rural relationships which are to be enhanced. The process of developing the Tuscan Landscape Plan was highly interactive and was guided by a report by the special commissioner on communication (Regione Toscana 2013). Since 2011, a public-access GIS facility on the landscape plan has been accessible on the Internet (Figure 6.4), and numerous meetings have been held in the many different landscapes, including special sessions with environmental and civil society NGOs (www502.regione.toscana.it/geoscopio/pianopaesaggistico.html)

Although there was a lively debate on the specific elements of the plan within the various Tuscan professional, commercial and citizens' communities in the years before the approval, the main issues, based on sound scientific evidence, have been retained:

- The basic ambition of the Tuscan Landscape Plan, as previously expressed in the PIT (2007), remains focussed on long-term return on investment rather than on the mere goal of profit making. It promotes the integration of the Tuscan urban pattern in the form of 'Tuscany as a polycentric city' and it aims to safeguard industry in Tuscany, which comprises all manufacturing, servicing and research activities that characterise the region in a global competitive and innovative way. Finally, it plans infrastructural projects of regional importance to stimulate planning agreements.
- The key role of the concept of 'territory' reflects the definition of 'territory' as a common good and an essential public heritage and, at the same time, a virtual arena in which the decisions on the future of Tuscany should be made. The 'territory' is considered a production space, which, on one hand, represents an environment for the collective realisation of fine local produce and, on the other, the motor for such production so that it is an essential production factor.
- The innovative governance approach of genuinely *public* governance of the territory – public in the sense of sharing interests among a territorial community based on a joint vision of future perspectives beyond simple economic profit making, but also in the sense of mutual collaboration between sectoral policies and spatial arrangements, which combine private economic initiatives with social, cultural and environmental values.
- The legislative integration of the PIT with the more economically oriented Tuscan Regional Development Plan 2014–2020 emphasises the relationship between spatial planning and regional development as the nexus of the regional social capital in all its aspects. The spatial designations defined in the PIT serve not only as boundary conditions for the regional development as identified in the Regional Development Plan, but also function as incentives for future perspectives.
- The holistic spatial planning approach – rather than continuing on the increasingly counterproductive route of zonation, the PIT focusses on the public goods that are considered indispensable for Tuscany, which represent values which go beyond the summation of values of constituent elements, and which should be cherished for their crucial value of identity and quality of life in the Tuscan region. Moreover, the richness of the polycentric urban pattern is recognised as a crucial Tuscan characteristic, as well as the heritage of the age-old rural countryside, which is irreversibly connected to the urban culture (and vice versa).

The principle of *active conservation* fosters conservation by finding a proactive modus between public authorities and private entrepreneurship to safeguard the long-term characteristics of the environmental, landscape, social, economic and cultural heritage as a common good. In particular, the concrete spatial planning strategy, which touches all kinds of vested interests, has generated discussion and still involves some challenges for the Tuscan government and community. However, the basic planning framework has now been legally established.

These six key points represent basic boundary conditions for spatial planning, something which various authors have put forward, though in varying combinations and order. The concept of 'territory', for example, is less prevalent in Anglo-Saxon spatial planning literature since the term 'landscape' often takes this role and combines the spatial (production space) and the local (heritage; place) aspects of 'territory' (Harvey 1993; Healey 2009). It is paramount to consider both aspects in spatial planning as is becoming increasingly apparent from the literature (Davoudi 2012; Primdahl et al. 2013a; Van Rij and Korthals Altes 2010).

Public participation, which is one of the principal requirements of the European Landscape Convention, is addressed in a sophisticated way in the Tuscan Landscape Plan. This is surely because it is understood as a logical component of genuine public governance of the landscape, which is propagated under the second issue. Michels and De Graaf (2010, p. 481) assert that there is a close relationship of the governance mode with the specific type of democracy adopted through various modes of participation (Table 6.2). Many examples of public participation have been reported (see e.g. Jones and Stenseke 2011), but

TABLE 6.2 *Aspects of citizen participation and democracy according to Michels and De Graaf (2010)*

Aspects	Clarification	Theoretical Perspective
Inclusion	Allows individual voices to be heard (openness; diversity of opinions)	Social capital Deliberative democracy
Civic skills and virtues	Civic skills (debating public issues, running a meeting) and civic virtues (public engagement and responsibility, feeling a public citizen, active participation in public life, reciprocity)	Participatory democracy Social capital
Deliberation	Rational decisions based on public reasoning (exchange of arguments and shifts of preferences)	Deliberative democracy
Legitimacy	Support for process and outcome	Participatory democracy

it is often unclear which mode was adopted in the specific project. The Tuscan Landscape Plan seems to have adopted various modes of participation with an emphasis on process governance and, thus, on legitimacy, which seems to be a prerequisite for a complex landscape approach (Sayer et al. 2013, p. 8353). For this purpose, at least all basic cartographical documentation remains available to all interested stakeholders in the public-access GIS facility on the Internet mentioned earlier, which facilitates genuine coordination with institutional stakeholders and local interest groups alike.

The integration of structural regional development and spatial planning in the Tuscan Landscape Plan is novel. Although the principle has been propagated in various countries, as far as we are aware, a real vision-led open and integrated spatial planning has not yet been realised. This is probably due to the fact that spatial planning usually follows social and economic development (Van der Valk and Van Dijk 2009). The combination of these two planning instruments is promising, and an evaluation of its success after a few years will be very interesting.

Furthermore, the holistic planning approach seems to be an innovation, although it is still difficult to judge the impact of the approach. The associated ambition of the further development of the polycentric *città toscana* in par. 5.2 of the *Documento del Piano* is not futile: it is reminiscent of the statement "*la cité est devenue l'État*" about the Roman Empire as a world order that Marguerite Yourcenar imagines as part of 'Hadrian's Mémoires' (Yourcenar 1951, p. 125). Whether the Tuscan lifestyle can be compared to that of the Roman Empire is an open question, but the explicit emphasis on polycentric development is very interesting because the concept is the subject of much debate (Brezzi and Veneri 2015; Finka and Kluvánková 2015). It seems that Tuscany is a region where the concept may be suitable because of the existing intricate urban-rural pattern. Polycentric development may lead to a new geography of 'functional urban areas', settlement systems, kept together by a multitude of material and immaterial networks (Marson 2015), which would strengthen the regional and local economy, make the settlements more attractive for people and business development and put more cities and towns in a better position to provide or sustain services of general interest closer to the citizens (Ulied 2014, p. 15).

The principle of 'active conservation' resembles the conservation-through-development concept which has recently been propagated for spatial planning in heritage areas (Van der Valk 2014), but which has also been contested in many conservationist circles (Palang and Fry 2003). The associated risks of compromises threatening cherished values are often poorly balanced with the advantages of a living heritage (Janssen et al. 2014).

It is clear that regional landscapes result from decisions made by a multitude of actors. The planning challenge is to guide these decisions, grounded in a proper understanding of the processes that foster the vitality of the countryside, acting through planning that actively supports rather than merely protects (Van der Valk 2014, p. 34). Process design for strengthening regional landscapes rather than instant solutions should be promoted. In today's decentralised governance context, market-based planning mechanisms must be allowed to contribute and experiment, but at the same time, they need clarity regarding the future development of urban and non-urban land use to succeed. This requires a planning style which is responsive to actors and their aims and open to complementary informal processes, which implies much more than just aesthetics, access to recreational space or increasing the value of adjacent dwellings (Legacy 2010, p. 106). The Tuscan Landscape Plan demonstrates that an effective framework for such planning can be designed and made accessible to all parties concerned, including citizens. It remains to be seen whether the Tuscan society will make use of this opportunity and implement a landscape policy that safeguards the unique values of a multilayered and recognisable landscape for the future, not only for the sake of tourists, but primarily for the Tuscan inhabitants and their children and their sense of belonging. If they do, the Tuscan Landscape Plan may become a pioneering piece of regional policy and an example for other regions both within Italy and in a wider European context.

Agricultural Policy

The single most significant policy area that affects European agricultural landscapes is without doubt agriculture, and within the EU that mean the CAP. Historically, before the establishment of the EU, agricultural policy was also extremely influential for the simple reason that the basic conditions for agriculture – the farm properties and purchase rights as well as the associated user rights – were core issues in agricultural policies. This also means that, traditionally, agricultural policy was not only market policy, but also land-use and social policy and even occasionally security policy. Incentives were a key instrument for promoting agricultural expansion and intensification from the nineteenth century until the end of the twentieth century, as discussed in Chapter 4. Support was offered to farmers who wanted to carry out projects such as the reclamation of heathlands, wetlands or other uncultivated areas, drainage and irrigation in many parts of Europe during the nineteenth and twentieth centuries. Within the EU, most of these types of land-use incentives were either stopped because they were seen as irregular national or regional supports of agriculture or they were made part of EU schemes within various development programmes. Also subsidies for fertilisation, chemical spraying and soil improvement have been widespread, as have subsidies for the production and export of specific products.

However, since World War II, agricultural policy has increasingly become a 'market policy' (Tarrant 1992), which is why the general trend in recent decades in the EU has been towards deregulation and the streamlining of legislation concerning rural property ownership. However, national legislation in European countries including EU Member States still maintains rules of significance for landscape change. An owner of a Danish farm property, for instance, must live on the farm. Although this rule has been eased recently, and possibly violates EU legislation, it has over time prevented farms being converted into second homes and has, in combination with other rules, maintained a living, productive landscape in attractive coastal regions, for example.

The CAP evolved during the 1960s, not as a coherent streamlined policy, but rather as a result of disagreements and compromises between the six Member States, which had quite different interests regarding agriculture – some being large exporters (including France and the Netherlands), while Germany was a major importer. The overall aim was to develop a common market and the evolving CAP should be seen as the result of this, rather than the goal behind it (Fearne 1997). In the beginning, there is no doubt that increased food production was the most crucial goal, while guaranteeing incomes, market stabilisation, food supply and reasonable prices were also included in the stated aims (Roederer-Rynning 2010). Subsidies, especially price subsidies, was the main instrument applied and this was doomed to failure, as the Dutch agricultural commissioner, Sicco Mansholt, warned from the very beginning.

In the late 1960s, it became clear that the CAP had become too expensive and overproduction was looming on the horizon. Mansholt launched the famous 'Agriculture 80-Plan', the aim of which was to dramatically reduce not only the price subsidies, but also both the agricultural area (by 12.5 million hectares) and the farm population (by 5 million) between 1970 and 1980 and all within the six Member States (Fearne 1997). However, this did not occur – at least not during the 1970s. As became clear, such a thing as the CAP cannot be changed overnight. Mansholt's farsighted proposal was rejected and it was not until the early 1990s that an Irish commissioner, MacSharry, succeeded in reforming the CAP, which has continued until now. At that time the costs of the CAP had become more and more problematic. The overall budget had increased; the EU countries as a whole had become more than self-sufficient with regards to food, and the overproduction of many subsidised products incurred significant costs for storage and export support (Cunha and Swinbank 2011). Furthermore, it became increasingly clear that the CAP was not only an economic burden on the expanding European Union; it was also causing severe environmental problems due to the expansion of agricultural land and the intensification of crop and husbandry farming. Habitats, especially semi-natural grasslands, had been removed or were fragmented, and the aquatic system from groundwater to larger coastal water bodies was becoming increasingly polluted by nitrate and phosphorous, causing eutrophication and contamination by pesticides – often through practices directly linked to CAP support (Lowe and Baldock 2000).

Since the MacSharry reform, production support has been reduced dramatically as part of a number of reforms. As 'compensation' for withdrawn subsidies, farmers are now offered direct income support, which is connected with an range of requirements to comply with various environmental measures and specific good farming practices. Only a few exceptions which involve payment being coupled to production are permitted. These direct payments, which correspond to what is called Pillar I of the CAP, account for the majority of the total CAP budget and whether they can still be termed 'incentives' is debateable. An innovative instrument has been applied in recent versions of the CAP, namely the introduction of cross-compliance measures which link payments to specific environmental requirements such as a minimum level of maintenance of permanent grasslands or compliance with specific environmental regulations such as the implementation of buffer strips along watercourses. The decoupling of payments from production and the cross-compliance measures may mean that the time when CAP subsidies were directly detrimental to the European landscape is over. Indeed, it would be fair to say that the direct payments have had some overall positive effects in terms of maintaining valuable European landscapes.

Even before the MacSharry reform, EU agri-environmental schemes were offered to farmers for agreements, which were typically of five years' duration, although agreements of 10 and 20 years' duration have been seen. Farmers were offered compensation payments for the costs connected with carrying out environmentally friendly farming practices – or for income foregone for *not* carrying out specific farming practices. Agreements typically contained prescribed practices such as extensive grazing, reduced levels of fertilisation, avoidance of pesticides, conversion of arable land into grassland, conversion of conventionally farmed land into organically farmed land etc. From the late 1980s, the EU co-financed such schemes and the Member States paid the remaining part, which was typically 25–50 per cent of the total.

In 1992, it became obligatory for Member States to establish and implement agri-environmental schemes, which have become widely used instruments within Member States for enhancing, maintaining and protecting various environmental values linked with agriculture. From accounting for only a few per cent of the total agricultural land covered by agreements, the prevalence of agri-environmental schemes has increased so that they now cover about 25 per cent of the EU's utilised agricultural area, while the cost for the EU was budgeted at 22.2 billion Euro for the period 2007–2013 (European Court of Auditors 2011). Similar schemes have been introduced in non-EU

Member States such as Norway and Switzerland. The agri-environmental schemes are now included in what is termed the Pillar II of the CAP, which includes several different axes, which also provide support for, e.g. farm modernisation, agricultural infrastructure and marketing. The funds for Pillar II represent a low proportion of total CAP funding, and they are still distributed between different types of goals. Nevertheless, there is no doubt that these schemes have had an effect on European landscapes concerning the protection as well as the enhancement of values linked to the biodiversity, cultural assets and aesthetic values of landscapes (Primdahl et al. 2003). However, just how effective these schemes are is unknown due to a lack of monitoring and well-designed evaluations. There are, however, indications that the agri-environmental schemes are relatively poorly designed and implemented (European Court of Auditors 2011; Hodge 2001; Kleijn and Sutherland 2003) and that there is a need for: (1) more thoroughly developed impact models of the causal relationship between farming practices and landscape values; (2) more targeted measures; (3) more locally adapted schemes; (4) improved monitoring and evaluation of schemes.

Despite these shortcomings, agri-environmental schemes have gradually grown in both expenditure and in coverage of land. Will this development continue? Will agri-environmental policies even become the major expense within the CAP? In 2012, the British agricultural economist Hill thought that such an outcome was likely.

> This is because the much more expensive Pillar 1 of the CAP – and in particular the Single Farm Payment – seems destined to be transformed, at least in part, into some form of agri-environmental payment when the present financial package under which it operates runs out at the end of 2013. Indeed, a dominant – perhaps the major – plank in the argument for continuing financial support to farmers is the contribution agriculture makes to the environment. (2012, p. 182)

However, this has not transpired. Once again we are reminded that the CAP is a large ship – expensive, complicated to run and difficult to turn around. To us, the authors of this book, it is probable that the agri-environmental dimension of the CAP will continue to grow, but there are uncertainties. For example, climate change and possible harvest disasters may change the focus to food security policies, or increased market fluctuations may result in re-coupling policies (for milk, for example) and continuous urbanisation may reinforce the social dimensions of rural development. In any case, we do not foresee the end of the CAP as an important policy domain in respect to European landscape, but more reforms are clearly needed as the next example illustrates.

Box 6.1 The Example of the Montado

The high-nature-value silvo-pastoral system in southern Portugal illustrates the internal contradictions in the way the CAP functions in relation to the stated objectives in the official discourse (Ribeiro et al. 2014). The Montado covers 1.2 m hectares in Portugal, while the very similar Spanish land-use system, the Dehesa, covers 3 m hectares on the other side of the border. Both were established as production systems adapted to the natural constraints, but are today both highly valued systems from the point of view of farmland biodiversity, landscape heritage, quality production and recreation support. Even though the discourse has changed and Pillar II is receiving increased funding, empirical studies show that the practice of policy implementation remains strongly focussed on intensive farming systems. This is creating tensions in the Montado management accelerating its decay, even though it is considered a highly valued system seen from many different perspectives: cork production, the sustainable use of resources, balance in ecosystem services and the provision of public goods. It has been shown how the construction of CAP payments applied to livestock production in the Montado is creating instability regarding farmers' income and leading to misbalances in the management strategies (Fragoso et al. 2011; Godinho et al. 2016b). From management strategies which were based on a profound understanding of the farm system itself, its capacities and its limitations, the CAP payments have led landowners to management strategies focussed on securing subsidies and best use. Such strategies are oriented on short-term results and often in conflict with the long-term balance of the system. In particular in the Montado they have resulted in a significant increase in livestock intensity, and replacement of sheep by cattle, which is hardly compatible with the maintenance and renewal of the Montado tree cover. Using the national competences to decide upon certain aspects of the distribution of the national shares within the CAP payment structure, the Portuguese administration has kept livestock payments coupled so far, and decreased sheep payments in favour of increased cattle payments. Since the 1990s, cattle production in the Montado has increased significantly, while sheep production has decreased. The impact of this increased grazing intensity on the Montado balance is unavoidable, as too many grazing cattle impede the natural regeneration of trees, damage existing young tree shoots, exhaust the natural pastures and increase soil compaction (Fonseca et al. 2016; Pinto-Correia et al. 2011). Without the tree cover, the Montado would cease to exist, and recovery would be hard to achieve. Indeed, the Montado is presently in severe decline, with 5000 ha of the Montado disappearing annually in the Alentejo region (Godinho et al. 2016a; Godinho et al. 2016b). Assessments made on the whole Alentejo region or on local case studies have shown how the grazing intensity above 0.6 heads per ha is directly related to a significant decline in the Montado land cover (Almeida et al. 2013; Almeida et al. 2016; Guerra, Pinto-Correia and Metzger 2014). Most farm units with extensive grazing under the Montado now exceed this density, as explained previously. This results directly from the policy tools of Pillar I. And with this, the resilience capacity of the Montado is affected, with a decline of the heritage and cultural character, forest stability and cork production, which by all means the farmers also would like to preserve. This is in direct contradiction to the stated objectives of the CAP, i.e. to maintain the sustainable use of the natural resources, but also the goals and existing measures under Pillar II: the plantation of trees to increase the density of Montado tree cover, the preservation of biodiversity rich pastures and the preservation of shrub plots as habitats in Montado areas.

We now change gears, focussing on the environmental dimension of landscape policy.

Environmental Policy

An overall environmental policy domain has developed worldwide and can be seen as part of the sustainability agenda discussed in Section 6.1 (Clapp and Dauverge 2005; O'Riordan and Voisey 1998). Both European nation states and the European Union have been part of this development, and today there are 'environmental' ministries (or specific central environmental policy domains) in all countries, while the European Commission has a directorate for the environment. European environmental policy is organised into a number of action programmes, which in turn are framed by the Environment Action Programme, which contains strategic goals and policy initiatives. Since the late 1980s, a number of environmental directives have been developed, most based on EU-level regulatory measures functioning as frameworks for EU Member States, which in turn must establish national legislation to implement the directives. The most important environmental directives in relation to rural landscapes are presented in Table 6.3.

Most of these directives function as top-down command-and-control regulations. Over time, critical analyses have shown that they suffer from clear implementation problems (Beunen et al. 2009; EEA 2015; Knill and Lenschow 2000), also in relative terms compared to other domains. Both in terms of infringement cases opened by the Commission and Court cases (referrals and proceedings), there appears to be more environmental cases than cases from other policy areas such as the single market, industry or consumer affairs (Jordan and Lenschow 2010).

To claim, however, that EU environmental policies represent a genuine failure with insignificant effects on the environment, including rural landscapes, would be far from the reality. All the directives listed in Table 6.3 are having a clear and intended effect on European rural landscapes and will continue to in the future. The Water Framework Directive and the NATURA 2000 network (Bird Protection and Habitat Directives) are relatively broad, proactive and long term in their perspectives and focus not only on protection, but also on the enhancement and continued management of landscape resources. In recent years, these directives have established connections to the CAP both in respect to the requirement for cross-compliance and incentives especially through agri-environmental policies (Benneth et al. 2006; Keenleyside et al. 2011; Paracchini et al. 2008).

Despite progress being made, integrating the different ends and means in EU environmental directives remains a challenge. Based on a study of Dutch experiences, Beunen and colleagues conclude:

TABLE 6.3 *Examples of EU environmental directives of significance for rural landscapes. Based on information gained from www.ec.europa.eu and www.eea.europa.eu. Note: these are important examples: other EU directives also regulate the rural landscape*

Directive	Year initiated	Main objective(s)
Bird Protection Directive[1]	1979	Protect natural habitats for birdlife. So-called Special Protection Areas are designated by Member States
Environmental Impact Assessment Directive (EIA)	1985	Ensure that any potential environmental impacts of plans, programmes and projects are assessed
Strategic Environmental Assessment (SEA)	2001	
Nitrate Directive	1991	Reduce nitrate pollution from agriculture. Regulate storage and use of fertiliser and manure for all land or within discrete nitrate vulnerable zones.
Habitat Directive[1]	1992	Protect natural habitats and the wild flora and fauna. So-called Special Areas of Conservation are designated by Member States
Water Framework Directive	2000	Classification and monitoring of all water sources including inland (surface and groundwater) and coastal waters
Soil Framework Directive[2]	(2006)	Not yet initiated. Ensure sustainable use of soils and protection of their functions in a comprehensive manner.[2]

[1] *Designated areas are part of the so-called NATURA 2000 network*

[2] *The Commission withdrew the proposed directive in 2014 because the Council would not accept it. However, the 7th Environmental Action Programme approved in 2014 stipulates that by 2020 soils must be adequately protected and remediation of contaminated sites must have been initiated.*

Although integration is aimed for at the European Union level, the current practices in The Netherlands show that this can be difficult. Due to the strong focus on formal compliance and limited possibilities for discretion, it is difficult for the involved actors to link the multiple objectives from the different EU directives with each other and with their own objectives. This is even more complicated because the implementation of different directives is done by different actors and often follows a sectoral approach.

…

> In many areas in The Netherlands an integrative approach has become the current practice. Water boards, for example, already successfully adopted such an integrative approach years ago. However, the implementation of the European Directives might frustrate such an integrative approach, because too much focus is put on formal compliance, and separate goals for surface water, groundwater status and protected areas have to be set by different authorities. A rigid and static interpretation of the Birds and Habitats Directives for example conflicts with the management of highly dynamic ecosystems. (Beunen et al. 2009, pp. 65–66)

Although such problems with integrating objectives and the development of new, more governance-oriented policy have been recognised for some years (Holzinger, Knill and Schäfer 2006), it has been difficult to resolve them. Without doubt there is a long way to go before EU environmental policy becomes well integrated across political-administrative levels and horizontally across policy sectors.

Landscape and Food

Against the background of global concerns over food security, as well as concerns regarding the sustainability of the globalised food chain, there has been renewed interest in recent years on the localisation of food chains (Hinrichs 2014). The rural development pillar of the CAP 2014–2020 emphasises the 'development of short supply chains and local markets' as a priority. A local market is defined as a 'supply chain involving a limited number of economic operators, committed to co-operation, local economic development, and close geographical and social relations between producers, processors and consumers'. Supporting short food chains involves supporting local activities that may improve the sustainability of the holding and local processing and marketing businesses (OECD/FAO/UNCDF 2016). Short supply chains and the activation of local markets create new economic opportunities for farmers who cannot compete on the global markets, but they also create job opportunities, new markets based on local products (McManus et al. 2012) and social benefits as a result of new opportunities for connections between consumers and producers. Finally, reducing food miles also reduces emissions, which supports the achievement of European objectives regarding combating climate change (Hinrichs 2014). Short food chains also help maintain peri-urban agriculture. Besides the EU level, several national policies have emerged recently, the aim of which is to support these developments. For example, in 2014, the French government adopted a law that supports short food supply chains, which reinforce the connection between local producers and local consumers (LegiFrance 2014).

From the landscape perspective, these new trends regarding small-scale farming connected to local markets and short supply chains are creating a new dynamic in the countryside and leading to the maintenance and renewal of small-scale mosaic farming, mainly in peri-urban locations. Public interventions have not been the drivers for this development. The new dynamics in small-scale agricultural production, urban farming, short supply chains and local markets, as well as community-supported agriculture, have emerged as bottom-up processes, especially in Southern Europe and mostly after the global financial crisis after 2008. Public policies and other forms of public intervention, such as market, sanitary and tax regulations, are crucial in creating or not the conditions that can support these processes and reinforce local initiatives (Loeber et al. 2011).

6.4 The Need for Policy Integration and New Forms of Landscape Governance

Coming to the conclusion of this chapter, we return to the challenge of integration and of managing conflict and making places. As has been shown, spatial planning, agricultural policy and environmental policy are poorly integrated within the different political-administrative levels. Neither is coherence across the levels – from the EU to the municipal level – functioning well. It is also clear the place-making dimension of planning is poorly developed for rural areas. However, as we will outline in this final section, promising examples point towards pathways for landscape governance.

Overall the coordination and integration of European landscape policies at all levels represents a major challenge, as does the inclusion of environmental and social concerns in economic decisions and ensuring that policies work together within the spatial scale in question. Comprehensive spatial planning has traditionally played a role in coordinating sectors and comprehensive landscape plans on a regional scale have played this role for many years in Germany (Haaren 2004). At the national and EU levels, a network strategy such as the NATURA 2000 network and the Water Framework Directive, together with cross-compliance measures, will ensure some degree of policy integration and will in effect deal with the issue of land sharing and land sparing discussed in Chapter 1.7.

At the local landscape level, the policies will be meeting in any case, no matter how well they may or may not be integrated. The European Landscape Convention has a clear focus on this and the promotion of Landscape Character Assessments is clearly part of the solution towards better integration. Another pathway towards policy integration, which is also dealt with in the ELC, is to develop new ways of landscape governance.

At the regional and local levels throughout Europe, public policy and planning agencies – together with stakeholders of many kinds – are currently gaining experience with the integration of rural landscape governance (Breman 2015) either as part of the rural development programme of the CAP, or as part of landscape-planning activities linked to spatial planning systems. To our knowledge, no systematic overview of such activities exists. In the following, we briefly refer to two, quite different approaches to collaborative and integrating landscape policy.

We find the development of Dutch environmental cooperatives – now often named 'territorial cooperatives' – a useful source of inspiration for what could be termed 'rural landscape governance'. Over the years, more than 100 so-called environmental cooperatives' (ECOs) have evolved in the Netherlands. They vary in size and organisational setup, but fundamentally they are all associations of farmers and non-farmers concerned about agri-environmental issues linked to public policy, first of EU regulations (Renting and Ploeg 2001). These ECOs are organised and run by farmers at the grassroots level and, over time, some of them have grown to cover large territories with a high number of members, a secretariat with staff and a broad range of landscape policy issues on the agenda from agri-environmental schemes to spatial planning issues such as housing, tourism development and new recreational paths and roads. An example is the Northern Friesian Woodlands (NFW) territorial cooperative, which has more than 1000 members (850 farmers) managing 50,000 ha of farmland and nature conservation land (Swagemaker and Wiskerke 2011). For several years, the NFW has also been involved in designing tailor-made agri-environmental programmes for its region through experimental EU programmes. Thus, the NFW was substantially involved in designing landscape policies for its own area as well as being the institution which negotiates with individual farmers and draws up contracts which guide farmers' management. This has been termed the 'front-back door principle' as shown in Figure 6.4. Overall, the ECOs have regained political power concerning public policy and may have fundamentally regained collective autonomy over the rural landscape (Milone and Ventura 2015). Together with other ECOs, the NFW is currently involved in a much larger and more formalised programme within the context of the EU Rural Development Programme 2014–2020.

Danish experiences with 'rural landscape strategy making' represent an approach which to date has been developed through a number of experimental projects on the regional and local scales (Kristensen and Primdahl 2015, Primdahl et al. 2013, 2016). See Case H for an example of a landscape strategy made for a rural parish.

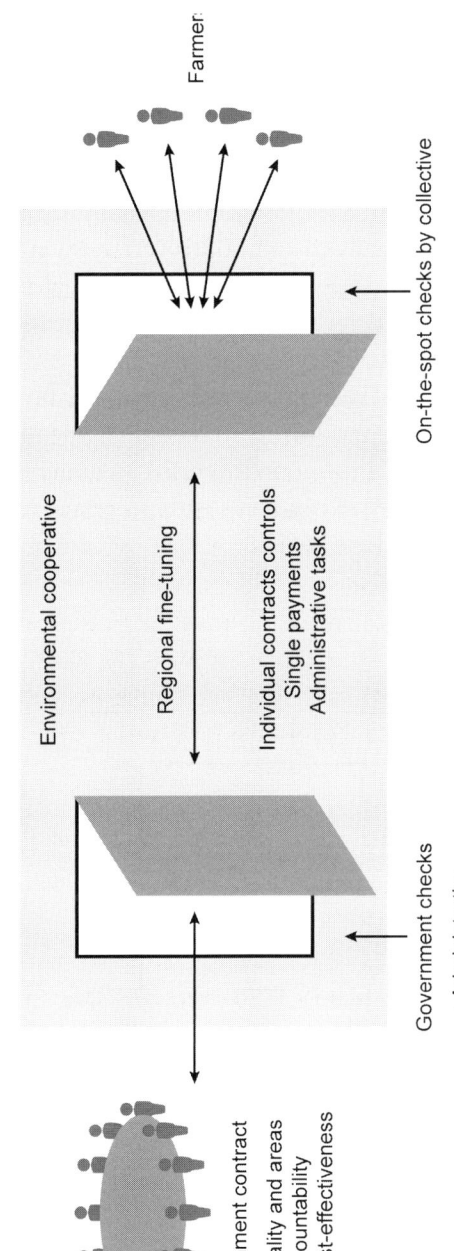

FIGURE 6.4
The Dutch Environmental Cooperative and the front door-back door approach (adapted from Milone and Ventura 2015, p. 73)

Inspired by spatial strategy making for cities and urban regions (Albrechts 2006; Healey 2009), a framework for strategy making for rural landscapes has been developed (Figure 6.5). Through inclusive processes usually involving the municipal administration (planning, nature conservation, and agri-environmental sections), individual farmers and local community groups develop a strategy with shared municipal and community ownership. Through lectures, excursions, seminars and workshops: (1) a common interest for and understanding of the landscape in question is established; (2) a vision and concrete objectives are agreed upon; (3) knowledge and ideas (from 'internal' as well as external resource persons) are mobilised and; (4) an overall strategy is produced. A typical strategy consists of a vision and objectives, a spatial framework for the future landscape and a number of so-called strategic projects that are to be implemented.

In the experiments carried out so far, the strategy-making process from the first meeting to a landscape strategy document has taken one to two years. In particular, the phase of creating a common interest and understanding takes time, and it is probably unrealistic to conduct such a process in less than a year in an area where it has not been done before.

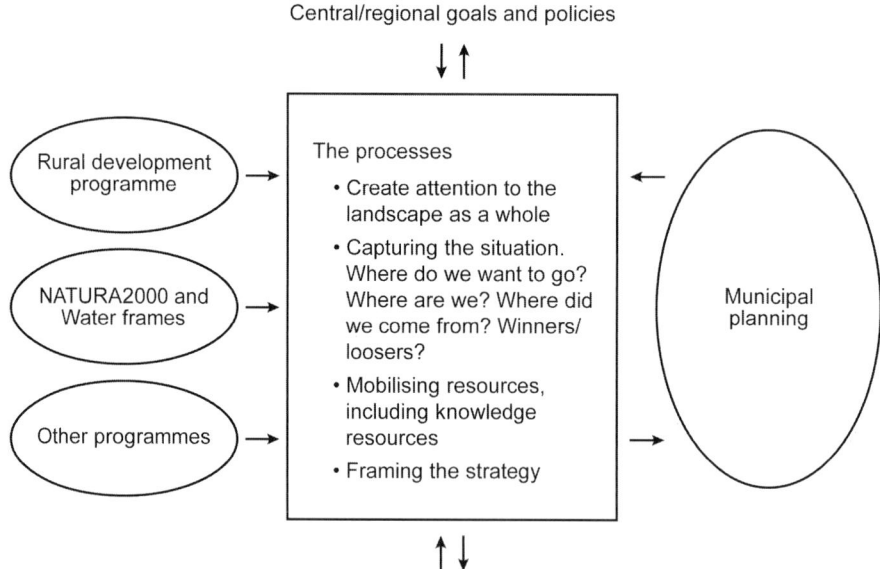

FIGURE 6.5
Dimensions and institutional context of rural landscape strategy making (adapted from Kristensen and Primdahl (2015, p. 16)

CASE H Landscape Strategy for Karby Parish, Northern Jutland, Denmark

Landscape Appearance

For thousands of years, the Karby parish in northern Jutland, Denmark, was farmed in a mixed farming system, including the extensively grazed salt marshes (Figure 6.H.1). The village is situated along the edges of the moraine plateau with fertile and intensively farmed arable land on the land side and large salt marsh areas on the coastal side. During most of the twentieth century, the population declined from more than 1000 to about 500, and most socio-economic services and businesses disappeared, including the school, the ferry connection and a number of small factories and shops.

Landscape Processes

During the twentieth century, agriculture became industrialised and concentrated, the salt marshes lost their agricultural function and the village declined in terms of social life and physical-aesthetic quality. A village renewal plan was implemented and a community-based landscape strategy was developed, defining a vision for Karby as a well-functioning village in a sustainable agri-environment. The strategy was a success and the local government decided that this kind of process should be the paradigm for countryside planning in the municipality.

Figure 6.H.1
Aerial photo of the parish taken in 2012 – today all the salt marshes to the right of the road have been fenced for extensive grazing

238 EUROPEAN LANDSCAPES IN TRANSITION

CASE H (*Cont.*)

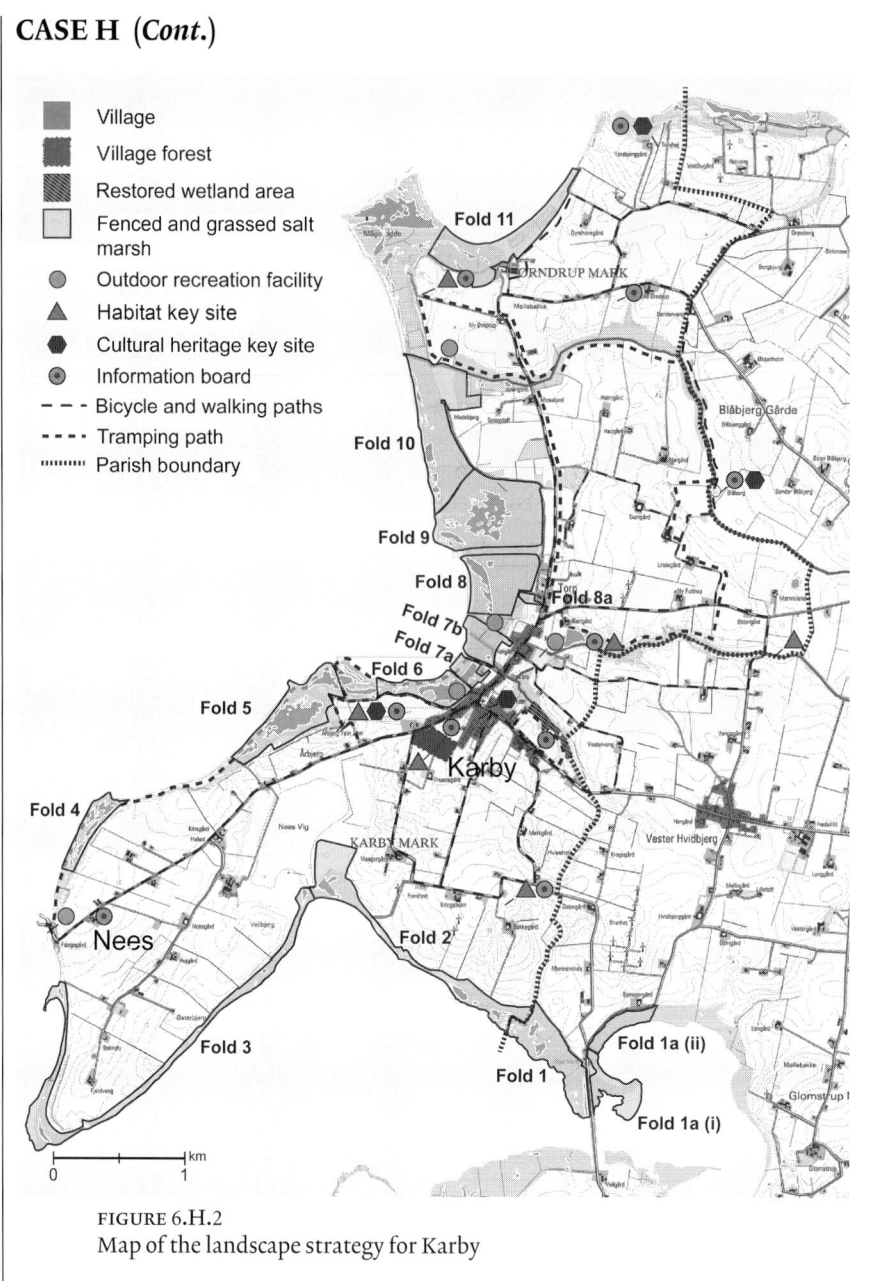

FIGURE 6.H.2
Map of the landscape strategy for Karby

> **CASE H** (*Cont.*)
>
> **Conclusion**
>
> Karby parish has benefitted from the landscape strategy-making process. Farmers and other citizens have gained positive experiences in working together in formulating common goals and initiating projects in cooperation with the municipal administration. The management of the salt marsh has been greatly improved through common fencing and grazing agreements and new walking trails, a bird-watching tower, wetland restoration within the village and other initiatives have enhanced Karby parish as a place to live and to visit.

In the next chapter, we summarise the main points made throughout this book, continue the discussion on policy and planning approaches to guide current change processes and conclude by presenting an example of a landscape strategy process for an imaginary European landscape.

7

Common Grounds for Colourful Futures

Layered Landscape
The sea is smooth. On the railing of the shipwreck, a cormorant is resting, its wings spread like a fallen black angel. A seal surfaces, its inquisitive snout emerging two or three times to watch me kindly, and then quietly disappear again. I stand on a huge sea dike, covered with asphalt and blocks of basalt. The sea can be mercilessly fierce here. Today cyclists and hikers are enjoying the pleasant autumn weather. Birdwatchers with telescopes are observing a multitude of different bird species migrating from their northerly breeding grounds to more southern ranges for wintering.

The grass on the land side of the dike has been neatly clipped by sheep and in the background, stretching into the distance, the wide Dutch polder land unfolds. I descend down to where a mighty sea arm once flowed far into the land. A row of rotating wind turbines, brilliant white in the October sun, draws my gaze to the horizon, to the cloudy sky, full of nuances of grey, white and pale blue, like in old Dutch paintings. Just in front, a small polder mill sits, its sails turning in their own traditional rhythm the other way around compared to the turbines. To the left and right are the pyramids of the red-tiled roofs of old farms, a single low treeline, some cows in the meadows, sheep as well. Closer, a band of dark yellow reeds waves above the steely blue of the broad pond. There are ducks, and a large flock of geese is settling down. Just in front of me, a wet meadow, colouring purple, where lapwings and a late redshank are foraging.

I find myself in a painting of a multi-coloured, layered landscape. Images of an eventful past stretch to represent this strangely peaceful ensemble of old and modern windmills. The longer I look, the more details I discern. Still, this painting reflects a landscape that lives, where the cattle slowly move across the meadow, and where birds find a welcome roosting place. As I stand, a cormorant lands on the pond.

The painting, that is us.

FIGURE 7.1
De Putten, landscape reserve behind the sea dike in North-Holland
(photo: Bas Pedroli)

This landscape image may stand for the common ground that we need to find for a colourful, multifaceted future of the European landscape. But it may also emphasise that landscape is a dynamic phenomenon that has developed through the centuries involving many different drivers and actors. Landscapes today are continuing to develop at an even faster pace, and with the involvement of a continuously changing community of actors. In this concluding chapter, we suggest pathways best able to cope with this uncertain future.

7.1 European Landscapes in Transition – Where Are We Headed?

In the previous chapters, we have shown how European rural landscapes are currently being affected by multiple processes of transition, in ways not seen before. These processes are partly based on factors which we have been aware of for long times, but also on novel and unforeseen factors. This has resulted in tensions and conflicts that demand new approaches for a comprehensive understanding of landscape and, particularly, for its management (Chapter 1). We have outlined the concepts that need to be mobilised in order to understand these changing processes in their multiple dimensions and beyond sophisticated assessments of the changing patterns or changing functions alone (Chapter 2). Subsequently, we introduced the conceptual frameworks that originate in different disciplinary constructs, which can help us grasp *what* is changing in the landscape, and *how* and *why*. These frameworks support the understanding of the multiple dimensions that need to be combined in the analytical approaches to study the ongoing and upcoming landscape transition (Chapter 3). Furthermore, we have presented how agriculture has evolved through time and how it has largely shaped the European rural landscapes to the present, but also how agriculture, in some parts of Europe, has ceased to be a driver of landscape use, or has ceased to be a driver of action in the landscape (Chapter 4). Subsequently, we discussed how rural and urban dynamics have interacted over time and how new rural–urban relationships are affecting current landscape changes (Chapter 5). Finally, we have discussed public policy and planning functions and how such interventions are organised in sectors and into different political-administrative levels with poor horizontal and vertical integration. Successes and failures, conflicts and opportunities, have been identified and discussed, and a framework for conflict management and place making has been outlined (Chapter 6).

For centuries, the rural has been the space which has provided natural resources and food, both for the people living in the rural itself and for those living in cities. This has evolved until today, when the food produced is mainly for urban areas, as the majority of the population now lives in cities (Mormont 1990; Woods 2011). As the producer of food and fibre, the rural has historically mainly functioned as the territorial basis for agriculture and forestry. Those who lived and worked in the rural were, until recently, directly or indirectly involved in these activities: there were close bonds between the land, the community and the activities of that community. In this way, management of the rural landscapes was a matter of practices intimately linked with agricultural and forestry production. This fundamentally changed during the second half of the twentieth century as environmental problems emerged on the public agenda, the scarcity of resources began to be understood and new uses of the rural space emerged. The differentiation of the rural has grown on a regional scale, reflecting different combinations of production, consumption and protection drivers, their organisation and hierarchy. This was also reflected in different categories of actors involved and affected on multiple governance scales (Robinson 2008).

New and uneven transition processes took place in the rural landscape, especially from the 1950s onwards. Reflecting on the five inherent landscape tensions we described at the start of this book reveals the following challenges we are currently facing in the European rural landscapes. (1) Landscape has no clear boundaries, but management arrangements depend on well-defined ones. How can specific landscape territories be guided by visions and aims which are open, in respect to both a dynamic surrounding world and to new visions? (2) Many policies have significant impacts on the landscape, but landscape competence itself is not a formal policy domain. What are the conditions for promoting a landscape approach within the public domain? (3) The demands for the rural landscape are increasingly multifunctional, but the landscape is still often structured as if agriculture were the only function. How can such different demands for rural land-use and landscape activities be combined in well-functioning rural landscapes? (4) Landscape is a common good, but it is managed individually. How can collaboration and collective actions be (re)developed as a landscape factor, and on which scale can this be done?

(5) Landscape heritage and biodiversity should be conserved, but a living landscape is a developing landscape. How can conservation and developments perspectives be brought together in landscape governance?

The recent transition processes have taken place without addressing these questions. The historical dominance of the farming and forestry sectors is still to a large extent defining the use of the land and natural resources and setting the agenda for the way land use is discussed in local communities. The causes of this are diverse, but there is no doubt that the long history of agricultural dominance in the rural landscape as well as the long-established and efficient organisation of the sector plays a significant role here (Bishop and Phillips 2004). New fields of public policy have emerged, such as environmental and rural development policies, but, as we saw in Chapter 6, this has often happened without reflection on or assessment of what should be kept under the realm of existing policy fields and what should be integrated in or coordinated with the new ones. There are many examples of effective measures and well-chosen landscape changes, but also cases of unintended and contested effects, conflicting drivers, loss of landscape values and tensions among different actors at play. So far, public intervention has generally been uncoordinated and reactive and has been driven by a response mechanism, often delayed in relation to the processes of change. More visionary and proactive interventions that pursue desirable societal goals have been rare. When they have been formulated, these strategic policies have often been weak in relation to market mechanisms and individual options (Ostrom 2007).

As we saw in Chapter 5, urbanisation appears in different forms – including traditional outmigration of rural areas, immigration from urban areas into rural areas and tourist development – all of which affects most European landscapes. In some areas, tensions from multiple uses are increasing, while maintaining the quality and character of the landscape is a constant challenge. When they can, people are increasingly considering landscape quality and attractiveness when choosing a place to live, but the increasing residential value of the rural space and the inherent changes in land use have led to changes in the structure of landscape and, thus, its character, which in turn has threatened the basis of this attractiveness. The effect of each individual action on the landscape is hardly considered and, thus, the values attached to the landscape may be lost without a joint awareness – and consequent action – being created.

On the other hand, populations are shrinking in remote and unattractive landscapes. This trend is likely to continue, leading to landscapes that will be either abandoned or managed through distant intervention. Many such remotely located landscapes will therefore, be prone to degradation, wildfires and erosion although they may at the same time support valued resources such as wildlife, water and air quality and heritage elements.

More globally, the demand for land everywhere in the world is set to rise in the coming decades given a global population that is increasing in number and wealth. Consumption of more food and energy, and a search for cleaner water, better-quality soils and protection from environmental hazards will receive greater attention as driving forces of landscape change. Migrations caused by military conflicts, climate change or simple poverty are set to increase the pressure on these resources, particularly in Europe. At the same time, society is demanding responses to climate change, conservation of cultural heritage and biodiversity and the development of a living countryside.

Given all the changes we have described in this book, what is the vision for the future of the rural landscape? Many of the landscapes we cherish today will most probably be subject to radical change or may even disappear. Not just what we call heritage landscapes will be difficult to preserve, but also everyday landscapes as we know them today will change. Meanwhile new landscapes are being created which are affected by drivers unknown until a few decades ago. Nevertheless, nowadays, communities on different scales rarely discuss the future of the rural landscape, or what is acceptable or desirable, what is sustainable or unsustainable. For the urban centres, the urban structure, the network of centres of different dimensions and different roles and future visions are often on the agenda, in the media and in the public and academic debate. But the same is not the case for the rural. This may be because the common image of the rural remains an image of the past as discussed in Chapter 5, or because it is changing in so many different directions that those concerned do not know how to tackle all changes, or maybe both factors interact. The reality is that communities and other users of the rural often remain spectators of a rural landscape that is changing extremely fast, while there are no governance mechanisms to address the associated problems and uncertainty.

7.2 Looking Forward – the Desired Future

These trends and tensions frequently imply potential conflicts and are, thus, often interpreted as threats, which they also are when seen from specific perspectives. However, it is more constructive to talk about the potential for innovation, entrepreneurship, new solutions and new narratives for a desired future. We are still learning about the processes of change. Predicting the future is a tricky affair because there are always unforeseeable changes and events causing tipping points and discontinuities in landscape development. With such change trajectories, we cannot rely on simply building on the experiences of the past. To achieve change in a positive and consciously chosen direction requires future-oriented thinking that goes beyond references to past experiences. This calls for a debate on which future developments are in focus and a clear vision of the desired goal shared by a large proportion of the members of the organisation or community, in this case those involved in each rural landscape (Costanza and Kubiszewski 2014).

The traditional strategies to cope with the unpredictability of the factors that influence landscape transitions are two: (1) react as adequately as possible on the external system developments (adaptation), building on proven methods to reduce negative effects and make optimal use of the opportunities offered with the 'optimum' usually defined from a narrow perspective; (2) develop policies which influence the external system developments, minimising negative influences (mitigation); this includes questioning the proven methods of (1) and opening a debate on cause–effect relationships.

There will be a need for more sophisticated strategies. If we address our twenty-first-century challenges with such reactive mind-sets that mostly reflect the realities of the nineteenth and twentieth centuries, we risk increasing frustration, cynicism and anger (Scharmer 2009). A clear boundary exists in these realities within the science policy interface: scientists tend to refer to the objectivity of their work to persuade policy makers to believe them, while policy makers legitimise their decisions by referring to democratic obligation to balance interests and values, which is something scientists are not supposed to do (Huitema and Turnhout 2009, p. 579). Based on the body of knowledge available from an analysis of the mentioned reactions (see e.g. Fuchs et al. 2014), it will be possible to identify ways forward to cross this boundary in a collaborative effort (Van Paassen et al. 2011b). Whereas traditional road-mapping methods are largely based on extrapolation, describing a sequence of measures to solve a specific problem (McDowall and Eames 2006), multiple – potentially conflicting – visions for a desirable future should be identified and analysed as a way to launch an awareness-raising and evolving debate, selecting goals and identifying the related pathways. There are already examples of such visions in multiple frameworks, e.g., in the VOLANTE project (Pedroli et al. 2015) and the ET2050 project of ESPON (Ulied 2014) on the European level, and in many examples at the local and regional scale (see e.g. Hirschi et al. 2013; Nainggolan et al. 2013). This will facilitate new creative approaches which cope with the expected emerging landscape management problems.

Developing visions is also about assessing opportunities and threats, and going further than avoiding the most negative impacts. It is complementary to more traditional forms of planning and serves as a way to determine shared desires and initiate a process towards change. It is a shared process, which means it needs to be inclusive and integrative and engage a range of actors at different governance scales. It is an ongoing process, which means that solutions can be identified, but also progressively adapted. Furthermore, it includes the identification of pathways for the vision: how to get from here to there. In a world of so much uncertainty and with so many new drivers at play, pathways cannot be defined at the start; they have to be continuously redefined and, therefore, the envisioning process needs to be continuous and adaptive. In this way, visions will comply with existing regulations and the sustainable use of resources, but will also facilitate creativity and new pathways to the future.

To cope with this need for new approaches, Sayer and colleagues (2013) propose 10 principles for a 'landscape approach'. The 10 principles include dimensions of learning, consensus building, raising awareness of multiple scales intersecting, multifunctionality and the associated trade-offs, engagement of multiple stakeholders, negotiated visions, clarifications of rights and duties, participatory and user-friendly monitoring, increased resilience by recognition of threats and vulnerabilities and, finally, strengthened stakeholder capacity (Sayer et al. 2013, pp. 8351–8352). These principles have been discussed and subsequently accepted by the Convention of Biological Diversity (CBD) and they do in our view represent a way forward in sustainable landscape governance. Guidance on how these principles are brought together in real-life governance processes should still be elaborated. We propose the landscape strategy making approach outlined in Chapter 6 and discussed at the end of this chapter could be used as the institutional framework to bring these principles at work in practice, in relation to both research and policy.

7.3 The Missing link: A Forward-Looking Rural Voice

As has been shown in this book, there are multiple interests in the rural landscapes of Europe, multiple expectations, multiple actors dealing with the rural space and the rural activities. However, when attempting to look ahead to the future and project the rural landscape for coming times, while acknowledging the particularities of each European rural landscape, it becomes apparent that a link is missing.

This missing link can be observed on at least three separate and highly relevant levels: the academic world, public policy intervention and societal engagement.

The Academic World

In relation to academia, what we observe today is that a large group of academics and professionals is dealing with the rural landscape, and this group increases even further in size when peri-urban landscapes are included. However, the vast majority of these academics focus on narrow aspects of this complex system and no one – or almost no one – is dealing with the whole picture. Furthermore, most academics working with rural landscapes are concerned with analysing and interpreting and not with the production of integrated solutions.

Until a few decades ago, agronomic journals addressed the improvement of the agricultural landscape, mainly from an agricultural production point of view, but they also discussed issues such as housing and gardening. This was still a time when the shaping of the rural landscapes was seen as mostly dependent on agriculture and agricultural expansion, regarding both land use and intensity. Agronomists, foresters, land surveyors, geographers, civil engineers etc. worked on this, while being sensitive to the territorial dimension of the land-use systems as part of their professions. However, since the 1980s, agricultural policy has been increasingly reduced to market policy formulated at the EU level, and such future-oriented approaches to agricultural landscapes have disappeared. The agricultural development discourse has, thus, progressively been reduced to questions of technological innovations and economic competiveness on the farm, field and livestock unit levels or in a single-farm business context. As a consequence of this professional discourse on the future, the rural landscape as such is rarely an issue for discussion in agricultural research.

From other sides, e.g., the growing environmental, socio-ecological and spatial planning research, landscape approaches have gained prominence in the search for solutions to reconcile conservation and development trade-offs. Through the insights from landscape ecology, early conservation theory promoted landscape-scale thinking (Burel and Baudry 1999; Forman and Godron 1986). Social processes and the people involved in the decisions were initially kept out of these analyses. Acknowledging this limitation, further development has come from the recognition of the need to address the priorities of people who live and work within and, ultimately, shape these landscapes. However, these priorities are often not aligned, and the proper integration of methods from social sciences into the landscape sciences to cope with this problem has thus far been limited. The trend towards the focus on ecosystem goods and services, strongly anchored in neoclassical economic principles, has gained momentum in recent years. By focusing on single services and benefits, it has even further led to a detachment from the place-based landscape perspective and it still does not solve the value plurality that underlies the different positions of the various societal groups (Bredin et al. 2015; Martín-López et al. 2014).

Public Policy Interventions

In relation to public policy and landscape planning, a similar division between sectors as the one between disciplines or disciplinary perspectives can be observed. These sectors have been described in Chapter 6.

Although attempts to achieve greater integration have been put forward and a sustainability agenda does occur in reality when it comes to the factual policy measures and their goals and instruments, the sectors remain segregated. Even within the same sector, specialisation and fragmentation are ongoing, as is clearly seen in the Common Agricultural Policy, where agricultural and rural development are increasingly being decoupled.

Spatial planning and strategy making represent a way to integrate the different activities and policies at play, as has clearly been demonstrated in recent urban renewal projects. Experiences from here can be used in rural landscape governance, but only to some extent. Patterns of consumption, public ownership of land, public discourses about place-making and conflict management are clearly very different in cities and in rural areas, and new governance approaches must, to some extent, emerge from within rather than being transferred from urban policy and planning experiences (Primdahl et al. 2013a).

When tensions and conflicts emerge, the single decision-maker at the local level, who is often the farmer when we are dealing with rural landscapes, has first to identify them and then to solve them. An integrated, cross-sectoral, place-based approach has been aimed at in different singular cases, but as a generalised approach it has still not been established. It would require a profound change in the policy process, which is not there yet. But setting up a locally based, landscape anchored approach to which different sectoral policies would converge is a fundamental condition for a novel public policy construction.

The Societal Engagement

This is perhaps where the most relevant link is missing: in the societal engagement in the rural, in what is going on and what will be the future for the rural. The discussion on futures is centred on the urban. This can be explained by the increasing degree of urbanisation in society. But it results in a wicked strategic perspective, as a large majority of this urban population relies on the rural as a source of food, natural vital resources, leisure amenities and identity bonds. Different social spaces meet and are superimposed on the same rural space, and often do not interact, or only to a limited extent (Ilbery 1998; Woods 2011). A nostalgic reference to a past rural does not help in shaping present-day relations or future-looking communities. At the same time, fewer and fewer people are involved in the production sector, farmers have become disconnected from rural services and progressively linked to global organisations and the stronger voices in the sector are no longer individual producers living in the rural, but investment companies, multinational concerns and agri-food cooperatives. The dominant voice of the rural, which was the agricultural sector, is no longer a voice of the rural, but a sectoral value-chain voice.

As suggested by McManus and colleagues (2012), the interactional community of place can be a conceptual anchoring. It is grounded on the principles of a community of place as suggested by Wilkinson (1991): a locality (geographic setting of the community defined by interactions and constructed meanings), a local society (people and organisations in the population, be it homogeneous or heterogeneous) and a community (social fields or processes of social interaction surrounding the organisation of a community). There is a strong need for a rural reappraisal if we want to continue to build upon the many assets of European landscape diversity. The rural community should be able to organise itself responsibly and address the deep-rooted conflicts over rights and resources in today's globalised world, and discuss visions for possible solutions which are tailored to each specific context. Only such a joint effort will enhance resilience by making it possible to cope with changing circumstances by defining visions, dealing with rights and constraints and resources and finding adaptive solutions to each different place. This would allow the missing voice of the rural to be acknowledged and respected by all parties involved, legitimately and equitably, and future oriented.

7.4 Local Futures – Towards a New Paradigm for Landscape Governance

An Imaginary Landscape Experiment

Imagine a local, agricultural landscape somewhere in Central Europe. Picture a mosaic landscape with a heterogeneous pattern of land use and vegetation cover, with a relatively small number of large, commercial farms and a larger number of small, part-time and hobby farmers, and with an even larger number of other residents living in the village and in houses scattered around the landscape. Imagine that the residents over the years have successfully worked together with the local municipality concerning the renewal of the village – improving the lighting and the village square and keeping the sports club alive. Even a local marketplace with facilities to store and sell local food products has been established by the residents in cooperation with municipal offices. The village community has developed through such initiatives into a lively society, and now a group of people is developing more ambitions for safeguarding the balanced future development of the community based on its cherished landscape. The group agrees with the municipality that a strategy for the overall future of the rural landscape should be drawn up by local groups of interested citizens (including farmers and other landowners) in cooperation with the relevant public offices.

The first phase of the process is to *create interest* in the rural landscape and its future, broadly among the local residents and other stakeholders and, equally important, among the different administrations within the municipality. Various initiatives are taken. A series of lectures is organised for winter evenings with experts from inside as well as outside the area, presenting perspectives on the history, the present state and the future of the landscape. Also an excursion is arranged with the participation of local resource persons, municipal officers and external experts. The themes of lectures and excursions include: (a) the natural history and ecological status of the landscape; (b) cultural heritage; (c) agricultural history and current production and; (d) the landscape character and recreational assets and needs.

The second type of activity runs partly parallel to the first and concerns discussions of *visions and concrete objectives* for the future landscape. Three questions are addressed in these discussions: Where do we want the landscape to be in 10 or 20 years? What is the current status of the landscape? And where did it come from, what is its historic background? Goals and objectives are compared to the current state of the landscape and gradually a common understanding – or more fundamentally – a common language evolves. Local and even individual histories are brought together with more comprehensive narratives on the landscape history – the old history as well as the more recent. Maybe the group and the municipality even manage to initiate open discussions about potential winners and losers in respect to the different envisaged landscape futures.

A third dimension of the process concerns the *mobilisation of ideas, knowledge and concrete resources*, part of which has already been dealt with in the first two phases. This receives more focus through more or less structured workshops. External experts and members of local groups from other landscapes where experience has been gained in comparable processes are invited to take part in the procedure and provide guidance and talk about relevant solutions. Possible resources available to be mobilised are identified and discussed. A small stream which was tubed decades ago may be opened up, thereby contributing to wildlife and scenery, large stable roofs may be used for solar panels by a cooperative of village residents, empty school buildings may be renovated and used for backpackers' accommodation and abandoned grasslands may be re-fenced and grazing reintroduced, this time managed by a small, local 'grazing association'. Such projects represent a mobilisation of resources which may contribute to making the area a better place to live (for both people and wildlife) and may also add to the local economy.

Finally, a *strategy* including visions, objectives, pathways, a spatial outline for the landscape future and specific projects is framed and formally agreed upon by the municipality and the local community.

The Experiment Interpreted

This example represents a relatively straightforward imaginary case of a spatial strategy as developed by Healy (2009) and discussed in Chapter 6. Such a combined spatial and strategic approach represents in our view a way forward in bringing the 10 principles for a 'landscape approach' (Sayer et al. 2013; see Section 7.3) together in the specific context.

Although our example refers to a Central European rural landscape, similar processes could take place in other European landscapes, differentiated according to specific context and culture. Imagine now that such a strategy together with other, similar examples of novel approaches to new forms of landscape governance and new examples of landscape patterns are analysed and discussed among landscape researchers and landscape professionals. Further, that the biophysical patterns of future landscapes become something that academics and professionals are actively occupied with – as urban researchers and landscape planners are discussing the future of urban landscapes. If such a culture of discussing needs and solutions to future landscape patterns developed, it would become much more likely that landscape research would be more closely linked to landscape policy and planning and more generally that the rural landscape as a public issue would be high on the environmental policy agenda.

As we are closing this book, let us continue imagining. Let us assume that the European community discussed here gets in direct contact with other rural communities in other parts of Europe and provides and receives inspiration concerning landscape governance and rural place-making. Such meetings could be organised within networks associated with the European Landscape Convention (ELC). Perhaps new forms of tourism or urban–rural interaction will evolve from this, and more fundamentally, new forms of local autonomy concerning the rural landscape, perhaps even a 'true landscape democracy' as it was phrased in the preamble to the ELC.

On the Road to Landscape Democracy

A social practice that includes: (1) traditional democratic processes concerning local cooperation on land use issues; (2) conflict management concerning rights and duties, and; (3) protecting, managing and changing the area as a living and visiting place may be termed 'landscape democracy' with reference to Arler (2006). The three key agents in such a landscape democracy are illustrated in Figure 7.1. As mentioned in Chapter 1, Arler distinguishes between the acts of the individual landowner and user (self-determination) and the acts of the collective in various forms (co-determination). These two different kinds of act may be seen as part of a local democracy in general. In some areas, well-functioning traditions of such cooperation, which are driven by civil society in general, have been maintained and are still very much alive or can be reinvented with the involvement of new actors, while in other areas, such traditions have vanished or have reduced to a few issues such as maintaining the common village green, the local pond etc., and here more effort is required to regain the sense of belonging among the residents.

The management of recreational access rights, mitigating environmental impact of agriculture, regulating new construction plans and similar types of practices are usually a matter for the individual landowner and user, on the one side, and the public authority, on the other. The former represent the pro-active agent while the latter act on behalf of the public good, the next generation and the regional interest. This dimension of landscape democracy is dynamic – rules are changing and so is the concrete management of such rules.

Finally, there is the cooperation between the civil society in various forms from the village association to the local nature conservation group and local entrepreneurs, on the one side, and public agencies from municipal spatial planers to regional or central nature conservation authorities, on the other. Conditions for such cooperative practice are, of course, very different from country to country, and we see no point in proposing one single model. The Dutch model of territorial cooperatives described in Chapter 6 is appealing and could probably be applied in many other regions and countries, although the competence distribution between public agency, local community and the individual landscape manager is likely to vary considerably between countries and regions. This is due to differences in political-administrative traditions, but also because different landscape contexts call for different approaches to conflict management and place-making. This also means that the different countries and regional landscape contexts will be located in different parts of the triangle in Figure 7.2. In any case the governance model that evolves from these cooperatives could be based on combinations of the back door–front door

FIGURE 7.2
Key agents in landscape democracy (inspired by Arler 2008 and moderated from Stahlschmidt et. al. 2017, p.178).

approach practised by the Dutch environmental cooperatives and a landscape strategy-making approach as presented in Chapter 6 and in the example cited here. Landscape democracy demands that the decisions of planning processes are credible (scientific adequacy of the technical evidence and arguments), salient (relevant to the needs of the decision-maker) and equitable (respectful of stakeholders' diverging values and beliefs, and fair in the treatment of opposing views and interests) (see e.g. Egoz, Makhzoumi and Pungetti 2011; Van Paassen et al. 2011a, pp. 31–32).

7.5 Conclusion

Due to the pressing need to integrate sectoral policies and to involve stakeholders more directly in conflict management and rural place-making, new forms of landscape governance as those outlined here are likely to be increasingly affecting European landscapes. No doubt this will happen through a wide range of new methodologies and such developments will challenge both landscape research and public policy making. Much more integrated approaches will be required to understand landscape dynamics and change. It is our conviction that the concepts and frameworks presented in this book will prove helpful in this context, and we also believe that landscape strategy making and the principles of landscape democracy have the potential to effectively deal with the major tensions characterising current landscape policy and planning practice.

We hope this book contributes to opening new directions in the agendas both within academia and in the various policy domains affecting European rural landscapes by enhancing landscape planning approaches that are grounded on the collective construction of place-based visions and strategies for their implementation. At the end of the day, it will be the engaged citizens who will need to find common ground for the colourful futures of the European landscape we are all longing for.

Bibliography

Aalen, F. H. A., Whelan, K. & Stout, M. (2011). *Atlas of the Irish Rural Landscape* (2nd edn.) Cork: Cork University Press.

Abercrombie, P. (1945). *Greater London Plan 1944*. London: His Majesty's Stationary Office.

Albrechts, L. (2006). Bridge the gap: from spatial planning to strategic projects. *European Planning Studies* 14, 1487–1500.

Alexander, P., Moran, D., Rounsevell, M. D. & Smith, P. (2013). Modelling the perennial energy crop market: the role of spatial diffusion. *Journal of the Royal Society Interface* 10 (88).

Allen, V. G., Batello, C., Berretta, E. J., Hodgson, J., Kothmann, M., Li, X., McIvor, J. Milne, J. Morris, C., Peeters, A. & Sanderson, M. (2011). An international terminology for grazing lands and grazing animals. *Grass and Forage Science* 66: 2–28.

Almeida, M., Azeda, C., Guiomar, N. & Pinto-Correia, T. (2016). The effects of grazing management in Montado fragmentation and heterogeneity. *Agroforestry Systems* 90 (1): 69–85.

Almeida, M., Guerra, C. & Pinto-Correia, T. (2013). Unfolding relations between land cover and farm management: high nature value assessment in complex silvo-pastoral systems. *Danish Journal of Geography* 113, 97–108.

Andersen, E., Baldock, D., Bennett, H. et al. (2003). *Developing a High Nature Value Indicator*. Copenhagen: European Environment Agency.

Andersen, E., Elbersen, B., Godeschalk, F. & Verhoog, D. (2007). Farm management indicators and farm typologies as a basis for assessments in a changing policy environment. *Journal of Environmental Management* 82, 353–362.

Andersen, E., Verhoog, A. D. V., Elbersen, B. S., Godeschalk, F. & Koole, B. (2006). *A multidimensional farming systems typology*. SEAMLESS Report n°12. Wageningen: Alterra Wageningen UR.

Andersson, S. I. & Floryan, M. (2005). *Great European Gardens: An Atlas of Historic Plans*. Copenhagen: The Danish Architectural Press.

Antrop, M. (1994). Landscapes of the urban fringe. *Acta Geographica Lovaniensia* 34, 501–513.

Antrop, M. (1997). The concept of traditional landscapes as a base for landscape evaluation and planning. The example of Flanders Region. *Landscape and Urban Planning* 38, 105–117.

Antrop, M. (2000). Changing patterns in the urbanized countryside of Western Europe. *Landscape Ecology* 15, 257–270.

Antrop, M. (2004a). Landscape change and the urbanization process in Europe. *Landscape and Urban Planning* 67, 9–26.

Antrop, M. (2004b). Rural–urban conflicts and opportunities. In R. H. G. Jongman (Ed.), *The New Dimensions of the European Landscape* (pp. 83–89). Dordrecht: Springer.

Antrop, M. (2013). A brief history of landscape research. In P. Howard, I. Thompson & E. Waterton (Eds.), *The Routledge Companion to Landscape Studies* (pp. 12–22). London/New York: Routledge.

Antrop, M., Brandt, J., Loupa-Ramos, I., Padoa-Schioppa, E., Porter, J., Van Eetvelde, V. & Pinto-Correia, T. (2013). How landscape ecology can promote the development of sustainable landscapes in Europe: The role of the European Association for Landscape Ecology (IALE-Europe) in the twenty-first century. *Landscape Ecology* 28, 1641–1647. doi: 10.1007/s10980-013-9914-9

Arler, F. (2008). A true landscape democracy. In S. Arttzen and E. Brady (Eds.) Humans in the land: *The ethics and aesthetics of the cultural landscape* (pp. 75–99). Oslo: Unipub.

Arnalte-Alegre, E. & Ortiz-Miranda, D. (2013). The 'Southern Model' of European agriculture revisited: continuities and dynamics. In D. Ortiz-Miranda, A. Moragues-Faus & E. Arnalte-Alegre (Eds.), *Agriculture in Mediterranean Europe. Between Old and New Paradigms* (pp. 37–74). Bingley: Emerald.

Arnaud, P. (1998). The classical world. In R. A. Butlin & R. A. Dodgshon (Eds.), *An Historical Geography of Europe* (pp. 26–53). Oxford: Clarendon Press.

Avillez, F. & Carvalho, M. (2015). A importância de uma gestão sustentável do solo para o crescimento da agricultura portuguesa. *Cultivar* 2, 27–40.

Badami, M. G. & Ramankutty, N. (2015). Urban agriculture and food security: a critique based on an assessment of urban land constraints. *Global Food Security* 4, 8–15.

Baldeschi, P. (2007). Chianti beyond its wine. In B. Pedroli, A. Van Doorn, G. D. Blust, M. L. Paracchini, D. Wascher & F. Bunce (Eds.), *Europe's Living Landscapes. Essays on Exploring Our Identity in the Countryside* (pp. 351–366). Zeist: KNNV Publishing.

Barbati, A., Bastrup-Birk, A., Baycheva, T., Bonhomme, C., Bozzano, M., Bücking, W. et al. (2011). State of Europe's forests, 2011: status & trends in sustainable forest management in Europe. In *Ministerial Conference on the Protection of Forests in Europe*. Forest Europe, Liaison Unit Oslo.

Barbati, A., Corona, P. & Marchetti, M. (2007). A forest typology for monitoring sustainable forest management: the case of European forest types. *Plant Biosystems* 141, 93–103.

Barbieri, C. & Valdivia, C. (2010). Recreation and agroforestry: examining new dimensions of multifunctionality in family farms. *Journal of Rural Studies* 26 (4): 465–473.

Bastian, O., Grunewald, K., Syrbe, R. U., Walz, U. & Wende, W. (2014). Landscape services: the concept and its practical relevance. *Landscape Ecology* 29, 1463–1479.

Bastian, O. & Steinhardt, U. (2002). *Development and Perspectives of Landscape Ecology*. London: Kluwer Academic Publishers.

Bateman, I. J., Harwood, A. R., Mace, G. M., Watson, R. T., Abson, D. J., Andrews, B. et al. (2013). Bringing ecosystem services into economic decision-making: land use in the United Kingdom. *Science* 341, 45–50.

Baudry, J., Poggio, S. L., Burel, F. & Laurent, C. (2010). Agricultural landscape changes through globalisation and biodiversity effects. In J. Primdahl & S. Swaffield (Eds.), *Globalisation and Agricultural Landscapes. Change Patterns and Policy Trends in Developed Countries* (pp. 57–72). Cambridge: Cambridge University Press.

Beldman, A. C. G., Daatselaar, C. H. G. & Prins, A. M. (2014). Developments in dairying worldwide, from a dairy farmer's perspective. In A. Kuipers, A. Rozstalnyy & G. Keane (Eds.), *Cattle Husbandry in Eastern Europe and China* (pp. 61–70). Wageningen: Wageningen Academic Publishers.

Benneth, H., Osterburg, B., Nitsch, H., Kristensen, L., Primdahl, J. & Verschuur, G. (2006). Strengths and weaknesses of cross-compliance in the CAP. *EuroChoices* 5, 50–57.

Benoît, M., Rizzo, D., Marraccini, E., Moonen, A. C., Galli, M., Lardon, S. et al. (2012). Landscape agronomy: a new field for addressing agricultural landscape dynamics. *Landscape Ecology* 27, 1385–1394.

Bergman, I., Ostlund, L., Zackrisson, O. & Liedgren, L. (2008). Varro Muorra: the landscape significance of Sami sacred wooden objects and sacrificial altars. *Ethnohistory* 55, 1.

Bernstein, H. & Byres, T. J. (2001). From peasant studies to agrarian change. *Journal of Agrarian Change*, 1 (1): 1–56.

Bessou, C., Basset-Mens, C., Benoist, A., Biard, Y., Burte, J., Feschet, P. et al. (2016). Life cycle assessment to understand agriculture-climate change linkages. In E. Torquebiau (Ed.) *Climate Change and Agriculture Worldwide* (pp. 263–275). Dordrecht: Springer Netherlands.

Beunen, R., Knaap, W. G. M. v.d. & Biesbroek, G. R. (2009). Implementation and integration of EU environmental directive. Experiences from the Netherlands. *Environmental Policy and Governance* 19(1): 57–69.

Bishop, K. & Phillips, A. (2004). *Then and Now: Planning for Countryside Conservation*. London/New York: Routledge.

Blacksell, M. (2010). Agriculture and landscape in 21st century Europe: the post-communist transition. *European Countryside* 2, 13–24.

Blake, A., Sinclair, M. T. & Sugiyarto, G. (2003). Quantifying the impact of foot and mouth disease on tourism and the UK economy. *Tourism Economics* 9, 449–465.

Blume, H. P., Brümmer, G. W., Fleige, H., Horn, R., Kandeler, E., Kögel-Knabner, I. et al. (2016). Land evaluation and soil protection. In H.-P. Blume, G. W. Brümmer, R. Horn, E. Kandeler, I. Kögel-Knabner, R. Kretzschmar, P. Schad, K. Stahr & B.-M. Wilke (Eds.) *Scheffer/Schachtschabel Soil Science* (pp. 561–585). Berlin/Heidelberg: Springer.

Bohn, U., Neuhäusl, R., Gollub, G., Hettwer, C., Neuhäuslová, Z., Raus, T. et al. (2003). Karte der natürlichen Vegetation Europas/Map of the natural vegetation of Europe. *Maßstab/Scale* 1: 2 500 000.

Bohnet, I. (2002). *Exploring Landscape Character: A Socio-ecological Analysis in the High Weald Area of Outstanding Natural Beauty*. London: University of London.

Bohnet, I. (2008). Assessing retrospective and prospective landscape change through the development of social profiles of landholders: a tool for improving land use planning and policy formulation. *Urban Planning* 88, 1–11.

Bomans, K., Dewaelheyns, V. & Gulinck, H. (2011). Pasture for horses: an underestimated land use class in an urbanized and multifunctional area. *International Journal of Sustainable Development and Planning* 6, 195–211.

Bosco, C., de Rigo, D., Dewitte, O., Poesen, J. & Panagos, P. (2014). Interactive comment on 'Modelling soil erosion at European scale: towards harmonization and reproducibility'. *Natural Hazards and Earth System Sciences Discussion* 2, 2639–2680.

Bowler, I. R. & Ilbery, B. W. (1987). Redefining agricultural geography. *Area* 19(4): 327–332.

Brandt, J. (1987). En regional analyse af bæreevnens udvikling I de færøske hauger. *Fródskaparrit* 33, 19–41.

Brandt, J. (1989). Planlægning af det åbne land i 1700-tallet. En inspirationskilde fra oplysningstiden. *Landskab* 2, 25–30.

Brandt, J., Primdahl, J. & Reenberg, A. (1999). Rural land-use and landscape dynamic – analysis of 'driving forces' in space and time. In R. Krönert, J. Baudry, I. Bowler & A. Reenberg (Eds.), *Land-Use Changes and Their Environmental Impact in Rural Areas in Europe* (pp. 81–102). Paris: Parthenon Publishing Group.

Brandt, J. & Vejre, H. (2004). *Multifunctional Landscapes, Vol. 1: Theory, Values and History*. Southampton: WIT Press.

Bredin, Y. K., Lindhjem, H., Van Dijk, J. & Linnell, J. D. (2015). Mapping value plurality towards ecosystem services in the case of Norwegian wildlife management: AQ analysis. *Ecological Economic* 118, 198–206.

Breman, B., Vihinen, H., Tapio Biström, M. L. & Pinto-Correia, M. T. (2010). Meeting the challenge of marginalization processes at the periphery of Europe. *Public Administration* 88 (2): 364–380.

Breman, G. (2015). *Territorial collaboration, self-governance, social capital and policy making in the European agricultural sector*. Research paper. Leiden University.

Brezzi, M. & Veneri, P. (2015). Assessing polycentric urban systems in the OECD: country, regional and metropolitan perspectives. *European Planning Studies* 23, 1128–1145.

Brouwer, F. & Van der Heide, M. (2009). *Multifunctional Rural Land Management: Economics and Policies*. London: Routledge.

Brouwer, F. M. & Van Ittersum, M. (2010). *Environmental and Agricultural Modelling: Integrated Approaches for Policy Impact Assessment*. Dordrecht: Springer Science & Business Media.

Bruckmeier, K. & Tovey, H. (2009). *Rural Sustainable Development in the Knowledge Society*. Farnham: Ashgate.

Brunn, S. D., Williams, J. F. & Zeigler, D. J. (Eds.) (2003). *Cities of the World: World Regional Urban Development*. Lanham, MD: Rowman & Littlefield.

Brunori, G., Malandrin, V. & Rossi, A. (2013). Trade-off or convergence? The role of food security in the evolution of food discourse in Italy. *Journal of Rural Studies* 29, 19–29.

Bryden, J. M. (1993). *Farm Household Adjustment in Western Europe, 1987–1991*. Office for Official Publications of the European Communities; UNIPUB.

Bryden, J., Arandia, A., Dunne, L. & Knickel, K. (2011b). A new approach to multifunctionality, sustainable rural development, and the outcomes of policy. In J. Bryden (Ed.), *Studies in Development and Society* (pp. 1–21). New York: Routledge.

Bryden, J., Efstratoglou, S., Ferenczi, T., Knickel, K., Johnson, T., Refsgaard, K. et al. (2011a). *Towards Sustainable Rural Regions in Europe. Exploring Relationships between Rural Policies, Farming, Environment, Demographics, Regional Economies and Quality of Life Using System Dynamics*. New York: Routledge.

Buijs, A. E., Pedroli, B. & Luginbühl, Y. (2006). From hiking through farmland to farming in a leisure landscape: changing social perceptions of the European landscape. *Landscape Ecology* 21, 375–389.

Buller, H. (2005). *Evaluation of Policies with Respect to Multifunctionality of Agriculture: Observation Tools and Support for Policy Formulation and Evaluation* (UK national report [WP 6], EU Multagri Project). Exeter: University of Exeter.

Burel, F. & Baudry, J. (1999). *Ecologie du Paysage. Concepts, methods et applications*. Paris: Ed. Tec&Doc.

Bürgi, M., Hersperger, A. M. & Schneeberger, N. (2004). Driving forces of landscape change – current and new directions. *Landscape Ecology* 19, 857–868.

Burkhard, B., Crossman, N., Nedkov, S., Petz, K. & Alkemade, R. (2013). Mapping and modeling ecosystem services for science, policy and practice. *Ecosystem Services* 4, 1–3.

Butler, R. (1980). The concept of a tourist area cycle of evolution: implications for management of resources. *The Canadian Geographer* 24, 5–12.

Butler, R. (2004). The tourism area life cycle in the twenty-first century. In A. A. Lew, C. M. Hall & A. M. William (Eds.), *A Companion to Tourism* (pp. 159–169). Oxford: Blackwell.

Buttimer, A. (2001a). Concluding reflections. In A. Buttimer (Ed.), *Sustainable Landscapes and Lifeways. Scale and Appropriateness* (pp. 367–387). Cork: Cork University Press.

Buttimer, A. (2001b). *Sustainable Landscapes and Lifeways: Scale and Appropriateness*. Cork: Cork University Press.

Cabannes, Y. & Ross, P. (2015). *21st Century Garden Cities of To-morrow. A Manifesto*. Letchworth Garden City: New Garden City Movement.

Cairol, D., Coudel, E., Knickel, K., Caron, P. & Kröger, M. (2009). Multifunctionality of agriculture and rural areas as reflected in policies: the importance and relevance of the territorial view. *Journal of Environmental Policy and Planning* 11 (4): 269–289.

Cancela d'Abreu, A., Pinto-Correia, T. & Oliveira, R. (2004). *Contributos para a Identificação e Caracterisação da Paisagem em Portugal Continental*. Vols. I–V, Colecção Estudos 10.

Carey, P. D., Short, C., Morris, C., Hunt, J., Priscott, A., Davis, M. et al. (2003). The multi-disciplinary evaluation of a national agri-environment scheme. *Journal of Environmental Management* 69, 71–91.

Carolino, J., Primdahl, J., Pinto-Correia, T. & Bojesen, M. (2011). Hunting and the right to landscape: comparing the Portuguese and Danish traditions and current challenges. In S. Egoz, J. Makhzoumi & G. Pungetti (Eds.), *The Right to Landscape. Contesting Landscape and Human Rights* (pp. 99–112). Farnham: Ashgate.

Carvalho-Ribeiro, S., Madeira, L. & Pinto-Correia, T. (2013). Developing comprehensive indicators for monitoring rural policy impacts on landscape in Alentejo, southern Portugal. *Danish Journal of Geography* 113, 87–96.

Casey, E. S. (2001). Between geography and philosophy: what does it mean to be in the place-world? *Annals of the Association of American Geographers* 91, 683–693.

Castells, M. (2000a). *The Rise of the Network Society* (2nd edn.). Oxford: Blackwell Publishers.

Castells, M. (2000b). Materials for an exploratory theory of the network society 1. *The British Journal of Sociology* 51, 5–24.

Castells, M. (2010). The culture of real virtuality: the integration of electronic communication, the end of the mass audience, and the rise of interactive networks. *The Rise of the Network Society: With a New Preface, Volume I, Second edition*. Wiley Online doi: 10.1002/9781444319514.ch5, pp. 355–406.

CEC (1999). *European Spatial Development Perspective: Towards Balanced and Sustainable Development of the Territory of the EU*.

Champion, T. (2001). Urbanization, suburbanization, counterurbanization and reurbanization. In R. Paddison (Ed.), *Handbook of Urban Studies* (pp. 143–161). London: Sage Publications.

Chavez-Tafur, J. & Zagt, R. J. (2014). *Towards Productive Landscapes*. Wageningen: Tropenbos International.

Cheshire, P. (1995). A new phase of urban development in Western Europe? The evidence for the 1980s. *Urban Studies* 32, 1045–1063.

Christensen, A. A., Svenningsen, S. R., Lommer, M. S. & Brandt, J. (2014). New multifunctional hunting landscapes in Denmark. *Danish Journal of Geography* 114, 25–40.

Christians, C. (1979). L'évaluation des paysages et sites ruraux. Essais de méthodes et resultats dans quelques régions wallones. *Bulletin de la Societé Géographique de Liège* 15, 167–208.

Cinà, G. & Dansero, E. (2015). *Localizing urban food strategies. Farming cities and performing rurality*. 7th International Aesop Sustainable Food Planning Conference Proceedings Turin, Italy 7–9 October 2015. 610.

Clapp, J. & Dauverge, P. (2005). *Paths to a Green World. The Political Economy of the Global Environment*. Cambridge, MA: MIT Press.

Clark, J. (2010). Geographies of multifunctional agriculture: developing governance explanations. *Geography Compass* 4, 803–818.

Committee on Spatial Development (2016). *European Spatial Development Perspective ESDP*. Luxembourg: European Commission.

Copus, A., Courtney, P., Dax, T., Meredith, D., Noguera, J., Talbot, H. et al. (2011a). European Development Opportunities for Rural Areas (EDORA), Final Report, Applied Research 2013/1/2, ESPON & UHI Millenium Institute, 2nd revised version, August 2011, Luxembourg. 101 pp.

Copus, A. & Hörnström, L. (2011). *The New Rural Europe: Towards Rural Cohesion Policy* Stockholm: Nordregio.

Copus, A., Shucksmith, M., Dax, T. & Meredith, D. (2011b). Cohesion policy for rural areas after 2013. *Studies in Agricultural Economics* 113, 121–132.

Cosgrove, D. E. (1984). *Social Formation and Symbolic Landscape*. London: Crom Helm.

Cosgrove, D. E. (1985). Prospect, perspective and the evolution of the landscape idea. *Transactions of the Institute of British Geographers* 10 (1): 45–62.

Costanza, R., De Groot, R., Sutton, P., Van der Ploeg, S., Anderson, S. J., Kubiszewski, I. et al. (2014). Changes in the global value of ecosystem services. *Global Environmental Change* 26, 152–158.

Costanza, R. & Kubiszewski, I. (2014). *Creating a sustainable and desirable future: insights from 45 global thought leaders*. Singapore: World Scientific Publishing.

Council of Europe (2000). *European Landscape Convention (Florence Convention)*, Treaty Series Nr. 176. Strasbourg: Council of Europe.

Countryside Commission (1998). *Countryside Character: The Character of England's Natural and Man-Made Landscape (Volume 1)*, Cheltenham: Countryside Commission CA 10–14; 535–53.

Crivits, M., Prové, C., Block, T. & Dessein, J. (2016). Four perspectives of sustainability applied to the local food strategy of Ghent (Belgium): need for a cycle of democratic participation? *Sustainability* 8, 55.

Cronon, W. (1991). *Nature's Metropolis. Chicago and the Great West*. New York: W.W. Norton and Company.

Crossman, N. D., Burkhard, B., Nedkov, S., Willemen, L., Petz, K., Palomo, I. et al. (2013). A blueprint for mapping and modelling ecosystem services. *Ecosystem Services* 4, 4–14.

Cunha, A. & Swinbank, A. (2011). *An Inside View of the CAP Reform Process: Explaining the MacSharry, Agenda 2000, and Fischler Reforms*. Oxford: Oxford University Press.

Curry, N. & Winter, M. (2000). European briefing: the transition to environmental agriculture in Europe: learning processes and knowledge networks. *European Planning Studies* 8, 107–121.

da Silva, D. S., Figueiredo, E., Eusébio, C. & Carneiro, M. J. (2016). The countryside is worth a thousand words – Portuguese representations of rural areas. *Journal of Rural Studies* 44, 77–88.

Daily, G. C. (1997). *Nature's Services: Societal Dependence on Natural Ecosystems*. Washington, DC: Island Press.

Dale, V. H., Efroymson, R. A, & Kline, K. L. (2011). The land use–climate change–energy nexus. *Landscape Ecology* 26, 755–773.

Dannenberg, P. & Kuemmerle, T. (2010). Farm size and land use pattern changes in postsocialist Poland? *The Professional Geographer* 62, 197–210.

Darnhofer, I. (2014). Resilience and why it matters for farm management. *European Review of Agricultural Economics* 41, 461–484.

Davoudi, S. (2012). The legacy of positivism and the emergence of interpretive tradition in spatial planning. *Regional Studies* 46, 429–441.

Davoudi, S. & Stead, D. (2002). Urban–rural relationships: an introduction and a brief history. *Built Environment* 28, 269–277.

De Groot, R. (2010). *Integrating the Ecological and Economic Dimensions in Biodiversity and Ecosystem Services Valuation. chapter 1, TEEF Report – The Economics of Ecosystems and Biodiversity*. Brussels: The Ecological and Economic Foundations.

De Groot, R. & Hein, L. (2007). The concept and valuation of landscape goods and services. In Ü. Mander, H. Wiggering & K. Helming (Eds.), *Multifunctional Land Use. Meeting Future Demands for Landscapes, Goods and Services* (pp. 15–36). Berlin/Heidelberg: Springer.

De Jong, A. & Van Duin, C. (2010). *Regionale prognose 2009–2040: Vergrijzing en omslag van groei naar krimp*. Den Haag, The Netherlands: PBL/CBS.

De Vries, J. (1984). *European Urbanisation, 1500–1800*. London: Methuen.

De Zeeuw, H. & Drechsel, P. (2015). *Cities and Agriculture: Developing Resilient Urban Food Systems*. London/New York: Routledge.

DeLeon, P. (1999). The stages approach to the policy process: what has it done? where is it going? In P. A. Sabatier (Ed.), *Theories of the Policy Process* (pp. 19–32). Oxford: Westview Press.

Department of the Environment (1989). *Planning Control in Western Europe*. London: HMSO.

Di Iacovo, F., Moruzzo, R., Rossignoli, C. & Scarpellini, P. (2014). Transition management and social innovation in rural areas: lessons from social farming. *The Journal of Agricultural Education and Extension* 20, 327–347.

Dijkstra, L. & Poelman, H. (2008). *Remote rural regions. How proximity to a city influences the performance of rural regions*. Regional Focus 1, Brussels: DG Regio, European Commission.

Doevendans, K., Lörzing, H. & Schram, A. (2007). From modernist landscapes to New Nature: planning of rural utopias in the Netherlands. *Landscape Research* 32, 333–354.

Domon, G. (2011). Landscape as resource: consequences, challenges and opportunities for rural development. *Landscape and Urban Planning* 100, 338–340.

Duby, G. (1961). *L'économie rurale et la vie des campagnes dans l'occident médiéval*. Aubier, Editions Montaigne.

Duby, G. (1976). *Le temps des cathédrales: l'art et la société (980–1420)*. Paris: Gallimard.

Dwiartama, A. & Piatti, C. (2016). Assembling local, assembling food security. *Agriculture and Human Values* 33, 153–164.

Dwyer, J. & Hodge, I. (2001). The challenge of change: demands and expectations for farmed land. In T. C. Smout (Ed.), *Nature, Landscape and People since the Second World War* (pp. 117–134). East Linton: Tuckwell Press.

EC (1999). Agriculture, environment, rural development. Facts and Figures. A Challenge for Agriculture. COM (1999) 22 final.

EC (2009). Adapting to climate change: towards a European framework for action. White Paper. COM(2009) 147 final, 1.4.2009.

EC (2011a). *Roadmap to a Resource Efficient Europe. Communication from the Commission to the European Parliament, the Council, the European Economic and Social Committee and the Committee of the Regions. COM 2011*, 517. Brussels: European Commission.

EC (2011b). What is a small farm? *EU Agricultural Economic Briefs*. 11.

EC (2013a). *Final Report on Greenhouse Production (Protected Cropping), egtop/6/2013*. Brussels: European Commission.

EC (2013b). Regulation on Rural Development (EU) No 1305/2013.

EEA (2005). *European Environmental Outlook*. Copenhagen: European Environment Agency.

EEA (2006). *Urban Sprawl in Europe. The Ignored challenge*. Copenhagen: European Environment Agency.

EEA (2007). Afforestation in Europe, 1990 – 2000. Available from http://dataservice.eea.europe.eu/atlas/viewdata/viewpub.asp?id=1899 2015.

Egnsplankontoret (1948). *Skitseforslag til egnsplan for Storkøbenhavn*. København: Egnsplankontoret.

Egoz, S., Makhzoumi, J. & Pungetti, G. (2011). *The Right to Landscape: Contesting Landscape and Human Rights*. Farnham: Ashgate.

Elkington, S. & Gammon, S. (2015). Reading landscapes: articulating a non-essentialist representation of space, place and identity in leisure. In S. Gammon & S. Elkington (Eds.), *Landscapes of Leisure: Space, Place and Identities* (pp. 1–7). London: Palgrave Macmillan UK.

Elson, M., Walker, S. & Macdonald, R. (1993). *The Effectiveness of Green Belts*. London: HMSO.

Elzen, B., Barbier, M., Cerf, M. & Grin, J. (2012). Stimulating transitions towards sustainable farming systems. In D. Gibbon & B. Dedieu (Eds.), *Farming Systems Research into the 21st Century: The New Dynamic* (pp. 431–455). Dordrecht: Springer.

Emanuelsson, U., Arding, M. & Petersson, M. (2009). *The Rural Landscapes of Europe: How Man Has Shaped European Nature*. Stockholm: Formas.

Enquist, P. O. (1988). On the art of flying backward with dignity. *Daedalus* 117, 485–498.

Eriksson, C. & Wästfelt, A. (2011). Är ett landskap enbart en utsikt? Två frågor i och med införandet av landskabskonventionen i Sverige. *Bebyggelsehistorisk Tidskrift* 61, 90–92.

European Court of Auditors (2011). *Is Agri-environment Support Well Designed and Managed?* Luxembourg: European Court of Auditors.

Eurostat (2013). *How Many People Work in Agriculture in the European Union? An Answer Based on Eurostat Data Sources. EU Agricultural Economics Briefs no 8 | July 2013* Brussels: DG Agriculture and Rural Development, Unit Economic Analysis of EU Agriculture.

Faludi, A. (2004). Spatial planning traditions in Europe: their role in the ESDP process 1. *International Planning Studies* 9, 155–172.

Faludi, A. & Van der Valk, A. (1994). *Rule and Order. Dutch Planning Doctrine in the Twentieth Century*. Dordrecht: Kluwer Academic Publishers.

Fang, X., Zhao, W., Fu, B. & Ding, J. (2015). Landscape service capability, landscape service flow and landscape service demand: a new framework for landscape services and its use for landscape sustainability assessment. *Progress in Physical Geography* 39, 817–836.

Fearrne, A. (1997). The history and development of the CAP 1945–1990. In C. Ritson & D. Harvey (Eds.), *The Common Agricultural Policy*, 2nd edn. (pp. 11–55). Wallingford: CABI International.

Fertner, C. (2012). Downscaling European urban-rural typologies. *Danish Journal of Geography* 112, 77–83.

Finka, M. & Kluvánková, T. (2015). Managing complexity of urban systems: a polycentric approach. *Land Use Policy* 42, 602–608.

Fisher, B., Turner, R. K. & Morling, P. (2009). Defining and classifying ecosystem services for decision making. *Ecological Economics* 68, 643–653.

Folke, C., Hahn, T., Olsson, P. & Norberg, J. (2005). Adaptive governance of social-ecological systems. *Annual Review of Environment and Resources* 30, 441–473.

Fonseca, A. M., Marques, C., Pinto-Correia, T. & Campbell, D. E. (2016). Emergy analysis of a silvo-pastoral system, a case study in southern Portugal. *Agroforestry Systems* 89, 1–21.

Forestry Commission Scotland (2006). *The Scottish Forestry Strategy*. Edinburgh: Forestry Commission Scotland, National Office.

Forman, R. T. T. & Godron, M. (1986). *Landscape Ecology*. New York: John Wiley & Sons.

Fragoso, R., Marques, C., Lucas, M. R., Martins, M. B. & Jorge, R. (2011). The economic effects of common agricultural policy on Mediterranean Montado/dehesa ecosystem. *Journal of Policy Modeling* 33, 311–327.

Frayn, J. M. (1993). *Markets and Fairs in Roman Italy*. Oxford: Oxford University Press.

Frederiksen, P. & Vesterager, J. P. (2013). *Policy Drivers of Land Use/Landscape Change and the Role of Institutions. VOLANTE EU FP7-env-2010–265104, Deliverable D2.2*. Wageningen: Alterra Wageningen UR.

Frey, W. H. & Zimmer, Z. (2001). Defining the city. In R. Paddison (Ed.), *Handbook of Urban Studies* (pp. 14–36). London: Sage Publications.

Friberg, E., Landström, B. & Schoolfield, G. C. (1998). *The Kalevala: Epic of the Finnish People*. Otava.

Fritzbøger, B. (1998). *Det åbne lands kulturhistorie 1680–1980*. Frederiksberg: DSR-Forlag.

Fuchs, R., Herold, M., Verburg, P. H. & Clevers, J. G. P. W. (2013). A high-resolution and harmonized model approach for reconstructing and analysing historic land changes in Europe. *Biogeosciences* 10, 1543–1559.

Fuchs, R., Herold, M., Verburg, P. H., Clevers, J. G. P. W. & Eberle, J. (2014). Gross changes in reconstructions of historic land cover/use for Europe between 1900 and 2010. *Global Change Biology* 21 (1): 299–313.

García-Ruiz, J. M. & Lana-Renault, N. (2011). Hydrological and erosive consequences of farmland abandonment in Europe, with special reference to the Mediterranean region – a review. *Agriculture, Ecosystems & Environment* 140, 317–338.

Gardi, C., Panagos, P., Van Liedekerke, M., Bosco, C. & De Brogniez, D. (2015). Land take and food security: assessment of land take on the agricultural production in Europe. *Journal of Environmental Planning and Management* 58, 898–912.

Gaski, H. (2011). Song, poetry and images in writing: Sami literature. *Nordlit* 27, 33–54.

Gaspar, J. (2006). Evolução e perspectivas do desenvolvimento territorial. In *Geografia de Portugal, Vol. 4: Planeamento e Ordenamneto do Território* (pp. 15–28). Lisboa: Circulos de Leitores.

Gasson, R. & Errington, A. J. (1993). *The Farm Family Business*. Wallingford: CABI International.

Geddes, P. (1915). *Cities in Evolution. An Introduction to the Town Planning Movement and to the Study of Civics*. London: Williams and Norgate.

Geels, F. W. (2011). The multi-level perspective on sustainability transitions: responses to seven criticisms. *Environmental Innovation and Social Transformation* 1, 24–40.

Geels, F. W. & Schot, J. (2007). Typology of sociotechnical transition pathways. *Research Policy* 36, 399–417.

Geels, F. W. & Schot, J. (2010). The dynamics of transitions: a socio-technical perspective. In J. Grin, J. Rotmans & J. Schot (Eds.), *Transitions to Sustainable Development. New Directions in the Study of Long Term Transformative Change* (pp. 9–101). New York: Routledge.

Geist, H. J. & Lambin, E. F. (2002). Proximate causes and underlying driving forces of tropical deforestation. Tropical forests are disappearing as the result of many pressures, both local and regional, acting in various combinations in different geographical locations. *Bioscience* 52, 143–150.

Gereffi, G., Humphrey, J. & Sturgeon, T. (2005). The governance of global value chains. *Review of International Political Economy* 12, 78–104.

Giddens, A. (1990). *The Consequences of Modernity*. Cambridge: Polity Press.

Gilg, A. W. (1996). *Countryside Planning: The First Half Century*. London: Psychology Press.

Gobster, P. H., Nassauer, J. I., Daniel, T. C. & Fry, G. (2007). The shared landscape: what does aesthetics have to do with ecology? *Landscape Ecology* 22, 959–972.

Godinho, S., Guiomar, N., Machado, R., Santos, P., Sá-Sousa, P., Fernandes, J. P., Neves, N. & Pinto-Correia, T. (2016a). Assessment of environment, land management and spatial interaction effects on recent changes in Montado land cover in southern Portugal. *Agroforestry Systems* 90 (1): 177–192.

Godinho, S., Guiomar, N., Neves, N. & Pinto-Correia, T. (2016b). A remote sensing-based approach to estimating Montado canopy density using the FCD model: a contribution to identifying HNV farmlands in southern Portugal. *Agroforestry Systems* 90 (1): 23–34.

Gómez-Baggethun, E., De Groot, R., Lomas, P. L. & Montes, C. (2010). The history of ecosystem services in economic theory and practice: from early notions to markets and payment schemes. *Ecological Economics* 69, 1209–1218.

Goudie, A. S. (2013). *The Human Impact on the Natural Environment: Past, Present, and Future*. New York: John Wiley & Sons.

Grigg, D. B. (1974). *The Agricultural Systems of the World. An Evolutionary Approach*. Cambridge: Cambridge University Press.

Grin, J. (2012). The politics of transition governance in Dutch agriculture. Conceptual understanding and implications for transition management. *International Journal of Sustainable Development* 15, 72–89.

Groenewegen, P. P., Van den Berg, A. E., De Vries, S. & Verheij, R. A. (2006). Vitamin G: effects of green space on health, well-being, and social safety. *BMC Public Health* 6, 1.

Grove, A. T. & Rackham, O. (2003). *The Nature of Mediterranean Europe. An Ecological History*. New Haven, CT: Yale University Press.

Guerra, C., Pinto-Correia, T. & Metzger, M. J. (2014). Mapping soil erosion prevention using an ecosystem service modeling framework for integrated land management and policy. *Ecosystems* 17, 878–889.

Haaren, C. V. (2004). Was ist Landschaftplanung. In C. V. Haaren (Ed.), *Landschaftsplanung* (pp. 20–31). Stuttgart: Ulmer.

Habermas, J. (1981). *Theorie des kommunikativen Handelns.*, Frankfurt am Main: Suhrkamp.

Haigh, B. (2008). Horsiculture and planning. *Journal of the Royal Agricultural Society of England* 169, 101–107.

Halfacree, K. (2007). Back-to-the-land in the twenty-first century – Making connections with rurality. *Tijdschrift voor economische en sociale geografie* 98 (1), 3–8.

Hall, P. (1996). *Cities of Tomorrow*, 2nd edn. Oxford: Blackwell Publishers.

Hall, P. & Tewdwr-Jones, M. (2011). *Urban and Regional Planning*, 5th edn. London: Routledge.

Hansen-Møller, J. (2006). The meaning of landscape: a diagram for analysing the relationship between culture and nature, based on CS Peirce's semiotics. *Studies in Environmental Aesthetics and Semiotics* 5, 85–108.

Harvey, D. (1985). *Consciousness and the Urban Experience. Studies in the History and Theory of Capitalist Urbanization*. Oxford: Basil Blackwell.

Harvey, D. (1993). From space to place and back again: reflections on the condition of postmodernity. In J. Bird, B. Curtis, T. Putnam, G. Robertson & L. Tickner (Eds.), *Mapping the Futures: Local Cultures, Global Change* (pp. 3–29). London: Routledge.

Harvey, D. (2005). *A Brief History of Neoliberalism*. Oxford: Oxford University Press.

Hassink, J., Grin, J. & Hulsink, W. (2013). Multifunctional agriculture meets health care: applying the multi-level transition sciences perspective to care farming in the Netherlands. *Sociologia Ruralis* 53, 223–245.

Hayami, Y. & Ruttan, V. W. (1985). *Agricultural Development: An International Perspective (revised and expanded edition)*. Baltimore, MD: Johns Hopkins University Press.

Hazeu, G., Elbersen, B., Andersen, E., Baruth, B., Van Diepen, K. & Metzger, M. (2010). A biophysical typology in agri-environmental modelling. In F. M. Brouwer & M. van Ittersum (Eds.) *Environmental and Agricultural Modelling* (pp. 159–187). Dordrecht: Springer.

Hazeu, G. W., Metzger, M. J., Mücher, C. A., Perez-Soba, M., Renetzeder, C. H. & Andersen, E. (2011). European environmental stratifications and typologies: an overview. *Agriculture, Ecosystems & Environment* 142, 29–39.

Healey, P. (1998). Collaborative planning in a stakeholder society. *Town Planning Review* 69 (1): 1–21.

Healey, P. (2004). The treatment of space and place in the new strategic planning in Europe. *International Journal of Urban and Regional Research* 28, 45–67.

Healey, P. (2009). In search of the 'strategic' in spatial strategy making. *Planning, Theory & Practice* 10 (4): 439–457.

Hedberg, C. & do Carmo, R. M. (Eds.) (2012). *Translocal Ruralism: Mobility and Connectivity in European Rural Spaces*. Dordrecht: Springer.

Hediger, W. & Knickel, K. (2009). Multifunctionality and sustainability of agriculture and rural areas: a welfare economics perspective. *Journal of Environmental Policy and Planning* 11, 291–313.

Helming, K. & Wiggering, H. (2003). *Sustainable Development of Multifunctional Landscapes*. Dordrecht: Springer Science & Business Media.

Herod, A. (2011). *Scale (Key Ideas in Geography)*. New York: Routledge.

Hersperger, A. M., Gennaio, M. P., Verburg, P. H. & Bürgi, M. (2010). Linking land change with driving forces and actors: four conceptual models. *Ecology and Society* 15, 1.

Hill, B. (2012). *Understanding the Common Agricultural Policy*. London: Earthscan.

Hinrichs, C. C. (2014). Transitions to sustainability: a change in thinking about food systems change? *Agriculture and Human Values* 31, 143–155.

Hirschi, C., Widmer, A., Briner, S. & Huber, R. (2013). Combining policy network and model-based scenario analyses: an assessment of future ecosystem goods and services in Swiss mountain regions. *Ecology and Society* 18, 42.

Hodge, I. (2001). Beyond agri-environmental policy: towards an alternative model of rural environmental governance. *Land Use Policy* 18, 99–111.

Hodge, I. (2016). *The Governance of the Countryside. Property, Planning and Policy*. Cambridge: Cambridge University Press.

Hodge, I. & Adams, W. M. (2012). Neoliberalisation, rural land trusts and institutional blending. *Geoforum* 43, 472–482.

Hodge, I., Hauck, J. & Bonn, A. (2015). The alignment of agricultural and nature conservation policies in the European Union. *Conservation Biology* 29, 996–1005.

Hoggart, K., Black, R. & Buller, H. (1995). *Rural Europe: Identity and Change*. London: Hodder Arnold.

Hogwood, B. W. & Gunn, L. A. (1984). *Policy Analysis for the Real World*. Oxford: Oxford University Press.

Holden, A. (2008). *Environment and Tourism*, 2nd edn. Oxon: Routledge.

Holmes, J. (2006). Impulses towards a multifunctional transition in rural Australia: gaps in the research agenda. *Journal of Rural Studies* 22, 142–160.

Holmes, J. (2012). Cape York Peninsula, Australia: a frontier region undergoing a multifunctional transition with indigenous engagement. *Journal of Rural Studies* 28, 252–265.

Holzinger, K., Knill, C. & Schäfer, A. (2006). Rhetoric or reality? 'New governance' in EU environmental policy. *European Law Journal*, 12, 403–420.

Howard, E. (1965). *Garden Cities of To-morrow. Edited, with a Preface by F. J. Osborne*, 23rd edn. Cambridge, MA: MIT Press.

Huitema, D. & Turnhout, E. (2009). Working at the science–policy interface: a discursive analysis of boundary work at the Netherlands Environmental Assessment Agency. *Environmental Politics* 18, 576–594.

Hunter, C. & Green, H. (1995). The environmental impacts of tourism. In C. Hunter & H. Green (Eds.), *Tourism and the Environment. A Sustainable Relationship?* (pp. 10–51). London: Routledge.

Hägerstrand, T. (2001). A look at the political geography of environmental management. In A. Buttimer (Ed.), *Sustainable Landscapes and Lifeways. Scale and Appropriateness* (pp. 35–58). Cork: Cork University Press.

Häkli, J. (1999). Cultures of demarcation: territory and national identity in Finland. In G. H. Herb & D. H. Kaplan (Eds.), *Nested Identities – Nationalism, Territory and Scale* (pp. 123–149). Lanham/Boulder/New York, Oxford: Rowman & Littlefield.

IEEP (1994). *The Nature of Farming: Low Intensity Farming Systems in Nine European Countries*. London: Report from the Institute for European Environmental Policy (IEEP) Retrieved from www.ieep.eu.

IEEP (2007). *Guidance Document to the Member States on the Application of the HNV Impact Indicator*. London: Report from the Institute for European Environmental Policy (IEEP). Retrieved from www.ieep.eu.

Ilbery, B. (Ed.) (1998). *The Geography of Rural Change*. London: Routledge.

Ilbery, B., Chiotti, Q. & Rickard, T. (1997). *Agricultural Restructuring and Sustainability. A Geographical Perspective. Sustainable Rural Development Series n°3*. Wallingford: CABI International.

Ingold, T. & Kurttila, T. (2000). Perceiving the environment in Finnish Lapland. *Body and Society* 6, 183–196.

Jacobs, M. (2004). Metropolitan matterscape, powerscape and mindscape. In G. Tress, B. Tress, B. Harms, P. Smeets & A. Van der Valk (Eds.), *Planning Metropolitan Landscapes – Concepts, Demands, Approaches* (pp. 26–38). Wageningen: Alterra Wageningen UR.

Janssen, J., Luiten, E., Renes, H. & Rouwendal, J. (2014). Heritage planning and spatial development in the Netherlands: changing policies and perspectives. *International Journal of Heritage Studies* 20, 1–21.

Jensen, F. S. & Koch, N. E. (2004). Twenty-five years of forest recreation research in Denmark and its influence on forest policy. *Scandinavian Journal of Forest Research* 19, 93–102.

Jensen, F. S. & Tvedt, T. (2012). Skovene på førstepladsen som mål for friluftsliv. *Geografisk Orientering* 4, 586–591.

Jensen, K. M. & Reenberg, A. (1984). *Dansk landbrug: udvikling i produktion og kulturlandskab*. Brenderup: Geografforlaget.

Jepsen, M. R., Kuemmerle, T., Müller, D., Erb, K., Verburg, P. H., Haberl, H. et al. (2015). Transitions in European land-management regimes between 1800 and 2010. *Land Use Policy* 49, 53–64.

Jepsen, M. R., Reenberg, A., Kümmerle, T., Müller, D., Erb, K., Haberl, H. et al. (2013). *Technological, Institutional and Economic Drivers of Land Use Change. Volante (fp7-env-2010-265104) Deliverable D4.3.*

Jepson, P. (2015). A rewilding agenda for Europe: creating a network of experimental reserves. Forum Ecography 10.1111/ecog.01602.

Jollivet, M. (1997). *Vers un rural post industriel. Rural et Environment dans huit pays européens*. Paris: L'Harmattan.

Jones, M. (1988). Land-tenure and landscape change in fishing communities on the outer coast of Central Norway, c. 1880 to the present. Methodological approaches and modes of explanation. *Geografiska Analer* B 1, 197–204.

Jones, M. (2006). Landscape, law and justice – concepts and issues. *Norwegian Journal of Gepgraphy* 60, 1–14.

Jones, M., Howard, P., Olwig, K. R., Primdahl, J. & Sarlöv-Herlin, I. (2007). Multiple interfaces of the European Landscape Convention. *Norwegian Journal of Geography* 61, 207–216.

Jones, M. & Stenseke, M. (Eds.) (2011). *The European Landscape Convention: Challenges of Participation*. Dordrecht: Springer.

Jones, N., De Graaff, J., Rodrigo, I. & Duarte, F. (2011). Historical review of land use changes in Portugal (before and after EU integration in 1986) and their implications for land degradation and conservation, with a focus on the Centro and Alentejo regions. *Applied Geography* 31, 1036–1048.

Jordan, A. & Lenschow, A. (2010). Environmental policy integration: a state of the art review. *Environmental Policy and Governance* 20, 147–158.

Kafkoula, K. (2013). On garden-city lines: looking into social housing estates of interwar Europe. *Planning Perspectives* 28, 171–198.

Kaplan, J. O., Krumhardt, K. M., Ellis, E. C., Ruddiman, W. F., Lemmen, C. & Goldewijk, K. K. (2011). Holocene carbon emissions as a result of anthropogenic land cover change. *The Holocene* 21, 775–791.

Keay, S. J. (2012). *Rome, Portus and the Mediterranean*. London: British School at Rome.

Keenleyside, C., Allen, B., Hart, K., Menadue, H., Stefanova, V., Prazan, J., Herzon. I., Clement, T., Povellato, A., Maciejczak, M. & Boatman, N. (2011) *Delivering environmental benefits through entry level agri-environment schemes in the EU*. Report Prepared for DG Environment, Project ENV.B.1/ETU/2010/0035. Institute for European Environmental Policy: London.

Kempen, M., Elbersen, B. S., Staritsky, I., Andersen, E. & Heckelei, T. (2011). Spatial allocation of farming systems and farming indicators in Europe. *Agriculture, Ecosystem & Environment* 142, 51–62.

Kitchen, L. & Marsden, T. (2006). *Assessing the Eco-economy of Rural Wales*. Wales Rural Observatory.

Kitto, H. D. F. (1996). 'The polis'. In R. T. Le Gates & F. Stout (Eds.), *The City Reader* (pp. 43–48). London: Routledge.

Kjaergaard, T. & Hohnen, D. (1994). *The Danish Revolution, 1500–1800: An Ecohistorical Interpretation*. Cambridge: Cambridge University Press.

Kleijn, D., Berendse, F., Smit, R. & Gilissen, N. (2001). Agri-environment schemes do not effectively protect biodiversity in Dutch agricultural landscapes. *Nature* 413, 723–725.

Kleijn, D. & Sutherland, W. J. (2003). How effective are European agri-environment schemes in conserving and promoting biodiversity? *Journal of Applied Ecology* 40, 947–969.

Knickel, K. (1990). Agricultural structural change: impact on the rural environment. *Journal of Rural Studies* 6 (4): 383–393.

Knickel, K. (2011). Multifunctionality in agriculture and rural development: an empirical analysis based on survey data from eleven European regions. In J. Bryden (Ed.), *Studies in Development and Society* (pp. 82–113). New York: Routledge.

Knickel, K., Kröger, M., Bruckmeier, K. & Engwall, Y. (2009). The challenge of evaluating policies for promoting the multifunctionality of agriculture: When 'good' questions cannot be addressed quantitatively and 'quantitative answers are not that good'. *Journal of Environmental Policy and Planning* 11, 347–367.

Knickel, K. & Peter, S. (2005). Amenity-led development of rural areas: the example of the regional action pilot programme in Germany. In G. P. Green (Ed.), *Amenities and Rural Development: Theory, Methods and Public Policy. New Horizons in Environmental Economics* (pp. 302–321). Northampton: Edward Elgar Publishing.

Knickel, K. & Renting, H. (2000). Methodological and conceptual issues in the study of multifunctionality and rural development. *Sociologia Ruralis* 40, 512–528.

Knickel, K., Renting, H. & Van der Ploeg, J. D. (2004). Multifunctionality in European agriculture. In F. Brouwer (Ed.), *Sustaining Agriculture and the Rural Environment. Governance, Policy and Multifunctionality* (pp. 81–103). Cheltenham: Edward Elgar.

Knickel, K., Zemeckis, R. & Tisenkopfs, T. (2013). A critical reflection of the meaning of agricultural modernization in a world of increasing demands and finite resources. In *6th International Scientific Conference on Rural Development* (pp. 28–29).

Knill, C. & Lenschow, A. (2000) (Eds.). *Implementing EU Environmental Policy. New Directions and Old Problems*. Manchester: Manchester University Press.

Kobler, A., Cunder, T. & Pirnat, J. (2005). Modelling spontaneous afforestation in Postojna area, Slovenia. *Journal of Nature Conservation* 13, 127–135.

Kovách, I. & Kučerová, E. (2006). The project class in Central Europe: the Czech and Hungarian cases. *Sociologia Ruralis* 46 (1): 3–21.

Kristensen, L. (2001). Agricultural change in Denmark between 1982 and 1989: the appearance of post-productivism in farming? *Danish Journal of Geography* 101, 77–86.

Kristensen, L. S. & Primdahl, J. (2015). Dialogbaseret planlægning i det åbne land – nye behov, nye planprocesser. In L. S. Kristensen, J. Primdahl & H. Vejre (Eds.), *Dialogbaseret planlægning i det åbne land – om strategier for kulturlandskabets fremtid* (pp. 10–45). København: Bogværket.

Kuemmerle, T., Kroisleitner, C., Plutzar, C., Erb, K., Estel, S., Müller, D. et al. (2013). *Report on recent land use transition hotspots in Europe*. VOLANTE FP7-ENV-2010–265104. Deliverable no: 3.2.

Kuemmerle, T., Levers, C., Erb, K., Estel, S., Jepsen, M. R., Müller, D., Plutzar, C., Stürck, J., Verkerk, P. J., Verburg, P. H. & Reenberg, A. (2016). Hotspots of land use change in Europe. *Environmental Research Letters* 11 (6), p. 064020.

Kuemmerle, T., Stürck, J., Levers, C., Müller, D., Erb, K., Gingrich, S. et al. (2014). *Interpretation of scenario results in terms of described and mapped land change trajectories and archetypes* VOLANTE FP7-ENV-2010–265104, Deliverable 11.2.

Küster, H. & Hoppe, A. (2010). *Das Gartenreich Dessau-Wörlitz: Landschaft und Geschichte*. München: CH Beck.

Kuusela, K. (1994). *Forest Resources in Europe 1950–1990*, 1st edn. Cambridge: Cambridge University Press.

La Rosa, D., Privitera, R., Martinico, F. & La Greca, P. (2013). Measures of safeguard and rehabilitation for landscape protection planning: a qualitative approach based on diversity indicators. *Journal of Environmental Management* 127, S73–S83.

Lambert, A. M. (1985). *The Making of the Dutch Landscape: An Historical Geography of the Netherlands*. London: Academic Press.

Lambin, E. & Geist, H. (2006). *Land-Use and Land-Cover Change. Local Processes and Global Change*. Berlin/Heidelberg/New York: Springer.

Lambin, E. F. & Meyfroidt, P. (2010). Land use transitions: socio-ecological feedback versus socio-economic change. *Land Use Policy* 27, 108–118.

Lambin, E. F., Turner, B. L., Geist, H. J., Agbola, S. B., Angelsen, A., Bruce, J. W. et al. (2001). The causes of land-use and land-cover change: moving beyond the myths. *Global Environmental Change* 11, 261–269.

Larsen, L. (1991). *Nature as therapy: an assessment of schizophrenic patients' visual preferences for institutional outdoor environments*. MSc thesis. University of Guelph, Guelph.

Larsen, L. & Harlan, S. L. (2006). Desert dreamscapes: residential landscape preference and behavior. *Landscape and Urban Planning* 78, 85–100.

Larsson, T. B., Angelstam, P., Balent, G., Barbati, A., Bijlsma, R. J., Boncina, A. et al. (2001). Biodiversity evaluation tools for European forests. *Ecological Bulletins* 50, 2001 www.jstor.org/stable/20113288.

Lassini, P., Monzani, F. & Pileri, P. (2007). A green vision for the renewal of the Lombardy landscape. In B. Pedroli, A. Van Doorn, G. De Blust, M. L. Paracchini, D. Wascher, & F. Bunce (Eds.), *Europe's Living Landscape. Essays Exploring Our Identity in the Countryside* (pp. 83–104). Wageningen/Zeist: Landscape Europe/KNNV Publishing.

Latour, B. (2012). *Enquêtes sur les modes d'existence*. Paris: La Découverte.

Lawton, J., Brotherton, P., Brown, V., Elphick, C., Fitter, A., Forshaw, J. et al. (2010). *Making Space for Nature: A Review of England's Wildlife Sites and Ecological Network*. Report to the Department of Environment, Food and Rural Affairs. London: DEFRA.

Lefebvre, H. (2000). *La Production de L'espace*, 4e éd. Paris: Coll. 'Ethnosociologie', Anthropos.

Legacy, C. (2010). Regional planning for open space. *Australian Planner* 47, 105–106.

Lionello, P. (2012). *The Climate of the Mediterranean Region: From the Past to the Future*. Amsterdam: Elsevier.

Lipsky, M. (1980). *Street-Level Bureaucracy*. New York: Russell Sage Foundation.

Lomba, A., Alves, P., Jongman, R. H. G. & McCracken, D. I. (2015). Reconciling nature conservation and traditional farming practices: a spatially explicit framework to assess the extent of High Nature Value farmlands in the European countryside. *Ecology and Evolution* 5, 1031–1044.

Lörzing, H. (2001). *The Nature of Landscape. A Personal Quest*. Rotterdam: 010 Publishers.

Louglin, J. (2014). The 'transformation' of governance: new directions in policy and politics. *Australian Journal of Politics and History* 50, 8–22.

Lowe, P. & Baldock, D. (2000). Integration of environmental objectives into agricultural policy making. In F. Brouwer & P. Lowe (Eds.), *CAP Regimes and the European Countryside* (pp. 31–52). Wallingford: CABI International.

Lowe, P., Marsden, T., Murdoch, J. & Ward, N. (2003). *The Differentiated Countryside*. London: Routledge.

Lowe, P. & Ward, N. (2007). Sustainable rural economies: some lessons from the English experience. *Sustainable Development* 15, 307–317.

Luginbühl, Y. (2012). *La mise en scène du monde: La construction du paysage européen*. Paris: CNRS Éditions.

MacDonald, D., Crabtree, J. R., Wiesinger, G., Dax, T., Stamou, N., Fleury, P. et al. (2000). Agricultural abandonment in mountain areas of Europe: environmental consequences and policy response. *Journal of Environmental Management* 59, 47–69.

Macnaghten, P. & Urry, J. (1998). *Contested Natures*. London: Sage Publications.

Mander, Ü., Helming, K. & Wiggering, H. (2007). *Multifunctional Land Use: Meeting Future Demands for Landscape Goods and Services*. Berlin: Springer.

Marin, L. E. M. & Russo, V. (2016). Re-localizing 'legal' food: a social psychology perspective on community resilience, individual empowerment and citizen adaptations in food consumption in southern Italy. *Agriculture and Human Values* 33, 179–190.

Marsden, T. (1999). Rural futures: the consumption countryside and its regulation. *Sociologia Ruralis* 39 (4): 501–526.

Marsden, T. (2003). *The Condition of Rural Sustainability*. Assen: Royal Van Gorcum.

Marsden, T. (2013). From post-productionism to reflexive governance: contested transitions in securing more sustainable food futures. *Journal of Rural Studies* 29, 123–134.

Marsden, T., Murdoch, J., Lowe, P., Munton, R. & Flynn, A. (1992). *Constructing the Countryside*. England: UCL Press.

Marsden, T. & Sonnino, R. (2008). Rural development and the regional state: denying multifunctional agriculture in the UK. *Journal of Rural Studies* 24, 422–431.

Marson, A. (2015). Introduzione. In C. Agnoletti, S. Iommi & P. Lattarulo (Eds.), *Rapporto sul territorio – Configurazioni urbani e territori negli spazi europei* (pp. 5–7). Firenze: IRPET.

Martín-López, B., Gómez-Baggethun, E., García-Llorente, M. & Montes, C. (2014). Trade-offs across value-domains in ecosystem services assessment. *Ecological Indicators* 37, 220–228.

Mather, A. S., Hill, G. & Nijnik, M. (2006). Post-productivism and rural land use: cul de sac or challenge for theorization? *Journal of Rural Studies* 22, 441–455.

Maye, D. & Kirwan, J. (2013). Food security: a fractured consensus. *Journal of Rural Studies* 29, 1–6.

Mazmanian, D. A. & Sabatier, P. A. (1989). *Implementation and Public Policy*. Lanham, MD: University Press of America.

McCarthy, L. & Danta, D. (2014). Cities of Europe. In S. D. Bruun, J. F. Williams & D. J. Zeigler (Eds.), *Cities of the World* (pp. 168–222). Lanham, MD: Rowman & Littlefield.

McDowall, W. & Eames, M. (2006). Forecasts, scenarios, visions, backcasts and roadmaps to the hydrogen economy: a review of the hydrogen futures literature. *Energy Policy* 34, 1236–1250.

McManus, P., Walmsley, J., Argent, N., Baum, S., Bourke, L., Martin, J. et al. (2012). Rural community and rural resilience: what is important to farmers in keeping their country towns alive? *Journal of Rural Studies* 28, 20–29.

Mees, H. L., Driessen, P. P. & Runhaar, H. A. (2014). Legitimate adaptive flood risk governance beyond the dikes: the cases of Hamburg, Helsinki and Rotterdam. *Regional Environmental Change* 14, 671–682.

Metzger, M. J., Bunce, R. G. H., Jongman, R. H. G., Mücher, C. A. & Watkins, J. W. (2005). A climatic stratification of the environment of Europe. *Global Ecology and Biogeography* 14, 549–563.

Metzger, M. J. & Schröter, D. (2006). Towards a spatially explicit and quantitative vulnerability assessment of environmental change in Europe. *Regional Environmental Change* 6, 201–216.

Michels, A. & De Graaf, L. (2010). Examining citizen participation: local participatory policy making and democracy. *Local Government Studies* 36, 477–491.

Milestad, R., Dedieu, B., Darnhofer, I. & Bellon, S. (2012). Farms and farmers facing change: the adaptive approach. In I. Darnhofer, D. Gibbon & B. Dedieu (Eds.), *Farming Systems Research into the 21st Century: The New Dynamic* (pp. 365–385). Dordrecht: Springer.

Millward, H. (2006). Urban containment strategies: a case-study appraisal of plans and policies in Japanese, British, and Canadian cities. *Land Use Policy* 23, 473–485.

Milone, P. & Ventura, F. (2015). Is multifunctionality the road to empowering the farmers? In P. Milone, F. Ventura & Y. Jingzhong (Eds.), *Constructing a New Framework for Rural Development* (pp. 59–88). Bingley: Emerald Group Publishing.

Miossec, J. M. (1976). *Éléments pour une théorie de l'espace touristique*. Aix-en-Provence: Centre des hautes études touristiques.

Mok, H. F., Williamson, V. G., Grove, J. R., Burry, K., Barker, S. F. & Hamilton, A. J. (2014). Strawberry fields forever? Urban agriculture in developed countries: a review. *Agronomy for Sustainable Development* 34, 21–43.

Mooney, P. & Hoover, R. (1996). The design of restorative landscapes for Alzheimer's patients. In W. Wagner (Ed.), *Proceedings 1996 Annual Meeting of the American Society of Landscape Architects* (pp. 50–55). Washington, DC: American Society of Landscape Architects.

Moreira, F., Ferreira, P. G., Rego, F. C. & Bunting, S. (2001). Landscape changes and breeding bird assemblages in northwestern Portugal: the role of fire. *Landscape Ecology* 16, 175–187.

Mormont, M. (1990). Who is rural? or, how to be rural: towards a sociology of the rural. In T. Marsden, S. Whatmore & P. Lowe (Eds.), *Rural Restructuring. Global Processes and Their Responses* (pp. 21–44). London: David Fulton.

Mücher, C. A., Klijn, J. A., Wascher, D. M. & Schaminée, J. H. J. (2010). A new European Landscape Classification (LANMAP): a transparent, flexible and user-oriented methodology to distinguish landscapes. *Ecological Indicators* 10, 87–103.

Müller, D., Kuemmerle, T., Rusu, M. & Griffiths, P. (2009). Lost in transition: determinants of post-socialist cropland abandonment in Romania. *Journal of Land Use Science* 4 (1–2), 109–129.

Müller, F., De Groot, R. & Willemen, L. (2010). Ecosystem services at the landscape scale: the need for integrative approaches. *Landscape Online* 23.

Mumford, L. (1970). *The Myth of the Machine. Vol. 2: The Pentagon of Power*. New York: Harcourt, Brace & World.

Munafò, M., Salvati, L. & Zitti, M. (2013). Estimating soil sealing rate at national level – Italy as a case study. *Ecological Indicators* 26, 137–140.

Murdoch, J., Lowe, P., Ward, N. & Marsden, T. (2003). *The Differentiated Countryside*. London [etc.]: Routledge.

Murdoch, J. & Marsden, T. (1994). *Reconstructing Rurality*. London: UCL Press.

Nainggolan, D., Termansen, M., Reed, M. S., Cebollero, E. D. & Hubacek, K. (2013). Farmer typology, future scenarios and the implications for ecosystem service provision: a case study from south-eastern Spain. *Regional Environmental Change* 13 (3), 601–614.

Nassauer, J. I. (1997). *Placing Nature. Culture and Landscape Ecology*. Washington, DC: Island Press.

Nassauer, J. I. (1995). Culture and changing landscape structure. *Landscape Ecology* 10, 229–237.

Nassauer, J. I. (2012). Landscape as medium and method for synthesis in urban ecological design. *Landscape and Urban Planning* 106, 221–229.

Navarro, L. M. & Pereira, H. M. (2012). Rewilding abandoned landscapes in Europe. *Ecosystems* 15, 900–912.

Naveh, Z. & Lieberman, A. (1984). *Landscape Ecology. Theory and Application*. New York: Springer.

Nellemann, V., Møller, K. M., Møller, P. G., Primdahl, J., & Øberg, A. (2016). Strategi for Karby Sogn – landskab og landsby. In L. S. Kristensen, J. Primdahl & H. Vejre (Eds.), *Dialognaseret planlægning i det åbne land – om strategier for kulturlandskabets fremtid* (pp. 66–85). København: Bogværket.

Neuray, G. (1982). *Des Paysages, Pour qui ? Pourquoi ? Comment ?* Gembloux, Belgium: Presses Agronomiques de Gembloux.

Nexö, M. A. (1913). *Pelle the Conqueror, volume II. The Great Struggle and Daybreak* (vols. 2). Glouster: Peter Smith.

Norgaard, R. B. (2010). Ecosystem services: from eye-opening metaphor to complexity blinder. *Ecological Economics* 69, 1219–1227.

O'Riordan, T. & Voisey, H. (1998). The political economy of the sustainability transition. In T. O'Riordan & H. Voisey (Eds.), *The Transition to Sustainability: The Politics of Agenda 21 in Europe* (pp. 5–30). London: Earthscan.

OECD (1993). *OECD Core Set of Indicators for Environmental Performance Reviews*. Paris: OECD Environmental Directorate Monographs No. 83.

OECD (1997). *Environmental Indicators for Agriculture. Volume 1 Concepts and Framework*, 1st edn. Paris: OECD.

OECD (2001). *Environmental Indicators for Agriculture: Methods and Results* (Vol. 3). Paris: OECD.

OECD/FAO/UNCDF (2016). *Adopting a Territorial Approach to Food Security and Nutrition Policy*. Paris: OECD. http://dx.doi.org/10.1787/9789264257108-en

Olwig, K. (1984). *Nature's Ideological Landscape*. London: George Allen and Unwin.

Olwig, K. R. (1996). Recovering the substantive nature of landscape. *Annals of the Association of American Geographers* 86, 630–653.

Olwig, K. R. (2002). *Landscape, Nature and the Body Politics: From Britain's Renaissance to America's New World*. Madison: University of Wisconsin Press.

Olwig, K. R. (2008). Has 'geography' always been modern? Choros, (non) representation, performance, and the landscape. *Environment and Planning* 40, 1843–1861.

Oostindie, H. & Broekhuizen, R. (2008). The dynamics of novelty production. In J. D. Van der Ploeg & T. Marsden (Eds.), *Unfolding Webs. The Dynamics of Regional Rural Development* (pp. 68–86). Assen: Royan van Gorcum.

Oppermann, R., Beaufoy, G. & Herzog, F. (2012). *High nature value farming in Europe. 35 European countries – experiences and perspectives*. Ubstad-Weiher, Germany: Verlag Regionalkultur/EFNCP.

Oreszczyn, S., Lane, A. & Carr, S. (2010). The role of networks of practice and webs of influencers on farmers' engagement with and learning about agricultural innovations. *Journal of Rural Studies* 26, 404–417.

Ortiz-Miranda, D., Moragues-Faus, A. & Arnalte-Alegre, E. (2013). Agriculture in Mediterranean Europe: challenging theory and policy. In D. Ortiz-Miranda, A. Moragues-Faus & E. Arnalte-Alegre (Eds.), *Agriculture in Mediterranean Europe. Between Old and New Paradigms*, 19th edn. (pp. 295–310). UK: Emerald.

Ostrom, E. (2007). A diagnostic approach for going beyond panaceas. *PNAS* 104, 15181–15187.

Paine, C. (1997). Creating and re-creating landscape for therapy and recreation. *Proceedings from the 34th IFLA Congress, Buenos Aires*, 56–68.

Palang, H. & Fry, G. (2003). *Landscape Interfaces: Cultural Heritage in Changing Landscapes* (Vol. 1). Dordrecht: Springer Science & Business Media.

Palang, H., Spek, T. & Stenseke, M. (2011). Digging in the past: new conceptual models in landscape history and their relevance in peri-urban landscapes. *Landscape and Urban Planning* 100, 344–346.

Paracchini, M. L., Pacini, C., Jones, M. L. & Pérez-Soba, M. (2011). An aggregation framework to link indicators associated with multifunctional land use to the stakeholder evaluation of policy options. *Ecological Indicators* 11, 71–80.

Paracchini, M. L., Petersen, J. E., Hoogeveen, Y., Bamps, C., Burfield, I. & Van Swaay, C. (2008). *High Nature Value Farmland in Europe. An Estimate of the Distribution Patterns on the Basis of Land Cover and Biodiversity Data*. Luxemburg: Office for Official Publications of the European Communities.

Parris, K. (2004). Measuring changes in agricultural landscapes as a tool for policy makers. *Multifunctional Landscapes* 1, 193–218.

Pearce, D. (1989). *Tourist Development*, 2nd edn. Essex: Longman Scientific & Technical.

Pedroli, B. (2000). *Landscape – Our Home / Lebensraum Landschaft. Essays on the Culture of the European Landscape as a Task*. Zeist: Indigo.

Pedroli, B. (2012). The arctic landscape as a living European heritage. In N. Raasakka & S. Sivonen (Eds.), *Implementation of the European Landscape Convention in the North Calotte Area Municipalities, 7–9 Sept. 2011* (pp. 72–77). Inari, Finland: Report Nr. 48, Centre for Economic Development, Transport and the Environment for Lapland, Rovaniemi.

Pedroli, B. (2016). Il piano toscano, riferimento per una politica integrata del paesaggio. In A. Marson (Ed.), *La struttura del paesaggio. Una sperimentazione multidisciplinare per il Piano della Toscana* (pp. 278–287). Bari-Roma: Laterza.

Pedroli, B., Elbersen, B., Frederiksen, P., Grandin, U., Heikkilä, R., Krogh, P. H., Izakovičová, Z., Johansen, A., Meiresonne, L. & Spijker, J. (2013a). Is energy cropping in Europe compatible with biodiversity? – Opportunities and threats to biodiversity from land-based production of biomass for bioenergy purposes. *Biomass and Bioenergy* 55 (0): 73–86.

Pedroli, B., Gramberger, M., Gravsholt Busck, A., Metzger, M., Paterson, J. S., Pérez-Soba, M. et al. (2015). *VOLANTE roadmap for future land resource management in Europe*. Wageningen: Alterra Wageningen UR.

Pedroli, B., Pinto-Correia, T. & Cornish, P. (2006). Landscape–what's in it? Trends in European landscape science and priority themes for concerted research. *Landscape Ecology* 21, 421–430.

Pedroli, B., Tagliasacchi, S., Van der Sluis, T. & Vos, W. (2013b). *Ecologia del paesaggio del Monte di Portofino / Landscape Ecology of the Monte di Portofino* (Vol. 2). Wageningen: Bilingual Italian-English Edition, FERGUS ON.

Pedroli, B., Van den Brink, T. & Bakker, M. M. (2018). European Landscape Character Areas – a first approximation. *in prep.*

Pedroli, B., Van Doorn, A., De Blust, G., Paracchini, M. L., Wascher, D. & Bunce, F. (2007a). *Europe's Living Landscapes. Essays on Exploring Our Identity in the Countryside*. Zeist: Landscape Europe/KNNV Publishing.

Pedroli, G. B. M., Van Elsen, T. & Van Mansvelt, J. D. (2007b). Values of rural landscapes in Europe: inspiration or by-product? *Journal of Life Sciences* 54, 431–447.

Pelosi, C., Goulard, M. & Balent, G. (2010). The spatial scale mismatch between ecological processes and agricultural management: do difficulties come from underlying theoretical frameworks? *Agriculture, Ecosystem & Environment* 139, 455–462.

Peneva, M., Draganova, M., Gonzalez, C., Diaz, M. & Mishev, P. (2015). High nature value farming: environmental practices for rural sustainability. In L. A. Sutherland, L. Zagata, I. Darnhofer & G. Wilson (Eds.), *Transition Pathways towards Sustainability in Agriculture* (pp. 97–112). Wallingford: CABI International.

Penker, M. (2015). Landscape governance for or by the local population? A property rights analysis in Austria. *Land Use Policy* 26, 947–953.

Pérez-Soba, M., Paterson, J. S. & Metzger, M. (2015). *Visions of Future Land Use in Europe. Stakeholder Visions for 2040*. Wageningen: Alterra Wageningen UR/VOLANTE.

Perfecto, I., Vandermeer, J. H. & Wright, A. L. (2009). *Nature's Matrix: Linking Agriculture, Conservation and Food Sovereignty*. London: Earthscan.

Petrarch, F. (1948). The ascent of Mont Ventoux. *Vives* 36–47.

Philo, C. (1997). Of other rurals? In P. Cloke & J. Little (Eds.), *Contested Countryside Cultures. Otherness, Marginalization and Rurality* (pp. 19–50). London: Routledge.

Pintar, M., Udovc, A., Istenic, M., Glavan, M. & Slavic, I. (2010). Goriska Brda (Slovenia) – sustainable natural resource management for the prosperity of a rural area. In H. Wiggering, H. P. Ende, A. Knierim & M. Pintar (Eds.), *Innovations in European Rural Landscapes*. Heildelberg: Springer.

Pinto-Correia, T. (2010). Assessing the future role of agriculture in peripheral rural landscapes. Application to Portugal. In J. Primdahl & S. Swaffield (Eds.), *Globalization and Agricultural Landscapes. Change Patterns and Policy Trends in Developed Countries* (pp. 127–148). Cambridge: Cambridge University Press.

Pinto-Correia, T., Almeida, M. & Gonzalez, C. (2016). A local landscape in transition between production and consumption: can new management arrangements preserve the local landscape character? *Danish Journal of Geography* 116 (1): 33–43.

Pinto-Correia, T., Barroso, F. & Menezes, H. (2010). The changing role of farming in a peripheric south European area – the challenge of the landscape amenities demand. In H. Wiggering, H. P. Ende, A. Knierim & M. Pintar (Eds.), *Innovations in European Rural Landscapes* (pp. 53–76). Dordrecht: Springer.

Pinto-Correia, T., Barroso, F., Surová, D. & Menezes, H. (2011). The fuzziness of Montado landscapes: progress in assessing user preferences through photo-based surveys. *Agroforestry System* 82, 209–224.

Pinto-Correia, T. & Breman, B. (2009). New roles for farming in a differentiated countryside: the Portuguese example. *Regional Environmental Change* 9, 143–152.

Pinto-Correia, T. & Fonseca, A. M. (2009). Historical perspective of Montados: the example of Évora. In J. Aronson, J. Santos Pereira & J. G. Pausas (Eds.), *Cork Oak Woodlands on the Edge: Ecology, Adaptative Management, and Restoration* (pp. 49–54). Washington, DC: Island Press.

Pinto-Correia, T., Gonzalez, C., Sutherland, L. A. & Peneva, M. (2015b). Lifestyle farming: countryside consumption and transition towards new farming models. In L. A. Sutherland, I. Darnhofer, G. Wilson & L. Zagata (Eds.), *Transition Pathways towards Sustainability in European Agriculture* (pp. 67–82). Wallingford: CABI International.

Pinto-Correia, T., Guiomar, N., Guerra, C. & Carvalho-Ribeiro, S. (2015c). Assessing the ability of rural areas to fulfill multiple societal demands. online first DOI 10.1016/j.landusepol.2015.01.031.

Pinto-Correia, T. & Kristensen, L. (2013). Linking research to practice: the landscape as the basis for integrating social and ecological perspectives of the rural. *Landscape and Urban Planning* 120, 248–256.

Pinto-Correia, T., Machado, C., Barroso, F., Picchi, P., Turpin, N., Bousset, J. P. et al. (2013). How do policy options modify landscape amenities? An assessment approach based on public expressed preferences. *Environmental Science and Policy* 32, 37–47.

Pinto-Correia, T., McKee, A. & Guimarães, H. (2015a). Transdisciplinarity in deriving sustainability pathways for agriculture. In L. A. Sutherland, I. Darnhofer, G. A. Wilson & L. Zagata (Eds.), *Transition Pathways towards Sustainability in Agriculture: Case Studies from Europe* (pp. 171–188). Wallingford: CABI International.

Pinto-Correia, T., Menezes, H. & Barroso, L. F. (2014). The landscape as an asset in Southern European fragile agricultural systems: contrasts and contradictions in land managers attitudes and practices. *Landscape Research* 39, 205–217.

Pinto-Correia, T. & Primdahl, J. (2009). When rural landscapes change functionality: examples from contrasting case-studies in Portugal and Denmark. In F. Brouwer (Ed.), *Multifunctional Rural Land Management, Economics and Policies* (pp. 255–277). London: Earthscan.

Pinto-Correia, T. & Vos, W. (2004). Multifunctionality in Mediterranean landscapes – past and future. In R. G. H. Jongman (Ed.), *The New Dimensions of the European Landscape* (pp. 135–164). Dordrecht: Springer.

Pitte, J. R. (1983). *Histoire du paysage français* (Vol. I). Paris: Talander.

Plieninger, T., Draux, H., Fagerholm, N., Bieling, C., Bürgi, M., Kizos, T. et al. (2016). The driving forces of landscape change in Europe: a systematic review of the evidence. *Land Use Policy* 57, 204–214.

Pointereau, P., Paracchini, M. L., Terres, J. M., Jiguet, F., Bas, Y. & Biala, K. (2007). Identification of High Nature Value farmland in France through statistical information and farm practice surveys. JRC Scientific and Technical Reports. EUR 22786.

Porsmose, E. (2008). *Danske landsbyer*. København: Gyldendal.

Potschin, M. & Haines-Young, R. (2013). Landscapes, sustainability and the place-based analysis of ecosystem services. *Landscape Ecology* 28, 1053–1065.

Potter, C. (2004). Multifunctionality as an agricultural and rural policy concept. In F. Brouwer (Ed.), *Sustaining Agriculture and the Rural Environment. Governance, Policy and Multifunctionality, Advances in Ecological Economics* (pp. 15–35). Cheltenham: Edward Elgar.

Potter, C. & Lobley, M. (1996). The farm family life cycle, succession paths and environmental change in Britain's countryside. *Journal of Agricultural Economics* 47, 172–190.

Pounds, N. J. (1990). *An Historical Geography of Europe Abridged Version*. Cambridge: Cambridge University Press.

Priestley, G. & Mundet, L. (1998). The post-stagnation phase of the resort cycle. *Annals of Tourism Research* 25, 85–111.

Primdahl, J. (2014). Agricultural landscape sustainability under pressure: policy developments and landscape change. *Landscape Research* 39, 123–140.

Primdahl, J., Andersen, E., Swaffield, S. & Kristensen, L. S. (2013a). Intersecting dynamics of agricultural structural change and urbanisation within European rural landscapes: change patterns and policy implications. *Landscape Research* 38, 799–817.

Primdahl, J., Bojesen, M., Vesterager, J. P. & Kristensen, L. S. (2012). Hunting and landscape in Denmark: farmers' management of hunting rights and landscape changes. *Landscape Research* 37, 659–672.

Primdahl, J. & Brandt, J. (1997). CAP, nature conservation and physical planning. In C. Laurent & I. Bowler (Eds.), *CAP and the Regions. Building a Multidisciplinary Framework for the Analysis of the EU Agricultural Space* (pp. 177–186). Paris: Institut National de la Reserche Agronomique.

Primdahl, J. & Kristensen, L. (2003). Danske erfaringer med det åbne lands planlægning. *Kungl. Skogs- och Lantbruksakademiens Tidskrift* 142, 11–24.

Primdahl, J. & Kristensen, L. S. (2011). The farmer as a landscape manager: management roles and change patterns in a Danish region. *Danish Journal of Geography* 111, 107–116.

Primdahl, J. & Kristensen, L.S. (2016). Landscape strategy making and landscape characterisation – experiences from Danish experimental planning processes. *Landscape Research* 41(2) 227–238.

Primdahl, J., Kristensen, L. S. & Busck, A. G. (2013b). The farmer and landscape management: different roles, different policy approaches. *Geography Compass* 7 (4): 300–314.

Primdahl, J., Kristensen, L. S., Busck, A. G. & Vejre, H. (2010c). Functional and structural changes of agricultural landscapes: how changes are conceived by local farmers in two Danish rural communities. *Landscape Research* 35, 633–653.

Primdahl, J., Kristensen, L. S., & Swaffield, S. (2013c). Guiding rural landscape change. Current policy approaches and potentials of landscape strategy making as a policy integrating approach. *Applied Geography* 42, 86–94.

Primdahl, J., Peco, B., Schramek, J., Andersen, E. & Oñate, J. J. (2003). Environmental effects of agri-environmental schemes in Western Europe. *Journal of Environmental Management* 67, 129–138.

Primdahl, J. & Swaffield, S. (2010a). Globalisation and the sustainability of agricultural landscapes. In J. Primdahl & S. Swaffield (Eds.), *Globalisation and Agricultural Landscapes. Change Patterns and Policy Trends in Developed Countries* (pp. 1–15). Cambridge: Cambridge University Press.

Primdahl, J. & Swaffield, S. (2010b). *Globalisation and Agricultural Landscapes: Change Patterns and Policy trends in Developed Countries*. Cambridge: Cambridge University Press.

Primdahl, J., Vesterager, J. P., Finn, J. A., Vlahos, G., Kristensen, L. & Vejre, H. (2010). Current use of impact models for agri-environment schemes and potential for improvements of policy design and assessment. *Journal of Environmental Management* 91, 1245–1254.

Pröbstl, U. (2016). Lessons learned, trends and strategies for the future. In U. Pröbstl, V. Wirth, B. Elands & S. Bell (Eds.), *Management of Recreation and Nature Based Tourism in European Forests* (pp. 287–298). Heidelberg: Springer.

Pröbstl, U., Elands, B. & Wirth, V. (2009). Forest recreation and nature tourism in Europe: context, history and current situation. In S. Bell, M. Simpson, L. Tyrväinen, T. Sievänen & U. Pröbstl (Eds.), *European Forest Recreation and Tourism. A Handbook* (pp. 12–32). London: Taylor & Francis.

Prokop, G., Jobstmann, H. & Schönbauer, A. (2011). Overview of best practices for limiting soil sealing or mitigating its effects in EU-27. *European Communities* 227.

Qviström, M. (2007). Landscapes out of order: studying the inner urban fringe beyond the rural–urban divide. *Geografiska Annaler B* 89 (3), 269–282.

Ramhøj, L. (2009). Hævd. *Landindspektøren* 51, 3–32.

Rasmussen, B. M. & Olsen, I. A. (1998). Byudviklingen i landskabet. *Byplan* 4–5, 142–145.

RCE (2015). *Atlas van de wederopbouw in Nederland, 28/30 Maas en Waal-West. Een landelijk gebied van nationaal belang.* Amersfoort: Rijksdienst voor het Cultureel Erfgoed.

Redwood, M. (2012). *Agriculture in Urban Planning: Generating Livelihoods and Food Security.* London: Routledge.

Regione Toscana (2013). Piano di indirizzo territoriale con valenza di piano paesaggistico. Rapporto del Garante della Comunicazione, a cura del garante della comunicazione per il Pit Massimo Morisi. Firenze, Assessorato urbanistica, pianificazione del territorio e paesaggio, Regione Toscana: 264.

Regione Toscana (2015). *Piano di indirizzo territoriale con valenza di piano paesaggistico.* Firenze: Assessorato urbanistica, pianificazione del territorio e paesaggio, Regione Toscana.

Relph, E. (1976). *Place and Placelessness*, 67th edn. London: Pion.

Renes, H. (2010). Grainlands. The landscape of open fields in a European perspective. *Landscape History* 31, 37–70.

Renes, H. & Piastra, S. (2011). Polders and politics: new agricultural landscapes in Italian and Dutch wetlands, 1920s to 1950s. *Landscapes* 12, 24–41.

Renting, H. & Ploeg, J.D. v.d. (2001). Reconnecting Nature, Farming and Society: Environmental Cooperatives in the Netherlands as Institutional Arrangements for Creating Coherence. *Journal of Evironmental Policy and Planning* 3, 85–101.

Renting, H., Rossing, W. A. H., Groot, J. C. J., Van der Ploeg, J. D., Laurent, C., Perraud, D. et al. (2009). Exploring multifunctional agriculture. A review of conceptual approaches and prospects for an integrative transitional framework. *Journal of Environmental Management* 90, Supplement 2, S112–S123.

Renwick, A., Jansson, T., Verburg, P. H., Revoredo-Giha, C., Britz, W., Gocht, A. & McCracken, D. (2013). Policy reform and agricultural land abandonment in the EU. *Land Use Policy* 30, 446–457.

Reynolds, K. (2015). Disparity despite diversity: social injustice in New York City's urban agriculture system. *Antipode* 47, 240–259.

Rhodes, R. A. (2007). Understanding governance: ten years on. *Organization Studies* 28, 1243–1264.

Ribeiro, P. F., Santos, J. L., Bugalho, M. N., Santana, J., Reino, L., Beja, P. et al. (2014). Modelling farming system dynamics in high nature value farmland under policy change. *Agriculture, Ecosystems & Environment* 183, 138–144.

Ringkamp, C. & Janssen, M. (2000), *Das Dessau-Wörlitzer Gartenreich: Inventarisation und Entwicklungspotentiale der historischen Infrastruktur.* Dessau: Kulturstiftung Desau-Wörlitz.

Rip, A. & Kemp, R. (1998). *Technological Change.* Columbus, OH: Battelle Press.

Ritter, J. (1962). Landschaft. Zur Funktion des Ästhetischen in der modernen Gesellschaft. Reprinted in G. Gröning & U. Herlyn (Eds.), *Landschaftswahrnehmung und Landschaftserfahrung* (pp. 28–68). Münster: Lit-Verlag.

Roberts, N. (2013). *The Holocene: An Environmental History*, 2nd edn. Hoboken, NJ: Wiley Blackwell.

Robinson, G. (2008). Sustainable rural systems: an introduction. In G. Robinson (Ed.), *Sustainable Agriculture and Rural Communities* (pp. 3–40). Farnham: Ashgate.

Roederer-Rynning, C. (2010). The Common Agricultural Policy. In H. Wallace, M. A. Pollack & A. R. Young (Eds.) *Policy Making in the European Union* (pp. 181–205). Oxford: Oxford University Press.

Röling, N. G. & Wagemakers, M. A. (1998). *Facilitating Sustainable Agriculture. Participatory Learning and Adaptative Management in Times of Environmental Uncertainty*. Cambridge: Cambridge University Press.

Romano, B. & Zullo, F. (2016). Half a century of urbanization in southern European lowlands: a study on the Po Valley (Northern Italy). *Urban Research & Practice* 9, 109–130.

Rossi, P. H. & Freeman, H. E. (1993). *Evaluation: A Systematic Approach*, 5th edn. London: Sage Publications.

Roth, M., Frixen, M., Tobisch, C. & Scholle, T. (2015). Finding spaces for urban food production – matching spatial and stakeholder analysis with urban agriculture approaches in the urban renewal area of Dortmund-Hörde, Germany. *Journal of Food and Society* 3, 79–88.

Rotmans, J. & Loorbach, D. (2010). Towards a better understanding of transitions and their governance. A systemic and reflexive approach. In J. Grin, J. Rotmans & J. Schot (Eds.) *Transitions to Sustainable Development. New Directions in the Study of Long Term Transformative Change* (pp. 105–198). London: Routledge.

Roturier, S. & Roué, M. (2009). Of forest, snow and lichen: Sami reindeer herders' knowledge of winter pastures in northern Sweden. *Forest Ecology and Management* 258, 1960–1967.

Rounsevell, M. D. A., Pedroli, B., Erb, K. H., Gramberger, M., Busck, A. G., Haberl, H. et al. (2012). Challenges for land system science. *Land Use Policy* 29, 899–910.

Sabatier, P. A. (1986). Top-down and bottom-up approaches to implementation research: a critical analysis and suggested synthesis. *Journal of Public Policy* 6 (1), 21–48.

Sanchez, B., Medina, F. & Iglesias, A. (2013). Typical farming system and trends in crop and soil management in Europe. Deliverable 2.2., Project Smart Soil, KBBE-2001–5. In (p. 64). Madrid.

Sanyé-Mengual, E., Anguelovski, I., Oliver-Solà, J., Montero, J. I. & Rieradevall, J. (2015). Resolving differing stakeholder perceptions of urban rooftop farming in Mediterranean cities: promoting food production as a driver for innovative forms of urban agriculture. *Agriculture and Human Values* 33, 101–120.

Saraiva, T. (2010). Fascist labscapes: geneticists, wheat, and the landscapes of Fascism in Italy and Portugal. *Historical Studies in Natural Sciences* 40, 457–498.

Sarda, R., Mora, J. & Avila, C. (2004). Tourism development in the Costa Brava (Girona, Spain). How integrated coastal zone management may rejuvenate its lifecycle. In J. Vermaat, W. Salomons, L. Bouwer, & K. Turner (Eds.), *Managing European Coasts: Past, Present and Future* (pp. 291–314). Berlin: Springer.

Sayer, J., Sunderland, T., Ghazoul, J., Pfund, J. L., Sheil, D., Meijaard, E. et al. (2013). Ten principles for a landscape approach to reconciling agriculture, conservation, and other competing land uses. *PNAS* 110, 8349–8356.

Schama, S. (1995). *Landscape and Memory*. New York: Alfred Knopf.

Scharmer, C. O. (2009). *Theory U. Leading from the Future as It Emerges*. San Francisco: Berret-Koehler Publishers.

Schneider, A. & Ingram, H. (1990). Behavioral Assumptions of Policy Tools. *The Journal of Politics* 52, 510–529.

Schröter, M., Van der Zanden, E. H., van Oudenhoven, A. P. E., Remme, R. P., Serna-Chavez, H. M., De Groot, R. S. et al. (2014). Ecosystem services as a contested concept: a synthesis of critique and counter-arguments. *Conservation Letters* 7 (6): 514–523.

Scott, A. (2011). Beyond the conventional: meeting the challenges of landscape governance within the European Landscape Convention? *Journal of Environmental Management* 92, 2754–2762.

Scott, A. J., Carter, C., Reed, M. R., Larkham, P., Adams, D., Morton, N. et al. (2013). Disintegrated development at the rural–urban fringe: re-connecting spatial planning theory and practice. *Progress in Planning* 83, 1–52.

Scott, A., Christie, M. & Midmore, P. (2004). Impact of the 2001 foot-and-mouth disease outbreak in Britain: implications for rural studies. *Journal of Rural Studies* 20, 1–14.

Scottish Government (2009). *Scotland's Climate Change Adaptation Framework*. Edinburgh, National Office.

Scottish Government (2011). *Getting the Best from Our Land – A Land Use Strategy for Scotland*. Edinburgh, National Office.

Scottish Natural Heritage (2009). *Loch Lomond and the Trossachs National Park Landscape Character Assessment* www.snh.org.uk/pdfs/publications/review/140.pdf.

Selman, P. H. (1988). Rural land use planning – resolving the British paradox? *Journal of Rural Studies* 4, 277–294.

Selman, P. (2008). What do we mean by sustainable landscape? *Science, Practice and Policy* 4, 23–28.

Selman, P. (2009). Planning for landscape multifunctionality. *Science, Practice and Policy* 5, 45–52.

Selman, P. (2012). *Sustainable Landscape Planning: The Reconnection Agenda*. London/New York: Routledge.

Senge, P., Scharmer, C. O., Jaworski, J. & Flowers, B. S. (2004). *Presence: Human Purpose, and the Field of the Future*. Cambridge, MA: Society for Organizational Learning.

Serra, P., Pons, X. & Saurí, D. (2008). Land-cover and land-use change in a Mediterranean landscape: a spatial analysis of driving forces integrating biophysical and human factors. *Applied Geography* 28, 189–209.

Seto, K. C., Reenberg, A., Boone, C. G., Fragkias, M., Haase, D., Langanke, T. et al. (2016). Urban land teleconnections and sustainability. *PNAS* 109 (20): 7687–7692.

Sharpley, R. & Craven, B. (2001). The 2001 foot and mouth crisis. Rural economy and tourism policy implications: a comment. *Current Issues in Tourism* 4, 527–537.

Short, C. (2008). Balancing nature conservation 'needs' and those of other land uses in a multi-functional context: high-value nature conservation sites in Lowland England. In G. Robinson (Ed.), *Sustainable Rural Systems: Sustainable Agriculture and Rural Communities* (pp. 125–144). Farnham: Ashgate.

Shucksmith, M. & Rønningen, K. (2011). The Uplands after neoliberalism? – The role of the small farm in rural sustainability. *Journal of Rural Studies* 27, 275–287.

Sikor, T. (2004). The commons in transition: agrarian and environmental change in Central and Eastern Europe. *Environmental Management* 34, 270–280.

Slee, B. & Pinto-Correia, T. (2014). Understanding the diversity of European rural areas. In L. A. Sutherland, I. Darnhofer, G. A. Wilson & L. Zagata (Eds.), *Transition Pathways towards Sustainability in Agriculture: Case Studies from Europe* (pp. 33–50). Wallingford: CABI International.

Smith, P. & Olesen, J. E. (2010). Synergies between the mitigation of, and adaptation to, climate change in agriculture. *Journal of Agricultural Science* 148, 543–552.

Snoodijk, D. (Ed.) (2011). Ruilverkaveling. *Gids Cultuurhistorie*, 8. Amersfoort: Rijksdienst Cultureel Erfgoed.

Spirn, A. W. (1998). *The Language of Landscape*. New Haven, CT: Yale University Press.

Stahlschmidt, P., Swaffield, S., Primdahl, J. & Nellemann, V. (2017) Landscape Analysis. *Investigating the Potentials of Space and Place*. London: Routledge.

Stanners, D. & Bordeaux, P. (1995). *Europe's Environment: The Dobříš Assessment*. Copenhagen: European Environmental Agency.

Steinitz, C., Parker, P. & Jordan, L. (2016). Hand-drawn overlays: their history and prospective uses. *Landscape Architecture* 9, 444–455.

Stoate, C., Boatman, N. D., Borralho, R. J., Carvalho, C. R., De Snoo, G. R. & Eden, P. (2001). Ecological impacts of arable intensification in Europe. *Journal of Environmental Management* 96, 337–365.

Stobbelaar, D. J. & Pedroli, B. (2011). Perspectives on landscape identity: a conceptual challenge. *Landscape Research* 36, 321–339.

Sturgess, I. & Dalton, G. (2000). The Agenda 2000 CAP reform and the 'Millennium' round: negotiations on agriculture. In S. Bilal & P. Pezaros (Eds.), *Negotiating the Future of Agricultural Policies: Agricultural Trade and the Millennium WTO Round* (pp. 97–111). The Hague: Kluwer Law International.

Surová, D., Pinto-Correia, T. & Marušák, R. (2014). Visual complexity and the Montado do matter: landscape pattern preferences of user groups in Alentejo, Portugal. *Annals of Forest Science* 71, 15–24.

Sutherland, L. A. (2010). Environmental grants and regulations in strategic farm business decision-making: a case study of attitudinal behaviour in Scotland. *Land Use Policy* 27, 415–423.

Sutherland, L. A., Peter, S. & Zagata, L. (2015a). On-farm renewable energy: a 'Classic case' of technological transition. In L. A. Sutherland, I. Darnhofer, G. Wilson & L. Zagata (Eds.), *Transition Pathways towards Sustainability in Agriculture. Case Studies from Europe* (pp. 113–126). Wallingford: CABI International.

Sutherland, L. A., Zagata, L., Darnhofer, I. & Wilson, G. (2015b). *Transition pathways towards sustainability in agriculture*. Wallingford: CABI International.

Svarstad, H., Petersen, L. K., Rothman, D., Siepel, H. & Wätzold, F. (2008). Discursive biases of the environmental research framework DPSIR. *Land Use Policy* 25, 116–125.

Swagemakers, P. & Wiskerke, S. C. (2011). Revitalizing ecological capital. *Danish Journal of Geography* 111 (2), 149–167.

Swanwick, C. (2004). The assessment of countryside and landscape character in England: an overview. In K. Bishop & A. Phillips (Eds.), *Countryside Planning: New Approaches to Management and Conservation* (pp. 109–124). London: Earthscan.

Swyngedouw, E. (2015). *Liquid Power: Contested Hydro-modernities in Twentieth-Century Spain*. Cambridge, MA: MIT Press.

Tarrant, J. (1992). Agriculture and the state. In I. Bowler & B. Ilbary (Eds.), *The Geography of Agriculture in Developed Market Economies* (pp. 239–274). London: Longman.

Termorshuizen, J. & Opdam, P. (2009). Landscape services as a bridge between landscape ecology and sustainable development. *Landscape Ecology* 24, 1037–1052.

Terres, J. M., Scacchiafichi, L. N., Wania, A., Ambar, M., Anguiano, E., Buckwell, A. et al. (2015). Farmland abandonment in Europe: identification of drivers and indicators, and development of a composite indicator of risk. *Land Use Policy* 49, 20–34.

Terwan, P., Ritchie, M., Van der Weijden, W., Verschuur, W. & Joannides, J. (2004). *Values of Agrarian Landscapes across Europe and North America. Centre for Agriculture and Environment, Renewing the Countryside*. Netherlands: Reed Business Information.

Tilzey, M. & Potter, C. (2008). Productivism versus post-productivism? Modes of agri-environmental governance in post-Fordist agricultural transitions. In G. Robinson (Ed.), *Sustainable Rural Systems: Sustainable Agriculture and Rural Communities* (pp. 41–66). London: Routledge.

Trauger, A. (2007). Un/Re-constructing the agrarian dream: going back-to-the-land with an organic marketing co-operative in south-central Pennsylvania, USA. *Tijdschrift voor economische en sociale geografie* 98, 9–20.

Tregear, A. & Cooper, S. (2016). Embeddedness, social capital and learning in rural areas: the case of producer cooperatives. *Journal of Rural Studies* 44, 101–110.

Trnka, M., Olesen, J. E., Kersebaum, K. C., Skjelvåg, A. O., Eitzinger, J., Seguin, B. et al. (2011). Agroclimatic conditions in Europe under climate change. *Global Change Biology* 17, 2298–2318.

Tscharntke, T., Clough, Y., Wanger, T. C., Jackson, L., Motzke, I., Perfecto, I. et al. (2012). Global food security, biodiversity conservation and the future of agricultural intensification. *Biological Conservation* 151, 53–59.

Turner, T. (1991). Pattern analysis. *Landscape Design*, October, 39–41.

Turok, I. & Mykhnenko, V. (2007). The trajectories of European cities, 1960–2005. *Cities* 24, 165–182.

Uhlig, H. (1961). Old hamlets with infield and outfield systems in western and central Europe. *Geografiska Annaler* 43, 285–312.

Ulied, A. (2014). *Making Europe open and polycentric. Vision and scenarios for the European territory towards 2050*. Luxemburg: ESPON 2013 Programme Unit.

Urry, J. (1990). *The Tourist Gaze*. London: Sage Publications.

Vallés-Planells, M., Galiana, F. & Van Eetvelde, V. (2014). A classification of landscape services to support local landscape planning. *Ecology and Society* 19, 1–44.

Van Berkel, D. B. & Verburg, P. H. (2011). Sensitising rural policy: assessing spatial variation in rural development options for Europe. *Land Use Policy* 28, 447–459.

Van-Camp, L., Bujarrabal, B., Gentile, A. R., Jones, R. J. A., Montanarella, L., Olazabal, C. et al. (2004). *Reports of the Technical Working Groups Established under the Thematic Strategy for Soil Protection. EUR 21319 EN/1-Environment*. Luxembourg pp. 872.

Van de Ven, G. P. (2004). *Man-Made Lowlands: History of Water Management and Land Reclamation in the Netherlands*, 4th edn. Utrecht: Matrijs.

Van den Berg, A. E., Hartig, T. & Staats, H. (2007). Preference for nature in urbanized societies: Stress, restoration, and the pursuit of sustainability. *Journal of Social Issues* 63, 79–96.

Van den Bergh, S. M. (2004). *Verdeeld land: De geschiedenis van de ruilverkaveling in Nederland vanuit een lokaal perspectief, 1890–1985*. Dissertation Wageningen University. Groningen/Wageningen: Nederlands Agronomisch Historisch Instituut.

Van den Bosch, M. A., Östergren, P. O., Grahn, P., Skärbäck, E. & Währborg, P. (2015). Moving to serene nature may prevent poor mental health – results from a Swedish longitudinal cohort study. *International Journal of Environmental Research and Public Health* 12, 7974–7989.

Van der Burg, M. (2002). *'Geen tweede boer': gender, landbouwmodernisering en onderwijs aan plattelandsvrouwen in Nederland, 1863–1968*. PhD thesis. Wageningen University, Wageningen.

Van der Ploeg, J. D. (1994). Styles of farming: an introductory note on concepts and methodology. In J. D. van der Ploeg & A. Long (Eds.), *Born from Within: Practice and Perspectives of Endogenous Rural Development* (pp. 7–30). Assen: Van Gorcum.

Van der Ploeg, J. D. (2009a). *The New Peasantries: Struggles for Autonomy and Sustainability in an Era of Empire and Globalization*. London: Routledge.

Van der Ploeg, J. D. (2009b). Transition: contradictory but interacting processes of change in Dutch agriculture. In K. J. Poppe, C. J. A. M. Termeer & M. Slingerland (Eds.), *Transitions towards Sustainable Agriculture and Food Chains in Peri-urban Areas* (pp. 293–307). Wageningen: Wageningen Academic Publishers.

Van der Ploeg, J. D., Long, A., & Banks, J. (2002a). Rural development: the state of the art. In J. D. van der Ploeg, A. Long & J. Banks (Eds.), *Living Countrysides. Rural Development Processes in Europe: The State of the Art* (pp. 8–17). Doetinchem: Elsevier.

Van der Ploeg, J. D. & Marsden, T. (2008). *Unfolding Webs: The Dynamics of Regional Rural Development*. Assen: Royal Van Gorcum.

Van der Ploeg, J. D., Roep, D., Renting, H., Banks, J., Mielgo, A. A., Gorman, M. et al. (2002b). The socio-economic impact of rural development processes within Europe. In J. D. van der Ploeg (Ed.), *Living Countrysides: Rural Development Processes in Europe* (pp. 180–191). Doetinchem: Elsevier.

Van der Ploeg, J. D., van Broekhuizen, R., Brunori, G., Sonnino, R., Knickel, K., Tisenkopfs, T. et al. (2008). Towards a framework for understanding regional rural development. In J. D. van der Ploeg & T. Marsden (Eds.), *Unfolding Webs. The Dynamics of Regional Rural Development* (pp. 1–28). Assen: Royal Van Gorcum.

Van der Ploeg, S. (2003). *The Virtual Farmer: Past, Present and Future of the Dutch Peasantry*. Assen: Royal van Gorcum.

Van der Sluis, T., Pedroli, B., Kristensen, S. B. P., Lavinia Cosor, G. & Pavlis, E. (2015). Changing land use intensity in Europe – recent processes in selected case studies. *Land Use Policy* 57, 777–785.

Van der Valk, A. (2014). Preservation and development: the cultural landscape and heritage paradox in the Netherlands. *Landscape Research* 39, 158–173.

Van der Valk, A. & Faludi, A. (1997). The green heart and the dynamics of doctrine. *Netherlands Journal of Housing and the Built Environment* 12, 57–75.

Van der Valk, A. & Van Dijk, T. (Eds.) (2009). *Regional Planning for Open Space*. London: Routledge.

Van Doorn, A. & Elbersen, B. S. (2012). *Implementation of High Nature Value Farmland in Agri-environmental Policies: What Can Be Learned from other EU Member States?* Wageningen: Alterra Wageningen UR.

Van Eetvelde, V. & Antrop, M. (2009). Indicators for assessing changing landscape character of cultural landscapes in Flanders (Belgium). *Land Use Policy* 26, 901–910.

Van Elsen, T., Günther, A. & Pedroli, B. (2006). The contribution of care farms to landscapes of the future: a challenge of multifunctional agriculture. In J. Hassink & M. van Dijk (Eds.), *Farming for Health. Proceedings of the Frontis Workshop on Farming for Health, 16–19 March 2005, Wageningen* (pp. 91–100). Dordrecht: Springer.

Van Eupen, M., Metzger, M. J., Pérez-Soba, M., Verburg, P. H., Van Doorn, A. & Bunce, R. G. H. (2012). A rural typology for strategic European policies. *Land Use Policy* 29, 473–482.

Van Huylenbroek, G. & Durand, G. (2003). Multifunctional agriculture. *A new paradigm for European agriculture and rural development*. Farnham: Ashgate.

Van Mansvelt, J. D. & Pedroli, B. (2003). Landscape – a matter of identity and integrity. In H. Palang & G. Fry (Eds.), *Landscape Interfaces* (pp. 375–394). Dordrecht: Kluwer.

Van Paassen, A., Opdam, P., Steingröver, E. & Van den Berg, J. (2011b). Landscape science and societal action. In A. Van Paassen, J. Van den Berg, E. Steingrover, R. Werkman & B. Pedroli (Eds.), *'Knowledge in Action'. The Search for Collaborative Research for Sustainable Landscape Development*. Mansholt Publ. series Vol. 11 (pp. 17–40). Wageningen: Wageningen Academic Publishers.

Van Paassen, A., Van den Berg, J., Steingrover, E., Werkman, R. & Pedroli, B. (2011a). *'Knowledge in Action'. The Search for Collaborative Research for Sustainable Landscape Development*. Mansholt Publ. series Vol. 11. Wageningen: Wageningen Academic Publishers.

Van Rij, E. & Korthals Altes, W. (2010). Looking for the optimum relationship between spatial planning and land development. *Town Planning Review* 81, 283–306.

Van Vliet, J., De Groot, H. L. F., Rietveld, P. & Verburg, P. H. (2015). Manifestations and underlying drivers of agricultural land use change in Europe. *Landscape and Urban Planning* 133, 24–36.

Vanslembrouck, I. & Van Huyenbroeck, G. (2005). *Landscape Amenities: Economic Assessment of Agricultural Landscapes*. Netherlands: Springer.

Vejre, H., Abildtrup, J., Andersen, E., Andersen, P. S., Brandt, J., Busck, A. G. et al. (2007a). Multifunctional agriculture and multifunctional landscapes – land use as interface. In Ü. Mander, H. Wiggering & K. Helming (Eds.), *Multifunctional Land Use, Meeting Future Demands for Landscape Goods and Services* (pp. 93–104). Berlin Heidelberg: Springer.

Vejre, H., Jensen, F. S. & Thorsen, B. J. (2010). Demonstrating the importance of intangible ecosystem services from peri-urban landscapes. *Ecological Complexity* 7, 338–348.

Vejre, H., Primdahl, J. & Brandt, J. (2007b). The Copenhagen Finger Plan. Keeping a green space structure by a simple planning metaphor. In B. Pedroli, A. Doorn, G. Blust, M. L. Paracchini, D. Wascher & F. Bunce (Eds.), *Europe's Living Landscapes. Essays Exploring Our Identity in the Countryside* (pp. 310–328). Zeist: KNNV Publishing.

Ventura, F., Brunori, G., Milone, P. & Berti, G. (2008). The rural web: a synthesis. In J. D. van der Ploeg & T. Marsden (Eds.), *Unfolding Webs. The Dynamics of Regional Rural Development* (pp. 149–174). Assen: Van Gorcum.

Verhoeve, A., Dewaelheyns, V., Kerselaers, E., Rogge, E. & Gulinck, H. (2015). Virtual farmland: grasping the occupation of agricultural land by non-agricultural land uses. *Land Use Policy* 42, 547–556.

Vlahos, G. & Schiller, S. (2015). Transition processes and natural resource management. In L. A. Sutherland, I. Darnhofer, G. Wilson & L. Zagata (Eds.), *Transition Pathways towards Sustainability in Agriculture. Case Studies from Europe* (pp. 113–126). Wallingford: CABI International.

Vos, W. & Meekes, H. (1999). Trends in European cultural landscape development: perspectives for a sustainable future. *Landscape and Urban Planning* 46, 3–14.

Vos, W. & Stortelder, A. (1992). *Vanishing Tuscan Landscapes. Landscape Ecology of a Submediterranean-Montane Area*. Wageningen: Pudoc Scientific Publishers.

Warren, E., Hawkesworth, S. & Knai, C. (2015). Investigating the association between urban agriculture and food security, dietary diversity, and nutritional status: a systematic literature review. *Food Policy* 53, 54–66.

Wascher, D., Kneafsey, M., Pintar, M. & Piorr, A. (2015). *Food Planning and Innovation for Sustainable Metropolitan Regions – Synthesis Report*. Wageningen: Wageningen UR.

Wascher, D. & Pedroli, B. (2008). *Blueprint for EUROSCAPE 2020. Reframing the Future of the European Landscape – Policy Visions and Research Support*. Wageningen: Landscape Europe.

Wattchow, B. (2013). Landscape and a sense of place: a creative tension. In P. Howard, I. Thompson & E. Waterton (Eds.), *The Routledge Companion to Landscape Studies* (pp. 87–96). London/New York: Routledge.

Werkcommissie Westen des Lands (1958). *De ontwikkeling van het Westen des Lands: toelichting*. The Hague: SDU.

Wiggering, H., Ende, H. P., Knierim, A. & Pintar, M. (2010). *Innovations in European Rural Landscapes*. Springer: Heidelberg.

Wilbur, A. (2013). Growing a radical ruralism: back-to-the-land as practice and ideal. *Geography Compass* 7, 149–160.

Wilkinson, K. P. (1991). *The Community in Rural America*. Santa Barbara, CA: Greenwood Publishing Group.

Williams, R. (1973). *The Country and the City*. Oxford: Oxford University Press.

Williams, R. (1984). Between country and city. In R. Mabey, S. Clifford & A. King (Eds.), *Second Nature* (pp. 209–219). London: Jonathan Cape.

Williams, R. H. (1996). *European Union Spatial Policy and Planning*. London: Paul Chapman Publishing.

Wilson, G. A. (2001). From productivism to post-productivism and back again? Exploring the (un)changed natural and mental landscapes of European agriculture. *Transactions of the Institute of British Geographers* 26, 77–102.

Wilson, G. A. (2007). *Multifunctional Agriculture: A Transition Theory Perspective*. Wallingford: CABI International.

Wilson, G. A. (2009). The spatiality of multifunctional agriculture: a human geography perspective. *Geoforum* 40, 269–280.

Winter, A. (2015). *Migrants and Urban Change: Newcomers to Antwerp, 1760–1860*. London: Routledge.

Winter, S. (1990). Integrating implementation research. In D. J. Palumbo & D. J. Calista (Eds.), *Implementation and the Policy Process. Opening up the Black Box* (pp. 19–38). New York: Greenwood Press.

Winter, S. & Nielsen, V. L. (2008). *Omplementing af politik*. København: Academica.
Woods, M. (2005). *Rural Geography*. London: Sage Publications.
Woods, M. (2007). Engaging the global countryside: globalization, hybridity and the reconstitution of rural place. *Progress in Human Geography* 31, 485–507.
Woods, M. (2011). *Rural*. London: Routledge.
World Commission (1987). *Our Common Future*. Oxford: Oxford University Press.
Wu, J. (2013). Landscape sustainability science: ecosystem services and human well-being in changing landscapes. *Landscape Ecology* 28, 999–1023.
Wylie, J. (2007). *Landscape*. London: Routledge.
Wylie, J. (2013). Landscape and phenomenology. In P. Howard, I. Thompson & E. Waterton (Eds.), *The Routledge Companion to Landscape Studies* (pp. 54–65). London/New York: Routledge.
Yourcenar, M. (1951). *Mémoires d'Hadrien*. Paris: Editions Gallimard, 1974.
Zasada, I. (2011). Multifunctional peri-urban agriculture – a review of societal demands and the provision of goods and services by farming. *Land Use Policy* 28, 639–648.
Zeigler, D. J., Brunn, S. T. & Williams, J. F. (2003). World urban development. In S. T. Brunn, J. F. Williams & D. J Zeigler (Eds.), *Cities of the World. World Regional Urban Development*, 3rd edn. (pp. 1–46).

Index

Aalen, 8
Abruzzi Apennines, 31
Afforestation, 18, 52, 75, 95, 143
Agent-based framework, 73
Agricultural policy, 12, 200, 225
Agriculture, 10, 51, 52, 111
  abandonment, 18, 129, 142
  agricultural production, 3, 11, 83, 166
  cropland, 21, 143
  first agrarian revolution, 165
  multifunctionality, 79, 85
  pasture, 21, 53, 127, 144
  production, 112
  second agrarian revolution, 166
  social farming, 192
Agri-environmental policy, 71, 153, 210, 213, 227
Agro-silvo-pastoral, 14, 132, 143
Alentejo, 30, 64, 128, 132, 229
Algarve, 189
Almeida, 229
Antrop, x, 5, 9, 23, 175
Arler, 35, 251
Arnaud, 162

Baltics, 12
Banská Štiavnica, 28
Baudry, 25
Beunen, 230
Biodiversity, 18, 151
Black Death, 164
Bocage, 112, 123
Brandt, 198
Broadening, 138
Brown fields, 23
Burgundy, 43
Butler, 190
Buttimer, 26

CAP. *See* Common Agricultural Policy
Carolino, 184
Castells, 56, 172
Celtic cross, 6
Champion, 174
Christianity, 6
Citizen, 34, 50, 74, 161, 162, 183, 223, 253
City state, 162
Classification
  ecosystem services, 59
  farming systems, 130
  forest types, 146
  HNV land use, 154
  land use intensity, 135
  landscape, 36
Climate change, 67, 148, 244
Commodification, 51, 61, 129
Commodities, 14, 79, 102
Common Agricultural Policy, 12, 143, 147, 204, 225, 248
Commons, ix, 114, 115, 206, 222
Community, 102, 116, 123, 139, 241, 242, 249, 252
  of place, 248
Conceptual framework, 67, 83, 87, 211
Conflict management, 200, 220, 251
Consumption landscape, 27
Convention
  European Landscape, 44
Copenhagen, 167, 176
  fingerplan, 171
Cosgrove, 47
Costa Brava, 189
Counter urbanisation, 158, 174
Countryside, 167
  consumption, 48, 81, 83, 102
  types, 82
Crete, 28
Cronon, 166
Czechoslovakia, 12

Deepening, 138
Denmark, 12, 142, 160, 179, 183, 237
Dessau-Wörlitzer Gartenreich, 156
Diversification, 85, 138
Dobříš Assessment, ix, 4, 36, 151
Drinking water, 26
Driving forces, 71, 142, 149, 151, 244
Duby, 10, 14, 115
Dwyer, 207

Eastern Europe, 23, 131, 143
Eastern Germany, 12
Ecological infrastructure, 26
Ecosystem services, 18, 30, 59, 66, 229, 247
Edam, 1
Egoz, 207, 253
Emanuelsson, 9, 165, 166
Enclosure, 115, 165
England, 16, 165, 167, 168, 218
Environment Action Programme, 230
Environmental co-operatives, 234
Environmental policy, 230
EU Environmental Directives, 231
European Court of Auditors, 210
European identity, 10
European landscape, 8
  agricultural landscape, 25
  classification, 36
  contexts, 23, 25
  current developments, 21
  diversity of, 4, 36
  feudal landscapes, 164
  functional basis, 193
  greening, 18
  heritage, 21, 27
  historic land changes, 18
  history, 13, 113
  map, 36
  medieval, 163
  peri-urban, 28
  policies, 217
  remote landscape, 31
  urbanisation, 174
European Landscape Convention, 3, 33, 205, 223, 251
European Spatial Development Perspective, 218
European Union
  agricultural policy, 12
  policy agenda, 207
  structure of agriculture, 108
  sustainability agenda, 226
Extensification, 78, 128, 136, 149

Farm
  broadening, 97
  dairy, 1, 40, 111, 126
  deepening, 97
  innovation, 97
  owner, 185, 192, 219
  owners, 73, 181
  re-grounding, 97, 140
Farmers, 68, 72, 74, 97, 99, 181
  full-time, 100, 139
  hobby, 141, 181
  lifestyle, 64, 101, 141
  occupational status, 181
  part-time, 100, 141
  strategies, 68, 72, 97, 139, 229
Farming, 52, 64
  family, 114, 116, 138, 140
  mixed, 126, 131
  small-scale, 66
  social, 192
  styles, 16, 74, 102
  system, 113, 122, 128, 129, 130, 135
Faroe Islands, 199
Faroe Sheep Letter, 198
Fearne, 226
Feuilla, France, 201
Fischler reform, 13
Fonseca, 219
Food policy, 232
Forest land, 18, 21, 131, 143
Forman, 45, 247
Fragmentation, 135, 152
Fuchs, 20, 245
Future landscapes, 8, 25, 37, 159, 221, 244

Garden city, 168, 195
Geddes, 171
Government and governance, 204, 242
Greenhouse gas, 68, 149
Grigg, 10, 124, 130, 136
Grove, 8, 11, 114, 122, 150

Habitat Directive, 231
Hägerstrand, 206
Hall, 171, 218
Healey, 35, 220, 221, 236
Hedgerow, 126, 218
Hellerau Garten Stadt, 170
High Nature Value, 55, 152
Hill, 228
Hodge, 207, 218
Hogwood, 208
Holland, 2
  Green Heart Plan, 171
  polder, 26, 123, 240
Holmes, 51, 83, 190
Hotspots of area changes, 22
Howard, 168
Hungary, 12
Hunting, 28, 57, 97, 99, 114, 184

Ideal rural area types, 87
Impact model, 210
Indicators, 71, 87, 91, 150
Industrial capitalism, 165
Infield, 9, 198
Innovation, 24, 84, 105, 114
Intensification, 128, 135, 136, 146, 150
Italy, 12, 23, 30, 111, 136, 162, 177, 221

Jepsen, 11, 166
Jones, 69, 72, 189, 217, 223
Jutland, 179, 237

Kent, 28
Kleijn, 210, 228

Land abandonment, 111, 128, 142
Land consolidation, 13, 15, 117
Land manager, 58, 72, 73
Land reform, 12, 15, 165, 166
Land use policy, 199, 217
Landscape
  appearance, 44, 49, 65
  as a common good, 34, 61, 242
  as a development factor, 35
  change, 22, 46, 65, 66, 114, 136, 142
  character, 45, 116
  commodification, 51, 61, 129, 213
  concept, 5, 32
  democracy, 251
  diversity, 3, 33, 129, 249
  ecology, 45, 60, 61, 247
  functions, 5, 45, 57, 60
  goods and services, 29, 58, 80, 103
  governance, 41, 199, 205, 233, 246
  manager, 74, 252
  pattern, 45, 74, 112, 122, 123, 124, 131, 151, 154
  services, 57, 60, 61
  structure, 45, 65
  territorial and spatial competences, 206
  transition, 3, 25, 50, 67, 241
  user, 58
  value, 60, 61, 79
Landscape character, 24, 36, 47, 93, 113, 206, 233
Landscape democracy, 35, 251
Landscape heritage, 34, 229
Landscape planning, 5, 39, 198, 220, 247
  conflict management and place making, 220, 253
Landscape policy, 33, 198, 242, 251
  design of, 208
  European Landscape Convention, 3, 33, 205, 223, 233, 251
  government and governance, 204
  impact model, 210
  implementation of, 214
  instruments, 211
  integration of, 233
  participation, 223
  policy agenda, 207
  policy process, 208
Landscape strategy making, 250, 253
Land-take, 113
Lipsky, 215
Lowe, 16, 51, 56, 227
Luginbühl, 11, 47, 115

Manchester, 165
Mansholt, 12, 171
Manure, 9, 115, 167
Market, 58, 61, 116, 128, 135, 137, 153
Marsden, 16, 50, 51, 56, 72, 80, 82, 85, 87, 98, 102, 103, 125
McSharry, 13
Mediterranean, 23, 129, 130, 143, 147, 150
Meeus, 36
Migration, 161, 244
Millward, 176
Montado, 64, 122, 146, 219, 229
Multifunctional landscape, 34
Multifunctionality, 28, 34, 59, 79, 81, 125, 138, 153
  concept, 85
Multi-level perspective, 106, 107
Mumford, 163, 164
Murdoch, 82, 90
Muslim influence, 9
Mussolini, 12
Mythical, 6

Narrative, 91, 244, 250
Nassauer, 45, 47, 48
NATURA 2000 network, 205, 230, 231, 233
Natural processes, 45, 67, 68, 74, 122
Nature conservation, 30, 52, 89, 117, 151, 153, 200, 234, 252
Naveh, 45
Netherlands, 2, 12, 29, 117, 234
New town, 170
Nitrate Directive, 231

O'Riordan, 230
OECD, 59, 71, 79, 208
Olwig, 5, 44, 47, 168, 217
Opdam, 57, 61, 174
Ostrom, 243
Outfield, 9

Paracchini, 37, 147, 152
Pathway, 41, 85, 221, 241, 245, 250
Peak District, 28, 188
Pearce, 190
Peatbog, 1, 2
Pedroli, 5, 37, 39, 48, 189, 221, 245
Perception, 44, 47, 61, 171, 211
Pérez-Soba, 37, 39

Peri-urban landscape, 28, 29, 196
Pinto-Correia, 11, 52, 80, 87, 92, 103, 146, 219, 229
Place, 48, 49, 62, 247
  making, 252, 253
Place and space, 7
Place making, 220, 248
Pluri-activity, 98
Polarising trend, 23
Polder, 2, 122
Policy implementation, 214, 215, 216
Policy instruments, 211
Policy integration, 233
Policy interventions, 247
Portugal, 12, 30, 65, 87, 128, 132, 146, 189, 217, 229
Post-industrial city, 172
Post-productivism, 80, 153
Pounds, 10, 113, 163
Primdahl, 50, 73, 87, 141, 178, 181, 207, 210, 228
Producer, 58, 74, 139, 164, 180, 186, 216, 232, 248
Productivism, 125
Public intervention, 11, 66, 152, 243
Public policy, 23, 35, 57, 66, 199, 201, 208, 234, 243, 247

Qviström, 176

Rackham, 8
Recreation, 177
Re-grounding, 138, 140
Renaissance cities, 164
Renewable energy, 140, 154
Renting, 234
Rhodes, 205
Roman Empire, 8, 115, 162
Romania, 12
Rounsevell, 21, 83
Rural, 50, 51
  actors, 73, 99, 102
  attributes, 89
  change, 82
  community, 56, 102
  development, 35, 81, 82, 107, 205, 232, 234
  drivers, 84
  occupance modes, 83
  population, 56
  representation, 51, 56
  trajectories, 85
  typologies, 88

Sámi culture, 6
Sayer, 224, 246
Scale, 62, 124, 138
Schama, 5
Schneider, 211
Scotland, 93

Scott, 176, 187, 188
Selman, 47, 212
Slovakia, 52, 53
Slovenia, 75
Slow landscape, 29
Social
  constructs, 57
  expectation, 85
  farming, 192
  learning, 50
  processes, 68
Society
  expectations, 57, 248
  network, 56
Soil degradation, 150
Southern European, 136, 168
Space, 47, 48, 50
Spatial planning, 217
Spirn, 35
Strategy making, 234, 248
Swaffield, 207

Teleconnection, 173, 197
Tenerife, 192
Termorshuizen, 174
Territorial and spatial competences, 206
Territorial perspective, 89, 135
Tewdwr-Jones, 171, 218
Tourism, 139, 187
Transhumance, 114, 122, 123
Transition, 67, 83, 135, 241
  processes, 44, 135
  theory, 104
Tuscany, 28, 177, 221
Typology, 90

Urban expansion, 23
Urban sprawl, 111, 112
Urbanisation, 15, 174, 243
Urban-rural relationship, 156
Urry, 188, 191

Van der Ploeg, 101, 141, 234
Vejre, 171
Village, 8, 9
Vision, 135, 140, 242, 244, 250
Vos, 11
Vries, 164

Water Framework Directive, 231
Williams, 158
Winter, 215
Woods, 50, 161

Zasada, 37